Psychoacoustic Foundations of M

Psychoacoustic Foundations of Major-Minor Tonality

Richard Parncutt

The MIT Press
Cambridge, Massachusetts
London, England

The MIT Press would like to thank the anonymous peer reviewers who provided comments on drafts of this book. The generous work of academic experts is essential for establishing the authority and quality of our publications. We acknowledge with gratitude the contributions of these otherwise uncredited readers.

This book was set in Stone Serif and Stone Sans by Westchester Publishing Services.
Printed and bound in the United States of America.

Library of Congress Cataloging-in-Publication Data

Names: Parncutt, Richard, 1957– author.
Title: Psychoacoustic foundations of major-minor tonality / Richard Parncutt.
Description: Cambridge, Massachusetts : The MIT Press, 2024. | Includes bibliographical references and index.
Identifiers: LCCN 2023007827 (print) | LCCN 2023007828 (ebook) | ISBN 9780262547352 (paperback) | ISBN 9780262377379 (epub) | ISBN 9780262377362 (pdf)
Subjects: LCSH: Tonality. | Psychoacoustics. | Chords (Music) | Musical intervals and scales.
Classification: LCC ML3811 .P37 2024 (print) | LCC ML3811 (ebook) | DDC 781.2/6—dc23/eng/20230606
LC record available at https://lccn.loc.gov/2023007827
LC ebook record available at https://lccn.loc.gov/2023007828

10 9 8 7 6 5 4 3 2 1

Contents

Terms and Abbreviations

aug; + augmented; augmented triad (in semitones: [048])

C/D consonance and dissonance

EDO equal divisions of the octave

dim; o diminished; diminished triad [036]

dim7; o7 diminished-7th chord [0369]

dominant $\hat{5}$, V, v

f0 fundamental frequency ("F-zero")

half-dim7; m7♭5; ø7; ø half-diminished-7th chord [036T]

JND just-noticeable difference

[L] eLeven (11) semitones; major-7th interval

M; maj major

m; min; - minor; minor triad [037]

M7; maj7; Δ7; Δ major-7th chord [047L]

m7; min7; -7 minor-7th chord [037T]

Mm7; 7 major-minor 7th chord [047T]

MmT major-minor tonality

mediant $\hat{3}$, III, iii

predominant chord that leads to the dominant (subdominant)

subdominant $\hat{4}$, IV, iv

submediant $\hat{6}$, VI, vi

subtonic $\flat\hat{7}$, ♭VII, ♭vii

supertonic $\hat{2}$, II, ii

sus; sus4 suspended; suspended triad [057]

[T] Ten (10) semitones; minor-7th interval

tonic $\hat{1}$, I, i

Preface

Like many of my research colleagues in the areas of music cognition and music theory, I have always been fascinated by the structure and emotional power of major-minor tonality (MmT). While this book is not directly about musical emotion, it is certainly motivated by it—as is probably all research on musical structure.

There can be little doubt about the emotional power of tonal music, including simple diatonic chord progressions. Where to begin? There are countless examples. Take the Latvian Song Festival in July 2018 in Riga, which celebrated Latvia's centenary. More than 1,600 choirs, dance ensembles, and music groups took part. As in most music festivals the world over, even today, most of the music was based on simple diatonic chord progressions. The climax of the festival was a performance by all participating choirs—about 11,000 singers—of *Pūt, vējiņi* (Blow, winds), an ancient wedding song that has become a symbol of Latvian national identity. Why do people love this music so much? The lyrics are rather weird, at least for outsiders. The simple tonal structure seems to play an important role, alongside the complex cultural context. Why is music of that kind so powerful?

This book is also motivated by a long-standing lack of academic consensus about seemingly trivial issues. Where do different diatonic scales and modes come from? Why are some chords and chord progressions more common than others? How and why do we perceive most music relative to tonal centers, and what are the defining features of a tonal center? How do these music-structural features contribute to music's emotionality?

Hovering above such questions are issues of nature versus culture. Are the listed aspects of MmT arbitrary? If not, to what extent? In what way might they depend on human biological or psychological universals or general perceptual principles? Moving to another level, we may ask: Is it politically incorrect to ask these questions? What about the colonial history of MmT?

Those are rather general questions, and there are also more specific ones. Why do most chords in MmT have three or four pitch classes and not some other number? Why are there two main types of trichord (major triad followed by minor triad), and why are major-minor 7th (Mm7) chords so common, despite the dissonant tritone interval between the major 3rd and the minor 7th above the root? Why are chords more often played in root position than in inversion? Why do most scales have seven pitch classes? Why are there two main types of scale in Western music, the other "modes" being less common? Why is the exact tuning of the piano so important to us in practice, although we understand it to be theoretically out of tune, and why do we tolerate much bigger intonational deviations in choral performance? Why are chord progressions with rising 4ths and falling 3rds between successive roots more common than progressions in the opposite direction? Why do rising leading tones tonicize better than falling ones?

In an attempt to answer questions of this kind, this book presents an evidence-based theory of MmT. The theory is founded on empirical psychological research and statistical corpus analyses while also drawing inspiration from the history of music theory and the author's musical practice and intuition. A novel combination of intellectual inputs—mainly from psychology, music theory, and psychoacoustics, but also from music information sciences, music history, and ethnomusicology—gives the theory of MmT a new foundation.

The book belongs to the academic discipline known as systematic musicology. The word "systematic" implies a combination of empirical observation, clear argumentation, and interdisciplinarity. As an academic label, systematic musicology dates to the late nineteenth century in Germany and Austria, although the discipline itself, one might argue, has ancient roots. Guido Adler (1885) divided musicology into two parts: historic and systematic. Under the heading of systematic musicology, he included music theory, music aesthetics, music education, and comparative musicology (non-Western music). He also listed parent disciplines: acoustics, mathematics, physiology, psychology, (philosophy of) logic, linguistics, education, and aesthetics. During the twentieth century, three of Adler's subdisciplines of systematic musicology became independent (music theory, music education, and comparative musicology, the latter transformed into ethnomusicology); they are no longer regarded as subdisciplines of systematic musicology.

Today's systematic musicologists tend to identify with (music) acoustics, neuroscience, psychology, sociology, computer science, and/or philosophy; with the exception of sociology, all these disciplines are relevant for the present book. In international English-speaking research, the term "music cognition" now seems equivalent to systematic musicology. Whichever term is used, the discipline is defined in practice by its leading peer-reviewed academic journals: *Music Perception, Psychology of Music, Musicae Scientiae, Music & Science, Jahrbuch Musikpsychologie* and *Journal of New Music Research*. The discipline is also defined by pan-national societies such as the European Society for the Cognitive Sciences of Music and the (North American) Society for Music Perception and Cognition.

My attempts to understand musical structure began in childhood. I diligently learned the piano while also singing in various choirs, including the chapel choir of Melbourne Grammar School. That school and I did not agree on everything, but singing in their chapel choir did teach me the foundations of MmT, for which I am grateful and without which my research in the area is unlikely to have happened. My musical explorations continued in the late 1970s as a student of music (piano) and science (physics) at the University of Melbourne and in the early 1980s as a doctoral candidate at the University of New England (Armidale, New South Wales). For decades, I performed regularly as a solo pianist, piano accompanist, or pop pianist, and sang in choirs. Today, I am writing and performing a cappella arrangements.

Some of my music-theoretical explorations led to dead ends. As a student, my background in physics inspired me to experiment with mathematical modeling in the style of pseudo-physical quasi-universal music theory, which for millennia has been an armchair pastime of the musically curious. But I came to understand a serious problem with that approach. There is something especially human about music that the quasi-universal discipline of physics cannot capture. That may seem obvious, but it is not so easy to incorporate that insight into a comprehensive theory of tonal music.

Around 1980, I remember attending a lecture-recital by Melbourne pianist Ronald Farren-Price in which he asked an interesting question. What should we put inside a time capsule as a relic of human civilization to allow future intelligent visitors from another galaxy to get some idea of who we were?

His answer: the notated scores of the thirty-two piano sonatas of Ludwig van Beethoven. The idea sounded convincing at the time, and the audience nodded in agreement. I only later realized how wrong it was. I love the music of Beethoven—no doubt about that. But it is a product of a specific cultural context whose track record has had its ups and downs to say the least. We can hardly sweep the biggest European white-male errors under the carpet. Those errors include the crusades, the Napoleonic wars, the slave trade, the scramble for Africa, two world wars, and the global climate and biodiversity crisis. Whereas Beethoven himself is surely innocent, at the same time, the idea that one white man's music is humanity's greatest achievement is surely sexist and racist, and it continues the familiar narrative according to which Western "man" is the pinnacle of creation or evolution and "his" culture is the pinnacle of culture (Ewell 2020).

While the focus of this book—like that of related books such as David Huron (2001, 2016), Carol Krumhansl (1990), and Fred Lerdahl and Ray Jackendoff (1983)—is on Western classical music, aspects of the theory may also shed light on the structure of diverse Westernized non-Western and/or nonclassical Western musics, especially if they use a twelve-tone chromatic scale. But the theory cannot claim to be cross-culturally valid, and when making intercultural comparisons, we need to avoid the implicit arrogance of some Western music-theoretical traditions. The focus on MmT in this and comparable books is not intended as a value judgment, nor is it intended to either include or exclude relevant non-European or non-Western styles. Intercultural issues are addressed in chapter 23.

If one day some intelligent aliens visit our planet, they will presumably have no idea what Beethoven's piano sonatas were for or what they were about. Music is inseparably linked to human biology and society and to the human environment. While Beethoven's most famous music may today be recognizable for people all over the planet, regardless of cultural background, that same music would surely mean nothing to visitors from another galaxy, even if they somehow managed to decode the relationship between notation and sound (something that most humans cannot do).

For those aliens would have no access to human subjectivity. They would have little idea how humans experience the world, let alone music. A color-blind philosopher could understand "red" better than those aliens could understand "music." We humans hardly understand our own subjectivity: the mind-body problem is not solved and may never be solved, although we have been wondering about it for millennia and although lately there have been big developments in neuroscience and philosophy (see chapter 4). Given our persistent failure to answer our biggest questions, those hypothetical aliens would have even less idea of what our music is about than we do. Nicholas Cook (1990) made a similar point, imagining "Martian musicologists of ten thousand years hence . . . attempting to reconstruct an authentic performance of Chopin's E minor Prelude" (122).

That being the case, it should be clear that physics alone can never fully explain the structure of any music, regardless of its cultural origin. Psychology and psychoacoustics are also necessary, but they too are insufficient; the culture specificity of Western music must be central to any explanation. In late-nineteenth-century colonialist Europe, German musicologist Hugo Riemann, in his *Natur der Harmonik* (1882), may not have fully grasped the cultural aspect of the problem, but he did understand its inherent interdisciplinary:

> Scientific investigation in the domain of music concerns itself, primarily, with ascertaining the laws which govern *sounding bodies*, and is thus a department of *physics*; i.e., the science of acoustics. Then pursuing tones still further, and inquiring into the effects they produce on the human ear, and the mode in which those effects are produced, it becomes a special department of *physiology*. Finally, concerning itself with tone-perceptions, with the *mental effects* of these acoustic and physiological phenomena, and with the mental connections and relations of the sensations produced by sound, it enters the domain of *psychology*. Out of the results of scientific investigation in all three fields of physics, physiology and psychology, we get the elements of an exact theory of the nature of harmony. (As cited in Mickelsen 1977, 45; italics in original)

Formulations of that kind are too positivist for modern tastes (especially the use of the word "exact" in the last line), and tacitly assume that Western society and culture are inherently superior. The following quote from US composer Norman Cazden (1954) comes closer to the truth on both accounts: "Few manuals of recent times enter into fundamental questions of music theory, and often one must deduce the implied axioms. Of theories deserving the name, we may distinguish three main types: those claiming to be founded on the laws of Nature, those claiming the priority of the arbitrary instinct of the composer, and those which rely on the observed practice of the art of music. We may term these the natural, the subjective and the empirical theories, and suggest that none of them fully satisfy our requirements" (288).

But again, this description is incomplete. First, there is still a Western cultural bias in Cazden's thinking. Second, he failed to notice a fourth type of music theory—one based on psychological listening experiments—that can be combined with the other three (physics, composition, and performance).

The omission was hardly surprising. In 1954, the academic discipline of music psychology barely existed, the German tradition having suffered terribly in the war. Things got started again in the 1960s with the international cognitive turn in psychology and linguistics (Pléh 2019). In 1973, the British journal *Psychology of Music* was founded. Between 1973 and 1985, Juan Roederer, an Italian-Austrian-Argentinian physicist at the University of Alaska Fairbanks, organized a series of international workshops on the "Physical and Neuropsychological Foundations of Music" at the Carinthian Music Festivals in Ossiach, Austria. Several key authors cited in this book took part in those workshops, including Jamshed Bharucha, Albert Bregman, David Butler, Robert Crowder, Lola Cuddy, Irène Deliège, Diana Deutsch, Adrian Houtsma, Carol Krumhansl, Fred Lerdahl, Stephen McAdams, Rudolf Rasch, Johan Sundberg, Ernst Terhardt, Joos Vos, and Dixon Ward. I had the good fortune to present my research at two of those meetings (1983 and 1985). In 1983, the *Deutsche Gesellschaft für Musikpsychologie* was founded by Klaus-Ernst Behne, Günter Kleinen, and Helga de la Motte-Haber; in the same year, Diana Deutsch founded the journal *Music Perception* in the USA (Spiller 1995). The first International Conference on Music Perception and Cognition was held in Japan in 1989. These developments were made possible by growth in the discipline of (cognitive) psychology in the late twentieth century, enabling research in music psychology to expand internationally, especially in North America.

As a result of these developments, we now understand more about MmT (music based on major and minor triads and scales) than we ever did before. But the same can be said of many other topics, and not enough has been done to bring together and unify the leading approaches to MmT in humanities and sciences, which is an aim of this book.

The second half of the twentieth century saw interesting new approaches. The best-known development in North American music theory was the organic, reductive approach to music analysis of Austrian music theorist Heinrich Schenker, made popular by US music theorists, including Allen Forte (Forte and Gilbert 1982). It plays an important role in this book, as does Forte's version of pitch-class set theory (Forte 1973), although it was originally intended for the analysis of post-tonal music. Parallel international developments in music psychology included Ernst Terhardt's (1974a) psychoacoustic theory of pitch perception and Carol Krumhansl's (1990) cognitive-structural approach to MmT. At first glance, these theories may seem incompatible; one of the goals of this book is to contradict that first impression.

When I started to think about the possible acoustical and psychological origins of MmT in the early 1980s, I was unaware of the richness of the epistemological environment that I was naively entering. For me at the time, there were merely interesting questions and answers, which I was seeing more or less in isolation. I only gradually discovered how much each cited author and paper had been influenced by academic traditions in different geographical locations and historical periods.

Terhardt's theory of pitch perception, which I learned about during a visit to the Technical University of Munich as a guest researcher and doctoral student in 1982–1983, overlaps in many ways with that of Brian Moore (1989). In their self-perception, however, these two scientists belonged to entirely different schools—one German, the other British; one engineering and cybernetic, the other psychological with a neurocognitive focus. Both nevertheless relied on empirical methods and modeling, and both developed models in which pitch depends on a combination of place of excitation the basilar membrane, timing information in the auditory signal, and the recognition of familiar patterns. Their approaches differed in many details, but those differences are not so important for the musical questions addressed in this book. For the purpose of building a music theory based on the psychoacoustics of pitch perception, one could equally refer to one or the other. I will focus on Terhardt's approach for the following reasons:

- Its general algorithmic formulation makes it convenient to apply to both everyday sound (including speech and music) and music-theoretical questions.
- The theory assumes an important role for the spontaneous learning of pitch patterns, consistent with both neuroscience (Hebbian learning in neural networks) and musicology (in particular, ethnomusicology).
- The theory clarifies the distinction between spectral and virtual pitch, which has interesting implications for both psychoacoustics and music theory.

This book focuses on simultaneous relationships among musical pitches, also called vertical aspects of musical scores. The vertical can hardly be separated from interacting horizontal aspects, and hence from counterpoint and voice leading, and underlying psychoacoustical and psychological principles. The leading source on the latter is Huron (2001, 2016), whose contribution relies on original computer-based analyses of databases of musical scores. Huron and I met after he reviewed Parncutt (1989)—a revision of my doctoral dissertation at the University of New England (Armidale, NSW, Australia) that presented a musically oriented interpretation of Terhardt's pitch theory. We collaborated during the 1990s, interpreting Terhardt's contribution in the context of North American music psychology and music theory.

This book is intended for researchers and graduate students in two main areas: music cognition and music theory. It may be too advanced for undergraduate or high school teaching, but I hope that some undergraduate teachers will become acquainted with its main ideas and adapt their teaching accordingly. Some readers (mainly in the humanities) will be disappointed by the paucity of musical examples, while others (mainly in the sciences) will have expected more empirical data. But the book is not intended as an introduction to relevant tonal repertoires or empirical methods. I am assuming that my readers are already acquainted with diverse tonal styles, on the one hand, and the foundations of psychological methodology, on the other. In particular, I am assuming that my readers can hear the pitch patterns that are discussed in the text—in the music that they listen to, perform, arrange, or compose themselves. Those who cannot do that will have difficulty evaluating the book's approach and claims.

The book may be received differently in North America and Europe due to persistent differences in the epistemology and pedagogy of music theory that emerged across the Atlantic in the late twentieth century. Post-secondary music students in North America may be better acquainted than their European colleagues with the details of theory that I am taking for granted in this book: Rameau, Riemann, Schenker, and Forte. They may also have better-trained ears and have received more detailed practical training in, for example, counterpoint in the style of Palestrina or harmonization in the style of J. S. Bach. Their Western European colleagues may be accustomed to a freer style of teaching and learning and may be more scholarly than their North American colleagues: better at constructing an original coherent argument and independently finding and evaluating literature.

These are clichés, of course, and the exceptions may be more common than I realize. The book may nevertheless be better received in North America if advanced music students there better understand its music-theoretical foundations: harmonic functions, Schenkerian analysis, and pitch-class set theory. But when it comes to critically evaluating the ideas that I am presenting and weighing different arguments up against each other, European students and colleagues may have the edge.

My main target readers will read a page or chapter here and there, depending on their interest or curiosity. Not many will read the book from start to finish, although that is also a possible strategy. With that in mind, I have aimed for a clear hierarchical structure on the contents page and consistent assignment of material within that structure. Given the complex web of connections within the material (that never stopped growing as I wrote), similar topics are sometimes addressed in different places. In those cases, the index will come in handy.

I should mention in closing that this book was written under a threatening cloud. The global climate and biodiversity crisis is steadily intensifying, along with a complex series of other, related global crises. This process is life-threatening for billions of people, and it also threatens to undermine academic research in all disciplines. This is not the place for a detailed explanation, but perhaps I should say this: Most readers of this book will have a high income and a high educational level relative to the average person in the world today—as I do. Many of us don't seem to have noticed how privileged we are, nor has it clicked with us that with privilege comes responsibility. If we care about young people and the future of humanity, it is up to us to send governments and corporations the right messages with the right level of urgency. It is also up to us to reduce our environmental footprints unilaterally, both individually and collectively. If we can't do that, one might reasonably ask why anyone else should; and if others don't do that, there is little point in writing books of this kind. I write these words knowing that most colleagues will ignore them, pretending not to understand. Thankfully, others will not, and some of us might have enough influence to make the necessary changes and avoid the worst.

Acknowledgments

This book would not have been possible without critical feedback of different kinds and at different times from countless colleagues. I can hardly mention them all, but the following lists give some idea of their diversity.

I am especially grateful to those who found the time to give me expert feedback on drafts of the book (or proposals for it). They included Josh Albrecht, Rytis Ambrazevičius, Diana Deutsch, Morwaread Farbood, Carol Krumhansl, Cheungkong Frederick Lau, Fred Lerdahl, Marcus Pearce, William Forde (Bill) Thompson, Barbara Tillmann, and four anonymous reviewers.

At the University of Graz, a series of student assistants and doctoral candidates helped me develop the ideas in diverse ways over several years. They include Zuzana Došeková (née Cenkerová), Helena Dukić, Isabelle Engel, Andreas Fuchs, Fabio Kaiser, Annekatrin Kessler, Manuela Marin, Nils Meyer-Kahlen, Lazar Radovanovic, Ana Rebrina, Gottfried Reichweger, Daniel Reisinger, Sabrina Sattmann, Theresa Schallmoser, Noemi Silvestri, and Evelyn Zemmel.

Jumping back a few decades, my PhD supervisors (Neville Fletcher †2017, physics; Catherine Ellis †1996, music; Willam G. Noble, psychology) and examiners (Howard Pollard, physics; Jeff Pressing †2002, music; Alan Costall, psychology) were important sources of support and inspiration. I especially thank Ernst Terhardt in Munich for his role as an informal additional doctoral supervisor, and my postdoc supervisors Johan Sundberg in Stockholm, Annabel Cohen in Halifax (Nova Scotia), Albert Bregman in Montreal, John Sloboda in Keele (UK), and Eric Clarke in Sheffield.

The book's content also depends on countless interactions with other academic colleagues over many years. I could hardly do justice to all of them, so allow me to present an alphabetical list that is bound to be incomplete:

Wolfgang Auhagen, Mario Baroni, Jamshed Bharucha, Emmanuel Bigand, Erica Bisesi, Bernd Brabec, William Caplin, Peter Cariani, Nicholas Cook, Ian Cross, Lola Cuddy, Irène Deliège, Johanna Devaney, Nicola Dibben, Roland Eberlein, Gerhard Eckel, Tuomas Eerola, Sean Ferguson, Andreas Fuchs, Anders Friberg, Christian Kaernbach, Douglas Keislar, Rudolf Flotzinger, Ichiro Fujinaga, Janina Fyk, Ali Cenk Gedik, Bruno Gingras, Robert Gjerdingen, Werner Goebl, Graham Hair, Robert Höldrich, David Huron, Stephan Koelsch, Reinhard Kopiez, Marc Leman, Daniel Levitin, Stephen McAdams, Neil McLachlan, Alan Marsden, Andrzej Miśkiewicz, Jeremey Poparad, Andrzej Rakowski, Bruno Repp, Robert Rowe, Frank Russo, Craig Sapp, Andrea Schiavio, Annemarie Seither-Preisler, Eleanor Selfridge-Field, Robert Stuckey, David Temperley, Sten Ternström, Petri Toiviainen, Caroline Traube, Anja Volk, Diana Weekes, Tillman Weyde, Frans Wiering, Geraint Wiggins, and Christian Utz.

I Interdisciplinary Introduction

1 Definitions

To avoid misunderstandings in an interdisciplinary approach, it helps to define some central terms, or at least to clarify their intended meanings. The definitions and explanations that I will propose are generally tentative and depend on cultural and historic context, which includes musical diversity (both globally and within Western or Westernized culture) and contrasting approaches in humanities and sciences. Music *is notoriously difficult to define because its origins and ultimate functions are shrouded in mystery.* Musicology *can be defined as the study of music,* psychology *as the study of human behavior and experience, and* music theory *as the study of musical structure.* Tonality *is about hearing music relative to a given reference pitch or pitch pattern (the "tonic"), and major-minor tonality* (MmT) *is tonality based on major and minor chords and scales. Musical* prolongation *happens when a musical sound or pattern is extended in time (e.g., by elaboration or decoration) such that a passage of music is heard relative to that original sound or pattern; prolongation may be the ultimate foundation of tonality, including MmT.* Salience *is the degree to which a musical event in the musical surface grabs our attention, whereas* stability *is how well the event is anchored in the musical structure.* Consonance *is how well (simultaneous or successive) tones are perceived to go together; dissonance is the opposite.* Psychohistory *is any combination of psychology and history, including the history of perception, the effect of psychological processes on history, and the effect of historical processes on psychology.*

Music

The average person devotes enormous amounts of time, energy, and money to music. Nonmusicians choose the music that they listen to, contribute actively as audience members, and discuss (often passionately) the music and the musicians. Musicians spend hours per day practicing, composing, performing, and recording. Music contributes to the construction of psychosocial (cultural, political) identities, particularly during adolescence. It provides stimulation, entertainment, and solace from infancy to old age.

The importance of music for most people justifies continuing research to try to understand how it works. After centuries of speculation, we still have remarkably little idea where music comes from or why all known human societies devote so much time and energy to it—at least not within today's dominant Darwinist evolutionary paradigm. The large and thriving literature on the origin of music (e.g., Honing 2018) can better explain *how* music works than *why* it is like it is (or how it originated). The task of explaining music's origin, function, emotion, and structure—while at the same time considering its cultural diversity—is further complicated by epistemological differences between researchers in humanities and sciences, as well as within humanities (e.g., ethnomusicology versus historical musicology) and within sciences (e.g., psychology versus physics). The continuing lack of agreement about musical origins means that the discipline of musicology lacks a stable foundation.

Because there is no clear objective criterion for the difference between music and nonmusic, definitions of music are often circular: we tend to define music in terms of music. We could for example say that music is a pattern of sound and movement that is perceived as music, as opposed to patterns of sound and music that are perceived as language or something else. Definitional circularity may be appropriate given that people in every musical culture have a good intuitive idea of what kinds of sound are perceived as music—at least for them. But circularity may prevent us from understanding music's origin and ultimate function.

In Western culture, music usually comprises artistic sound patterns that are based on relatively consonant chord progressions, played from tones taken from an almost equally spaced twelve-tone chromatic scale and dominated by tones from a diatonic scale (a subset of the chromatic). Whether that is a good or a bad thing is not the point: this book is inspired by the discipline of music cognition, in which the so-called average person is often the focus of attention. The focus is not usually on musical elites who perform and appreciate diverse post-tonal styles—although their experience is certainly relevant and sometimes crucial for understanding what MmT is and how it works.

The focus of this book is on music that is based on, or can somehow be reduced to, chord progressions (or harmonic progressions)—in particular, sequences of diatonic major and minor triads. In the key of C major or

A (natural) minor, diatonic triads are the chords C major, D minor, E minor, F major, G major, A minor, and B diminished. I will assume that Western listeners usually perceive progressions of such chords, or chords based on them, to be consonant—even if they are unsure what consonance means and may never use that word themselves. For the average person, the music may simply sound good (or like music), with the musical tones and chords going together in a musical way.

In classical harmony theory, a chord progression can be defined by three things: the root of each chord (the pitch, often ambiguous, relative to which the chord is often perceived and notated), the type of each chord (the intervals relative to the root, e.g., minor triad, diminished 7th; see chapter 3), and the inversion of each chord (which chord tone is in the bass). The voicing (which chord tones are in which voice, including the spacing and doubling of the voices) and the voice leading from one chord to the next are also important but are not necessarily implied when we say "chord progression," and they are not the focus of this book. If MmT can be reduced to harmonic progressions, as defined, we may ask a simple question: Why are some progressions more common than others? Much of the material in this book is intended as a foundation for answering this deceptively simple question.

In a global perspective, "music" is much more than MmT. In fact, one can argue that there is nothing particularly special about MmT. For many decades, the academic discipline of ethnomusicology (or comparative musicology) has been documenting the diversity of the world's musics. That diversity is one reason why scholars and scientists with different backgrounds have different ideas about the nature and origins of music; another reason involves underlying philosophical assumptions (see chapter 4). Given the importance of music for almost all humans, the apparent regularity (but also diversity) of its structure, the range of available explanations its function and existence, and the long history of trying to answer such fundamental musical questions, it is hardly an exaggeration to say that understanding music is one of humanity's great intellectual challenges.

Academic Disciplines

This book derives its material from a series of academic disciplines, including musicology, psychology, and music theory. What characterizes those disciplines, and how do they differ?

Musicology, understood to include ethnomusicology, addresses all phenomena called "music" (from either a cultural-insider "emic" or outsider "etic" perspective) in all its cultural and historical contexts. It includes all academic disciplines that successfully investigate musical phenomena and answer interesting musical questions ("all types of scholarship in music are, and properly ought to be, part of musicology"; Nettl 1999, 289). Some musicologists limit this broad vision in arbitrary ways or promote hierarchical structures in which some subdisciplines are central and others are auxiliary. Don Randel (1986) noted that "The great majority of scholars who describe themselves as musicologists (as distinct from ethnomusicologists) are students of Western art music" (520), implying a hierarchical structure within musicology in which Western art music is at the top and other forms of musicology (including systematic musicology, the psychology of music, the study of popular music and jazz, and so on) are subsidiary. But when one considers the entirety of published academic research on music and its interdisciplinary readership, musicology seems more like a collection of equally important and partly interacting subdisciplines (Parncutt 2007).

Psychology is the study of human behavior and experience. Perception (seen as an aspect of psychology) is a process of picking up information about the environment—usually while interacting with it—and taking advantage of previously acquired perceptual knowledge (perceptual memory). Cognitive psychology, in that view, is about how humans behave and what they experience when picking up and processing information from the environment. In the case of music, cognitive psychology involves behavioral and experiential aspects of complex interactions among musicians, listeners, social environments, and physical environments. What we know about the psychology of music is based mainly on experiments in which variables are independently varied and quantitative data collected. On that basis, possible causalities are investigated. Experiments often focus on specific responses to specific stimuli or situations. That approach differs markedly from the more qualitative and all-encompassing approach of musicology in the humanities.

Just as cognition is an important aspect of psychology, music cognition is an important aspect of music psychology. Cognition involves a long list of psychological and linguistic skills that are relevant for music, including perception, attention, thought, skill, knowledge, learning, memory, evaluation, reasoning, computation, problem solving, decision making, and the production and comprehension of language. The discipline of cognitive psychology and its subdiscipline, music cognition, tend to assume the existence of cognitive (or mental) structures (or representations) that correspond to an external reality and enable the organism to interact with that reality. Such cognitive representations, while possessing considerable explanatory power, can also be questioned; the interaction of the organism and its environment can instead be regarded as an emergent property of a complex system (e.g., Sterck and Begeer 2010).

According to English-language Wikipedia (checked on September 22, 2021), music theory is "the study of the practices and possibilities of music." That definition is at once too general and too vague. German-language Wikipedia, consulted on the same day, explained that music theory is both a subdiscipline of musicology and an independent artistic and academic discipline that involves harmony, counterpoint, form, and music analysis—which is closer to the definition assumed here. One could argue that music theory is a subdiscipline of musicology; it is music research that addresses musical structure, focusing on pitch (harmony), time (rhythm, form), and combinations of the two (melody, counterpoint). But in practice, music theory and musicology are often separate, with musicologists mainly studying music history.

Music theory has a long, dynamic history that dates to ancient times. As such, it represents an important chapter in the history of ideas. The Western music theory of recent centuries usually takes conventional music notation as its starting point. The focus tends to be on musical structure (melody, rhythm, harmony, phrasing, form, and so on) as represented in musical scores and their analysis (music analysis). At the same time, music theorists are keenly aware of the limitations of scores ("Notation conserves music . . . but it conceals as much as it reveals"; Cook 1998, 55), and ask broader questions about the context in which music is performed, composed, improvised, and received.

This book belongs to a long tradition of music theory, but it also departs from that tradition by juxtaposing contrasting perspectives on musical structure from different academic disciplines. Musical structure may be considered from the perspective of music notation, but that is not the only possibility. Music also has corresponding physical and psychological representations. All three, and their interrelationships, will be considered in depth.

In the late twentieth century, music theory changed in response to trends variously described as postmodernism or new musicology. Old certainties were undermined: there was no longer a correct analysis of a piece of music, nor was there a correct approach to analysis. The hermeneutic scope of analysis was expanded. The musical score was no longer regarded as the immutable and sacrosanct foundation upon which musicology was built. Instead, musical pluralism was embraced, along with performance and interpretation traditions, and the inherent subjectivity and context dependence of music-analytic decisions. Musicologists became more aware of the arbitrariness of social power and authority in determining academic opinion.

That development raised basic questions about the value and viability of music theory: "The history of musicology and music theory in our generation is one of loss of confidence: we no longer know what we know" (Cook and Everest 1999, v). How positivist, formalist, or purely musical should music theory be? To what extent should we construct or deconstruct existing music-theoretical concepts? What kind of questions should music theory answer? What is it for? Is music theory even a discipline? How valuable is an analysis of a piece of music that focuses on formalist elements (as this book does) and neglects socially mediated meanings? Similar questions can be asked of historical musicology: "What kind of truth can analysis of historical musicology reveal, and how might it relate to other kinds of truth about music? Should we be speaking of truth at all? Or does the act of engaging in analysis or the writing of history lock us into a predetermined epistemological stance?" (Cook and Everest 1999, xi).

These are serious questions, and as their formulation implies, clear answers are inherently lacking. A possible response to such a cascade of doubt is to try anew to articulate the discipline's scientific and formalist foundations, while at the same time remaining aware of the pitfalls of positivism. That is one of the goals of this book.

Tonality

Defining tonality is no easy task. Is it Western or cross-cultural? Is it part of the score, inherent in certain musical sounds, or a cognitive abstraction? Is it based on harmony and harmonic function (how chords progress from one to another), or melody and voice leading, or both? Is it based on a more general artistic or psychological principle, or is it unique?

Broadly speaking, any music is tonal, or has tonality, if some tones or pitches are perceived as psychological references by virtue of being repeated, accented, or sustained (cf. Hutchins and Palmer 2008). Those tones are called tonal centers, and the effect is called pitch centricity (cf. Kleppinger 2011; Lewin 1968; Perle 1972; W. E. Thomson 1999). Almost all pitch-based music, including innumerable non-Western tonal styles, features melodic tonality in the sense that some tones or pitches are more likely to act as psychological anchors or reference points for the others—usually because they are sounded more often. These references may also be hierarchically structured (Kessler et al. 1984; Kolinski 1967).

Tones may be perceived as references if they lie near the middle of a melody's pitch range or ambitus. Since melodic phrases tend to have an arch-like shape, starting and ending relatively low and reaching a peak in the middle (Huron 1996), tonal centers may also lie in the lower part of the range. In "atonal" or "pantonal" styles, attention may be drawn to specific pitches by repetition or different kinds of accentuation (Parncutt 2003), creating temporary centers. Tonal centers can also be predicted from the frequency, occurrence, or duration of individual tones (Lantz et al. 2020; Oram and Cuddy 1995). Additional cues to tonality include loudness, timbre, and learned patterns (Ross and Knight 2019, 387).

Repetition is a characteristic feature of all music, and it is one feature that distinguishes music from speech (Margulis 2014). Repetition can even turn nonmusical sound into music: when a short snippet of speech is recorded and repeated several times, we start to hear the pitches of the phonemes as if they were part of a musical scale (Deutsch et al. 2008). Repetition is also an important aspect of music learning; it is how music is normally passed from one generation to the next. Repetition is related to a series of music-structural features, including tonic return and implication-realization. Tonic return implies repetition of the tonic after intermediate elaboration. Implication often means starting from musical element A and moving to B, and realization means moving back to (or toward) element A (cf. intervallic reversal and registral return according to Narmour 1990). In such cases, element A is repeated—often with some variation.

Tonality involves pattern recognition (Krumhansl 1990, 140; Wertheimer 1938). Again, that is related to repetition: by favoring a limited set of familiar patterns, tonality encourages a certain kind of repetitiveness. Richard Norton (1984, x) regarded tonality as "a decision made against the chaos of audible nature, a bewildering abundance of sonic potential that perhaps only the contemporary composer at his electronic synthesizer can fully appreciate," implying that musical repetition reduces natural chaos.

More narrowly, MmT is about major and minor keys and subdominant-dominant-tonic progressions in Western music of the eighteenth and nineteenth centuries. It's what François-Joseph Fétis called *tonalité moderne*, Hugo Riemann *Funktionsharmonik*, and Carl Dahlhaus *harmonische Tonalität*. The latter rests on two assumptions: first, that a triad constitutes a primary, direct unit; and second, that the progression of chordal roots establishes the key (Dahlhaus 1968; as cited in W. E. Thomson 1999, 17).

Both these points are interesting, but both can also be challenged. The idea that harmonic tonality is based on the tonic triad is supported by evidence that the music is perceived relative to that triad. One form of evidence is the high correlation between Krumhansl's key profiles and the pitch-class salience profile of the tonic triad (Parncutt 2011; figure 1.1). Another is more intuitive: diatonic-tonal melodies can often be perceived relative to the tonic triad. Each tone in such a melody falls on one of seven scale degrees, of which three belong to the tonic triad ($\hat{1}$, $\hat{3}$, or $\hat{5}$) and four do not ($\hat{2}$, $\hat{4}$, $\hat{6}$, or $\hat{7}$). Those that do not are a step away from a tone in the tonic triad and so can be perceived as a neighbor tone. Typical neighbor decorations of tonic-triad tones include $\hat{3}-\hat{4}-\hat{3}$ and $\hat{5}-\hat{6}-\hat{5}$. In that way, a melody can be perceived relative to the tonic triad that is imagined in the background.

Dahlhaus also argued that progression of chordal roots establishes the key. Whereas that is usually true, it also contradicts the idea that harmonic tonality is based on the tonic triad alone. Sometimes, tonality can be established simply by prolonging the tonic triad in the sense of repeating it for a long time, such as in the opening of Richard Wagner's *Rheingold* or the entirety of Led Zeppelin's *Whole Lotta Love*.

The term "tonality" was first used by French musicologist Alexandre-Étienne Choron at the start of the nineteenth century. His ideas were adopted and elaborated by his more influential Belgian colleague Fétis. The French *tonalité* refers to *ton*, which among other things is the key of a piece of music. Early definitions focused on the scale in which a piece of music is written, and the distinction between church modes (which Fétis called *tonalités anciennes*) and major/minor keys (*tonalités modernes*; Simms 1975).

The tritone interval (augmented 4th; diminished 5th; half octave; six semitones) played an important role in Fétis's theory. He regarded it as a consonance when part of a diatonic scale, but also as appellative, given its tendency to resolve to a more consonant interval. The role of the tritone can be clarified in modern pitch-class set theory. The standard diatonic scale can be represented as [024579L], where the numbers are intervals in semitones relative to the tonic in a major key, and L = 11. The scale has an interval vector of <254361>: it includes two minor 2nds, five major 2nds, four minor 3rds, three major 3rds, six perfect 4ths, and one tritone. The relative rareness of the tritone means that a diatonic scale can be unambiguously defined by playing just three tones, provided two of them span a tritone. That rule can help listeners keep their bearings as tonal music modulates from one key to another (H. Brown and Butler 1981; Browne 1981; Butler and Brown 1984).

But listeners often spontaneously perceive the tonic without the help of a tritone. Annabel Cohen (1991) presented excerpts from the Preludes in J. S. Bach's *Well-Tempered Clavier*. Listeners could usually identify the tonic after only four tones, regardless of whether a chord was sounded or a tritone was included. Piet Vos (1999), noticing that pieces of tonal music often begin with an ascending 4th or a descending 5th between melodic tones or roots, showed that listeners easily identified the second sound as the tonic—again, in the absence of any tritone.

The *tonalité moderne* of Fétis included the tonic, dominant, and subdominant harmonies, as previously explained by Jean-Philippe Rameau, and later called *Funktionen* by Riemann (1893). According to Fétis, *tonalité moderne* was born when Monteverdi (in the madrigal *Cruda Amarilli*) sounded what later became known as unprepared dominant 7th chords and resolved them. The dominant 7th, according to Fétis, was natural to the extent that it represented ratios of small integers (4:5:6:7). Fétis also developed a theory of modulation that included remote key areas. Later, Louis Lambillotte spotted dominant 7th chords in Palestrina's *Pope Marcellus Mass* (Christensen 2019).

Fétis also used the term *tonalité* in a more general way to refer to relationships between scale degrees. For him, those relationships were ultimately mysterious, metaphysical, and culture specific—not acoustic or physiological in origin (Simms 1975). Each scale degree had its own character (an idea taken up later by Huron 2006) that was shared by the triad that could be built upon it; in that sense, melodic and harmonic functions were inseparable. *Tonalité* was determined by these tones and harmonies and by the relationships between them (Ceulemans 1990).

Jumping to the mid-twentieth century, Manfred Bukofzer (1947, 12; as cited in W. Thomson 1958, 37), having experienced various post-tonal styles and experiments, defined tonality as "a system of chordal relations based on the attraction of a tonal center. This tonic formed the center of gravitation for the other chords." He went on: "It is no mere metaphor if tonality is explained in terms of gravitation. Both tonality and gravitation were discoveries of the Baroque period made at exactly the same time." The idea of tonal gravitation comes from Rameau (Christensen 1987) and was inspired by Newton; it seems appropriate, given expectations that certain tones will move in certain directions. Needless to say, there is no actual gravitation here; in a scientific approach, music has nothing to do with the motion of the planets, despite what has been believed for millennia. Psychological expectation can nonetheless be studied scientifically in its own right, even if the reasons behind the expectations cannot be explained by analogy to gravitation in physics. It is true that we expect, based on our experience of the world, that a high object will fall to the ground. It is also true, based on our experience of music, that we expect a leading tone to rise by a semitone.

When Dahlhaus pointed out that the progression of chordal roots establishes the key, he was thinking of subdominant-dominant-tonic progressions. It is true that they define and reinforce the tonic common-practice tonal music from the eighteenth and nineteenth centuries. But today, it may be appropriate to broaden the definition of MmT to include music with harmonic progressions that do not conform to that mold. Christopher White and Ian Quinn (2018) applied hidden Markov modeling to corpora of notated tonal music; on that basis, they questioned the generality of Riemann's theory of harmonic function (in which all chords have a tonic, dominant, or subdominant function) and highlighted harmonic differences between different repertoires.

Given the lack of a simple, clear definition of tonality, we need to be open to somewhat vague and multifaceted definitions. Tonality can be split into quasi-objective and subjective aspects, also called structural versus cultural. Regarding the cultural aspect, Norton's (1984) concept of mass tonality or mass tonal consciousness is appropriate from a psychological or sociological perspective:

> In none of the current literature has anybody admitted that the phenomenon of mass tonality is the single most pervading component of those forms of mass culture which in any fashion whatsoever are involved with music: specifically radio, television, film, and the making of live and recorded music. This tonality is understood on both the unconscious and the conscious level by hundreds of millions of people around the world. This tonality easily accounts for *all* the popular music of

> the last two centuries—from the waltz, brass-band music, operetta, and the Victorian hymn in the nineteenth, to American band music, ragtime, jazz, blues, swing, rock, and the Broadway musical in the twentieth. There is no form of popular music in the modern, industrialized world that exists outside the province of mass tonal consciousness. It is the tonality of the church, school, office, parade, convention, cafeteria, workplace, airport, airplane, automobile, truck, tractor, lounge, lobby, bar, gym, brothel, bank, and elevator. Afraid of being within it while on foot, humans are presently strapping it to their bodies in order to walk to it, run to it, work to it, and relax to it. It is everywhere. It is music and it writes the songs. (271)

Today, decades after Norton's book, his claims and observations are still largely valid. If anything, the number of people across the world who are engaged in "mass tonal consciousness" has increased. Pop and rock, which seem to have become even more triadic (based on major and minor triads) in recent decades (or at least more conventional; Serrà et al. 2012), are surely just as "common practice" as classical music and just as worthy of analysis—even if "traditional constructs of scale-degree theory and harmonic functionality" must be "modified to address elements of rock music that do not conform to tonal norms" (Biamonte 2010, 95). Technology has become even more portable, reinforcing the social importance of recorded music by comparison to live performance. In the academic world, meanwhile, although schools of composition have become stylistically and aesthetically more diverse, moving away from postwar modernism (Guldbrandsen and Johnson 2015), Norton's general claims remain valid. Consequently, my goal in this book is similar to that of Philip Tagg (2009): to develop a "tonal theory of what most people hear"(i).

In the West today, and increasingly elsewhere, most music is based on major and minor triads and tonalities. That is remarkable, given the musical creativity of recent centuries and the number of brilliant and influential modernist twentieth-century composers who devoted their lives to expanding, undermining, and replacing MmT. Although those many composers achieved an impressive diversity of tonal styles—each of which departs more or less, and in different ways, from MmT—MmT remains the dominant style of musical pitch organization, at least in the everyday lives of most people.

In the world of popular music (e.g., indie pop), each generation seems to reinvent or rediscover the same old chord progressions and present them as if they were new. In May 2022, I listened to the top five hits in Austria according to radio station Ö3. For teaching purposes, I noted the opening or main chord progressions that I heard (denoting a major tonic triad as I, a minor tonic triad as i, and so on in the usual way):

- *We Made It* by t-low and Miksu/Macloud: i v VI iv
- *As It Was* by Harry Styles: ii V I IV
- *Sehnsucht* by Miksu/Macloud and t-low: i VI VII i
- *First Class* by Jack Harlow: i (VI III)
- *Heat Waves* by Glass Animals: ii I vi V (IV I vi V)

Were these chord progressions stolen from Mozart? Hardly. Did Mozart steal them from his contemporaries and predecessors? No. Triadic, diatonic chord progressions of this kind are a part of Western (or perhaps now global) culture and identity. For centuries, people have been performing, hearing, and loving these harmonic clichés. At the same time, musicologists (including music theorists and music psychologists) have been trying to explain their origin and appeal—with mixed success.

Some definitions of tonality are relatively broad, while others are narrower. I am taking a middle path by defining MmT as tonality that is based on major and minor triads but which does not necessarily conform strictly to the rules of tonality developed by nineteenth-century theorists, inspired by Bach, Mozart, Beethoven, and Brahms. Tagg (2009) defined tonality similarly broadly as "the system or set of norms according to which tones are configured in any musical culture" (54); "tonality includes such phenomena as key, melody, tonal polyphony, chords, and harmony" (61). In an approach of that kind, MmT includes:

- nonclassical Western cultures such as Renaissance polyphony, heard retrospectively as progressions of mainly major and minor triads within major and minor scales (cf. Long 2018), as well as all music composed in the (long) European nineteenth century;
- most jazz, pop, and rock from the twentieth century in all its diversity, from soft rock to heavy metal, from musical to video game music; all improvisations of Thelonius Monk, all songs by Adele, all Brazilian popular music, all K-Pop;
- most modern pop, including music that is confined to a diatonic scale and treats the three major and three minor triads in that scale (e.g., C, Dm, Em, F, G, Am) as equally important (cf. Clement 2019; Ferrandino 2022);

- urban contemporary music and Black genres, including hip-hop, contemporary R&B, quiet storm, electronic dance music (EDM), reggae, Latin music, Chicano rock, and brown-eyed soul; and
- (harmonic) tonalities in diverse non-Western musics, such as African choral and a cappella styles (Ofosu and Ofosu 2020).

MmT includes diverse tonal structures beyond classical chord progressions in major and minor keys. It includes planing in the music of Debussy—parallel chords conforming to a diatonic or pentatonic scale, or parallel Mm7 chords (for examples, see Uchida 1990) and similar developments in post-bop jazz (Pamies 2021). It includes passages that are based on or confined to standard pentatonic scales or quartal harmonies (chords created by stacking 4ths rather than 3rds; Day-O'Connell 2007), insofar as perceptible links to triadic or tertian (stacked-3rd) harmony remain (e.g., a 7th being perceived as a stack of 3rds although constructed from two 4ths).

In this way, MmT can be defined as any musical structure that is primarily based on major and minor triads and scales. This definition is broad in the sense of not insisting on a (clear) tonal center, but it is narrow in insisting on the role of major/minor triads or scales, or their extensions or derivatives.

The definition excludes music that is commonly described as atonal, pantonal, or post-tonal, insofar as the boundaries of those categories can be defined, while at the same time acknowledging a continuous transition between tonal and atonal. Relative to classical major and minor keys, the pitch structures of planing, pentatonicism, or quartal harmony tend to ambiguate tonality, but if they are based on relatively consonant interval patterns, as in tertian harmony and major/minor scales, they may be included in a broad definition of MmT. The definition also includes diverse non-Western styles that have been influenced by Western MmT and regards MmT as a phenomenon that—mainly for geopolitical reasons—is now widespread outside of the West (however defined).

Different scales can be classified as major or minor, given the size of the 3rd interval above tonic or reference pitch—even if the distinction is not entirely clear. MmT therefore includes diatonic modes (Dorian, Phrygian, and so on) and most (other) jazz scales. It might even include symmetrical scales such as the diminished scale, when the scale appears in tonal contexts with tonal implications.

Many borderline cases could be listed. One might argue, for example, that Indian classical music should be included because it is based on scales that have major or minor 3rds. But that music often does not imply major or minor triads—depending, perhaps, on who is listening. In any case, my knowledge of this wonderfully evocative and complex musical tradition is far too rudimentary to justify considering it here. Other similarly unclear examples could be provided, but it is beyond my scope to cover different stylistic possibilities systematically or to define a gray zone. Instead, my aim is to focus on basic principles that might allow the various stylistic possibilities within mainstream MmT to be explained.

As a psychologist, I could operationalize my definition of MmT by saying that for most Western or Westernized people, MmT practically means the same as music. That is because most of the music that most Western people hear is MmT, as broadly defined here. For that reason, many psychological studies of music are, on closer examination, studies of MmT. "Music" is preferred if it expresses values and identity (Schäfer and Sedlmeier 2009). "Music" preferences depend on gender and self-esteem (Shepherd and Sigg 2015) as well as personality (Dunn et al. 2012). "Music" that is somewhat unusual may be preferred (Chmiel and Schubert 2019; Miles et al. 2021). One "music" preference study found the following five genre groups (factors): Intense and Electronic; Devotional and Cultural; Emotional and Melodious; Spiritual and Reflective; and Contemporary and Rhythmic; all were examples of MmT, although the study was carried out at an Indian university (Upadhyay et al. 2016).

Nicholas Temperley (1987) argued that preference for tonality is a middle-class phenomenon. I beg to differ. In my experience, MmT is surprisingly independent of class, but different social classes may prefer different styles and genres. Adrian North and David Hargreaves (2007) found that "liking for 'high-art' music was indicative of a lifestyle of the upper-middle and upper classes, whereas liking for 'low-art' music was indicative of a lifestyle of the lower-middle and lower classes" (473); of course, the music was tonal in both cases. In addition, we can say that in Western and Westernized cultures, more or less everyone likes MmT.

For modernists and lovers of contemporary music, the continuing dominance of MmT in Western music (and increasingly outside the West) is regrettable. But it also gives basic questions about MmT a quasi-timeless character. I sympathize with those who yearn for more music-structural complexity and diversity, my own musical taste being rather eclectic. For the purpose of scientific research, however, it is not my task to evaluate the continuing preference for MmT. That is simply the way things are. A music psychologist should try to explain how music works and why it is the way it is, without judging.

Prolongation

Heinrich Schenker proposed that any tone or sonority (interval or chord) can be prolonged (*auskomponiert, verziert, ausgefaltet, fortgesponnen,* composed out, unfolded, unpacked, embellished, decorated, elaborated, spun out . . .) such that a passage of music is heard relative to that tone or sonority, whether it is actually sounding or not (Forte and Gilbert 1982). Prolongation is happening constantly in almost all the music we listen to. It can happen on different hierarchic levels simultaneously, such that a component of a prolongation is itself prolonged.

Simple prolongations can be divided into three parts: a statement of a pitch pattern A, an intermediate elaboration B, and a final repetition of A (Jonas 1982). The classical sonata form of exposition, development, and recapitulation is a high-level ABA pattern that can be regarded as thematic and tonal prolongation of A. The "organic unity" of a musical work of art, as appreciated by many music theorists, including Schenker, can be seen as a form of high-level prolongation.

MmT can be regarded or even defined as prolongation of the tonic triad. The idea can be illustrated with familiar examples of diatonic music in major keys. For better or for worse, we might start with some well-known national anthems:

- The melody of *God Save the King* starts like this: $\hat{1}\,\hat{1}\,\hat{2}\,\hat{7}\,\hat{1}\,\hat{2}$, $\hat{3}\,\hat{3}\,\hat{4}\,\hat{3}\,\hat{2}\,\hat{1}$, $\hat{2}\,\hat{1}\,\hat{7}\,\hat{1}$ (using commas to separate the phrases). In the first phrase, $\hat{2}$ and $\hat{7}$ ($\hat{7}$ being a semitone below $\hat{1}$) are perceived as upper and lower neighbors to $\hat{1}$; in the second phrase, $\hat{4}$ is perceived as an upper neighbor to $\hat{3}$, and $\hat{2}$ to $\hat{1}$; and the third phrase is like the first. Later in the melody, $\hat{6}$ is perceived as an upper neighbor to $\hat{5}$. In that way, the melody can be heard as a prolongation of the tonic triad $\hat{1}-\hat{3}-\hat{5}$.
- *La Marseillaise* starts with a flourish based on an arpeggiated tonic triad, like this: $\hat{5}\ \hat{1}\ \hat{2}\ \hat{5}\ \hat{3}\ \hat{1}$. In this context, $\hat{2}$ can be heard as either an upper neighbor to $\hat{1}$ or a passing note to $\hat{3}$. In the next phrase, a low $\hat{6}$ appears,

which becomes an upper neighbor to a low $\hat{5}$. The subsequent $\hat{4}$ can be heard an upper neighbor to $\hat{3}$, although its resolution to $\hat{3}$ is delayed to the start of the next phrase.

- At the start of the German national anthem *Einigkeit und Recht und Freiheit* ($\hat{1}\ \hat{2}\ \hat{3}\ \hat{2}\ \hat{4}\ \hat{3}\ \hat{2}\ \hat{7}\ \hat{1}$), $\hat{2}$ can be heard as an upper neighbor to $\hat{1}$ or a passing note to $\hat{3}$; $\hat{4}$ is an upper neighbor to $\hat{3}$; and $\hat{7}$ is a lower neighbor to $\hat{1}$. In the second phrase, the high $\hat{6}$ becomes an upper neighbor to $\hat{5}$.

Countless tonal diatonic melodies can be analyzed in this way, from themes of classical sonatas to popular songs. Every tone in such a melody can be perceived as either belonging to the tonic triad or related to it by stepwise neighbor relations. People do not necessarily hear melodies in MmT that way, but it is often reasonably possible to do so.

The concept of harmonic prolongation is embedded in a broader music-theoretical discourse. Steve Larson (1997) explained:

> Prolongation and musical forces are both related to "stability," a term whose operational definition depends on three more terms: "auralize," "trace," and "displace." To auralize means to hear internally sounds that are not physically present. A trace is the internal representation of a note that is still melodically active. However, a trace is more than a memory. A trace can be displaced by subsequent notes. If I play one note and then another that is a step away, you may be able to recall both notes. But there is some sense in which the second note will displace the trace of the first (Komar 1971). (104)

> In a melodic step, the second note tends to displace the trace of the first, leaving one trace in musical memory; in a melodic leap, the second note tends to support the trace of the first, leaving two traces in musical memory. (105)

Larson's "displacement" is related in psychological research to auditory scene analysis and in particular to the trill threshold, according to which two alternating tones one or two semitones apart create a trill (one auditory stream), whereas they create a shake (two streams) for intervals of three semitones or more. The idea is both intuitively clear in musical practice and psychologically testable (G. A. Miller and Heise 1950). It is also related to Jamshed Bharucha's (1996) melodic anchoring.

Larson (1997) also commented that "prolongation—and only prolongation—always determines which notes are heard as stable in a given context" (112) and "the act of *hearing* a passage of music *as* containing a prolongation I call (after Salzer 1952) 'structural hearing.' Accordingly,

Schenker's term *Fernhören* means global structural hearing" (115; italics in original). Moreover, "prolongation *is* embellishment; embellishment (and only embellishment) determines the relationships between tones that make some tones of lesser and greater structural weight than others" (130, italics in original).

Any musical tone or sonority can be prolonged, provided it is perceived to be *consonant.* In some early post-tonal music (e.g., Scriabin's *Vers la flame,* Berg's song Op. 2 no. 2, Debussy's *Voiles,* Weber's song Op. 3 no. 1), harmonies related to the harmonic series were prolonged (Väisälä 2002). Much medieval counterpoint can be understood as prolongations of single tones or perfect 5th intervals—with implications for tonality in a broad sense. Felix Salzer (1967) even claimed that "by the middle of the twelfth century, the basic principle of tonality had been created. This principle can be defined as directed motion within the framework of a single prolonged sonority" (54). He continued: "Tonality is chord prolongation. This means that the horizontalization of a sonority or the motion around a sonority is the unifying element of the contrapuntal voice-leading texture. This is the common denominator within the totality of a single language expressing itself in a rich variety of styles" (98). But there is a danger of overgeneralizing. The term "tonality" should be applied with caution to music before about 1600 (Sanders 2003).

Ernst Krenek (1940, cited by W. E. Thomson 1999, 18) argued that "a tone becomes the tonic only when the central triad is built over it." In other words, the tonic is a triad, not a tone. If MmT is a prolongation of the tonic triad, every tone in the diatonic scale is perceived relative to that triad. That idea is consistent with the high correlation coefficients ($r \approx 0.95$) that have repeatedly been found among the following three independently determined vectors of twelve numbers (Parncutt 2011; see figure 1.1):

- the pitch-class stability vector of a musical key (Krumhansl's key profiles),
- the pitch-class prevalence vector of a musical passage (how often each pitch class happens in typical musical scores in MmT), and
- the pitch-class salience vector of the tonic triad (according to Parncutt 1988, 1993).

The correlation coefficients are high for both major and minor keys but sometimes higher in major case, consistent with the minor key's greater ambiguity (Parncutt 1989, 1994).

Salience versus Stability

When examining the idea that MmT is a prolongation of the tonic triad, we will compare and contrast two concepts: tonal stability in music theory and pitch salience in psychoacoustics.

In music theory, stability is about tone-by-tone expectations. If a piece of music suddenly stops on a tone, that tone's stability is its closure or finality—the degree to which we perceive it as the end of the piece. Conversely, its instability is the degree to which we expect it to move up or down in subsequent music. In music theory, familiar examples of stability and instability include scale degrees $\hat{1}$ and $\hat{7}$ of the major or minor scale—the tonic and leading tone, respectively. The tonic is stable because it creates closure at the end of a passage, whereas the leading tone is unstable because we expect it to move (usually to rise through a semitone to the tonic).

The salience of a pitch in psychoacoustics is its ability to attract the listener's attention, or simply how important it sounds—how much it stands out in our sonic experience within the musical surface. It is the perceived clarity of a pitch or the probability of noticing a tone or directing attention to it. It depends on chord voicings, voice leadings, and accentuations. Musicians are constantly manipulating tone salience in performance. The easiest way to attract attention to a tone is to play it louder; other techniques include playing it more quietly (in a surprising or unexpected way, *subito piano*), changing its timbre, playing it out of time relative to its rhythmic context, varying its intonation, and so on. Musical expression is often about using performed accents (loudness, delayed onset, etc.) to attract attention to immanent accents (harmonic, melodic, grouping, etc.; Parncutt 2003).

The salience of a voice-leading pattern in music analysis is its likelihood of being noticed. But in hierarchical-reductive approaches to analysis, inspired by Schenker, the contents of the next-deeper level tend to depend more on stability than salience: events that are tonally stable at one level are more likely to appear at the next level down (Pellegrin 2013).

In performed music, there are big differences between salience and stability. Unstable tones can be salient when accented (e.g., appoggiatura and harmonic accent). Performers often expressively attract attention to unstable elements, making them salient. Stable tones are often not salient (e.g., scale degree $\hat{1}$ as an unaccented passing tone between $\hat{2}$ and $\hat{7}$).

If we take the expression (timing and dynamics) out of a music performance and imagine that all tones are played such that they would be equally salient if sounded alone (as in a well-controlled but ecologically invalid psychological experiment), we can ask how the stability of a tone depends on context in MmT—without distinguishing stability and salience. In standard tonal theory, there is a hierarchy of stability. The tonic ($\hat{1}$) is the most stable tone; then comes the dominant ($\hat{5}$), followed by the 3rd of the tonic triad ($\hat{3}$). After that, there are the other tones in the diatonic scale ($\hat{2}$, $\hat{4}$, and $\hat{6}$), then the leading tone ($\hat{7}$), and finally the remaining chromatic tones.

Carol Krumhansl and Edward Kessler (1982) encapsulated this hierarchy mathematically by measuring and quantifying the stability of each chromatic scale step to create key profiles with twelve elements. They did that by presenting three-chord key-defining cadences to listeners, followed by individual tones, and asking "how well, in a musical sense, each probe tone fit into or went with the musical element just heard" (342). Tones corresponding to stable tones in the tonality were perceived to go well, whereas those corresponding to unstable tones went poorly.

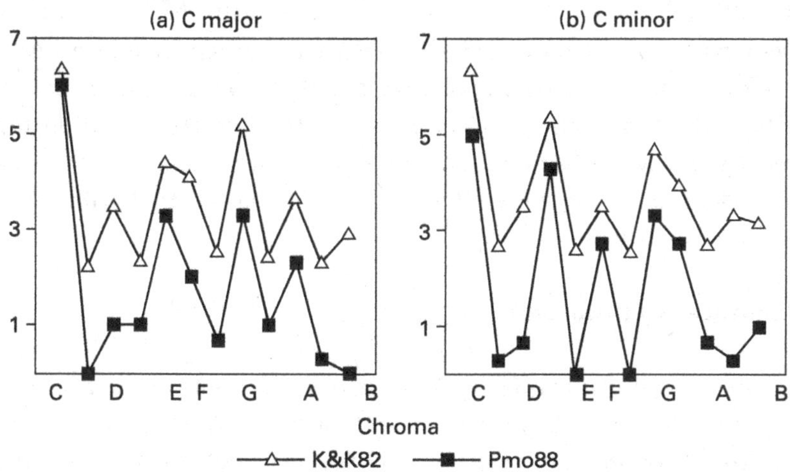

Figure 1.1
Comparison of the stability of scale degrees (key profiles; Krumhansl and Kessler 1982) with pitch salience within the tonic triad (pitch-class salience profiles of major and minor triads; Parncutt 1988 with revised root-support weights, divided by 3 for ease of comparison). From Parncutt (2011) by permission of University of California Press.

The concept of MmT presented in this book involves comparing (correlating) the stability of musical scale steps (Krumhansl's key profiles) with their salience within the tonic triad (the pitch-class salience profile of the tonic triad), as shown in figure 1.1. The pitch-salience profile is calculated using a simplified, octave-generalized version of the pitch algorithm of Terhardt et al. (1982b). The peaks in the profile are virtual pitches evoked by the tonic triad. They correspond to the fundamentals of harmonic patterns of audible partials (spectral pitches). The comparison is consistent with the Schenkerian idea that a passage of music in a major or minor key is a prolongation of its tonic triad.

The idea is not new. Some have argued that of seven pitch classes of a major or minor scale, the three that correspond to the tonic triad are consonances and the others are dissonances (Mickelsen 1977, 67; Riemann 1882; Tartini 1754). That is a good first approximation. Quantitatively, there may be little difference between the "consonance" of pitches in a major or minor scale, their tonal stability, and their salience within the tonic triad.

A triad can be performed in many different ways, including voicings (inversions, spacings, and doublings) and expressive interpretations (timbres and dynamics). When comparing the tonic triad with the key profile, we understand the comparison to involve an average or representative version of the triad, including typical voicings, and with all tones played equally loudly. Terhardt's pitch model implies that the harmonic relationships between the tones in the tonic triad cause some tones to sound more perceptually salient than others, even if all tones would have been equally salient if played separately.

Consonance and Dissonance

Consonance and dissonance (C/D) are central concepts in the history of Western music theory. If we want to understand MmT, we need to understand C/D. A sonority can only be prolonged, creating a tonal space, if it is perceived to be consonant.

The Latin roots of the word "consonance" refer to how different sounds sound together. That does not necessarily imply that the component sounds sound together in a particular way. Nor does it imply that the sounds are individual tones or groups of tones, or that the comparison is simultaneous or successive. Seen that way, the word is quite general. Nor does it necessarily apply to Western music. Although C/D may be important for some non-Western musics, it is irrelevant for others (McDermott et al. 2016). So, considerations in this book will be confined to the Western tradition. I will adopt a broad definition of C/D that includes simultaneous and successive pitch relations and encompasses the diverse approaches of different historical periods and academic disciplines.

In the humanities and the history of ideas, C/D is often understood to be categorical. Intervals are divided into consonances and dissonances, consonances are divided into different kinds of consonances, and so on. Scientists, including music psychologists, tend to think of C/D as a continuous variable or real number, which has the advantage of flexibility: no matter how consonant or dissonant a sound, it is possible to imagine a more consonant or dissonant one (depending on definition). If the terms "consonance" and "dissonance" are exact opposites, Joe Monzo (n.d.) argued that C/D should have a single name and proposed the term "sonance." But the two words can also be asymmetrical if, for example, consonance is promoted by one factor (harmonicity) and dissonance by another (roughness).

Both simultaneous and successive tones can be either consonant or dissonant. A major-2nd interval is normally considered consonant when the tones are successive (it is the most common melodic interval) and dissonant when they are simultaneous (due to roughness). In discussions of early monophonic (pre-polyphonic) music, C/D refers primarily to successive tones; in polyphony, it refers to simultaneous tones. The term "C/D" can be used in a more general sense, according to which an entire passage of music has C/D that depends among other things on its tonality and voice leading. In making that claim, I am assuming that music listeners tend to hear music holistically, drifting along with the music's emotion and imagery, allowing the music to change their state of consciousness (Herbert 2016), and without analyzing different aspects of C/D.

Researchers in psychology and psychophysics have often confused consonance with pleasantness. Sometimes, the two overlap, but not always. We may find complex jazz chords with bitonal upper structures "pleasant," although they are dissonant from a music-theoretical perspective. By the same token, we may find four-part harmony more "pleasant" than unison singing when listening to a choir, although chords are rougher than single tones according to psychoacoustic models.

Psychohistory

From a retrospective modern standpoint, music history can be seen as a process of collective creativity, in which composers approached aesthetic goals by a sophisticated form of trial and error: "The principles of sonata form, for example, which governed music of the late eighteenth and early nineteenth centuries were not enacted; they were arrived at by a trial and error process, reflected in the music of K. P. E. Bach, Johann Stamitz, Josef Haydn, and others during the period from 1750–1770" (Manns 1994, 88). Musicians were and still are trying out new sound patterns and evaluating the results. The final arbiter of what is good music is fundamentally subjective and may override existing conventions or rules, which, on the one hand, makes it possible for musical styles to evolve and, on the other, makes it difficult to study the underlying principles scientifically. One result of this historic process is that music perception changes as listeners become familiar with new or changing musical styles (Parncutt 2012).

In an attempt to do justice to this historic complexity, I will regard musical structures such as intervals, scales, and chords as psychohistoric entities. Although we can investigate their perception today in psychological experiments, and the results of those experiments depend on the physical properties of the sound, the results also depend on musical familiarity, and that depends in turn on cultural processes that span hundreds or thousands of years. Familiarity can often be evaluated by statistical analysis of musical databases on the assumption that listeners are familiar, either directly or indirectly, with the sound of the contents. In that sense, my explanations of musical elements and structures will often have a psychohistoric character, speculating about the history of music perception and its effect on the history of musical syntax. I will assume that the structure of music depends on unconscious or ineffable perceptual processes in the history of music perception, as well as the history of music theory and compositional convention.

This is not a book about the history of music theory, although there are many references to it. In any case, the history of music theory is not what I mean by psychohistory. I will not, for example, attempt to survey historic contributions to the theory of major and minor chords and tonalities systematically (cf. Parncutt 2011). Instead, I will engage with selected highlights in the history of music theory: the ancient idea that musical intervals are ratios of small integers, milestones in the history of C/D theory, and the emergence of concepts such as sonority, chord, root, progression, and tonality.

Questions about enharmonic spelling (such as G♯ versus A♭) can be seen from a psychohistoric perspective. To understand enharmonics, we need to look at both historic and psychological contexts, combining different ways of thinking in humanities and sciences. We need to ask whether and how enharmonics affect the real-time music experience of modern listeners and performers (see chapter 12).

A psychohistorical approach may be controversial. The attempt to combine history and psychology is problematic for psychologists because it is not possible to perform psychological experiments on people who died long ago. Conversely, any attempt to identify music-structural principles that apply equally to such diverse styles as medieval chant, Renaissance polyphony, common-practice tonality, jazz harmony, and even post-tonal music is problematic for humanities scholars, given the enormous stylistic differences and historic changes in both music and the history of ideas. It is nevertheless relevant that, at all times and in all places, humans have communicated acoustically using harmonic complex tones and fricatives produced by their vocal tracts. Humans also use linguistic cognitive structures that have global commonalities and are in some ways similar to musical cognitive structures. Such quasi universals in speech communication and song are consistent with a psychohistorical approach to music perception that considers historical changes in musical vocabularies, patterns, and fashions, and their relationship to unchanging perceptual universals.

2 Approaches

The task of explaining and understanding MmT can be approached in different ways. Positivism *is the confidence that central questions have discoverable answers (e.g., physical laws);* relativism *is the nagging doubt that positivist claims are impossible or problematic (cultural arbitrariness). Can we pursue a middle path? This book aims to identify psychoacoustic foundations of MmT, raising general issues about the foundations of academic disciplines. Relevant truths may be limited to domains such as the physical universe (the domain of physics), life on Earth (biology), Western culture (cultural studies), and MmT. The study of musical structure can be* descriptive *in the tradition of music psychology and cognition, describing structures while avoiding value judgments, or* prescriptive *in the tradition of music theory and analysis, with implications for good (or bad) music, or good (or bad) analysis. An* ecological *approach to music perception focuses on interactions between people and their physical and social environments, which involves* categorical perception *as the organism extracts useful information from incoming signals. Questions about MmT can appeal to either* nature *or* culture, *or a mixture of the two. To deal with this complex web of contrasting approaches, we need to combine insights from different relevant academic disciplines. Clarity can be achieved by preferring simpler explanations to more complex ones (principle of parsimony), whenever that is realistically possible.*

Questions

This book is motivated by general questions about MmT and the belief (perhaps naive) that a modern interdisciplinary approach is increasingly making convincing answers possible. The task is to understand how MmT works, where it comes from, and what general principles lie behind its familiar pitch-time patterns. While trying to address these issues, we will be confronted with more specific questions:

- Can musical subjectivity be measured? Can we learn something about musical structure by studying human behavior or the brain?
- Is there such a thing as a history of music perception or a psychohistory of music? What can statistical analysis of musical scores tell us about it?
- Did musicians and listeners of the past become less sensitive to the spectral structure of the musical stimulus as music became more complex due to the increased cognitive load? Did early listeners perceive music-structural elements or implications in musical passages that we no longer perceive today? Did psychohistoric effects change the statistical frequency of occurrence of relevant pitch-time patterns, which in turn changed how music is perceived today?
- Does the root of a chord have psychological reality, or is it merely a musical convention? Are the pitches predicted by psychoacoustic models at missing fundamentals, such as the missing root of a jazz voicing, perceptible in musical contexts? How might those pitches affect our perception of those contexts?
- Is it musically realistic to define the chromatic scale psychologically as twelve approximately equally spaced pitch categories within a slightly stretched octave? To what extent or in what way does such a scale have psychological reality? How old is this psychological schema?
- Is a musical interval a physical, mathematical, psychological, or cultural entity? What role, if any, do Just intonation and Pythagorean tuning play in music performance, past and present?
- To what extent does conventional music notation reflect the perception of a musical passage, beyond its primary role as instructions for performers? Are enharmonic spellings mere notational conventions or are they crucial for analyzing and understanding a musical passage?
- To what extent, and in what way, does Schenker's concept of prolongation have psychological reality? Do Krumhansl's key profiles represent expectations for musical continuation that are based on music experience, or do they represent the relative perceptual salience of chromatic scale steps within chords and chord progressions?

Questions of this kind can be answered in different ways. The following sections consider possible approaches.

Positivism versus Relativism

Many humanities scholars regard questions of the above kind as positivist—formulated as if clear answers exist. In their view, the music psychologists of today—or at least the author of this book—are naively continuing a millennia-old tradition, looking for nonexistent simplistic or scientific explanations for phenomena as complex and culture-dependent as musical structure and perception.

Musicologists and music theorists who reject or question positivist approaches may prefer instead to focus on the more realistic task of understanding the ideas and music of historical composers and music theorists in their social context. Music theories are not generally valid—they are valid for a specified style of music or stylistic range. With this insight, the question of whether historical ideas are true from a modern scientific viewpoint becomes unanswerable or irrelevant.

But in what sense might music theory approach truth at all? Nicholas Cook (2001) commented: "It seems to me that a qualified relativism provides a principled way to pass through the horns of the dilemma, in which we recognize the constructedness and contingency of our critical musical sounds (and sights, and other embodied experiments). In his *Harmonielehre,* Arnold Schoenberg speculated about the existence of immutable, exception-free laws governing music, but if there are such things, he said, 'I believe they will not be discovered very soon'" (185).

What, then, is the truth content of today's music theory? Is it purely intuitive, or is it empirically based in the scientific sense of music cognition research? Many past music theorists have attempted to explain MmT; their explanations make up a large part of the history of music theory. Assuming for a moment that their questions were answerable (an assumption that relativists regard as impossibly positivist), what if they failed because of misunderstandings transmitted from one generation to the next for hundreds or even thousands of years?

As the empirically founded discipline of music psychology (or music cognition) advanced and grew in the late twentieth century, it became increasingly reasonable to correct such misconceptions. But doing that seemed inappropriate in the intellectual climate of late-twentieth-century music theory because it suggested again that true answers to questions about musical structure and musical emotion might be possible. Music theorists had realized that music theory can seldom be true in the sense that physical theories might be true (in all times and all places). The mystical, positivist strivings of music theorists of the past, although understandable in their historical and social context, may have been just as wrongheaded as later scientific attempts to improve or replace them. There are also musical fashions: just as late-twentieth-century schools of composition unquestioningly promoted musical modernism, music theorists promoted analyses of the (quasi-)atonal repertoire that emphasized nontonal forms of musical organization, in which theories of the origin of consonant or tonal sounds had no place.

Relativism can be taken too far, as Alan Sokal (2010) scandalously demonstrated when he published a hoax article (Sokal 1996) in a postmodernist academic journal. The article ironically addressed the "transformative hermeneutics of quantum gravity." The reviewers who recommended publication were attracted by the idea of exposing the arbitrary cultural foundations of physics, thereby undermining the whole academic enterprise. What they failed to realize is that a lot of physics is in fact true in an absolute sense, and indeed it would be foolish to believe otherwise. If you jump off a ten-story balcony and hit the concrete pavement, the laws of physics inevitably mean that you will die—regardless of whether you believe in those physical laws, or how you have constructed their formulation in a subjective human cultural context. Certain positivistically formulated truths really are valid. If, for example, humanity continues to pump greenhouse gases into the atmosphere, we will eventually destroy ourselves, along with many other species. Unfortunately, there is no doubt that is true.

While positivist music theory may have fallen out of fashion, positivism remains an important motivator. Why devote years of your life to answering a question if, in the end, there will be no clear answer? Music theorists of the past often believed in clear answers to questions about musical structure, and some even suspected their answers would be valid in all times and all places. Their quest was to discover those principles. Those who realized this goal was unrealistic strove instead for an appropriate balance between positivism and relativism.

This book similarly acknowledges the importance of combining approaches from both the more relativist humanities (music history, music aesthetics, cultural studies) and the more positivist sciences (music psychology, neuroscience, acoustics, computer science). The discipline of cognitive ethnomusicology, understood as the psychological study of musical practice (Moisala 1995), can be an interesting model if it achieves an appropriate balance between positivism and relativism. Ethnomusicologists strive to understand music relativistically, in its cultural context and from the point of view of its practitioners rather than from a Western or quasi-universal viewpoint—as far as that is practically possible and remaining aware of the limitations. Many ethnomusicologists are also open to positivist approaches from the sciences (acoustics, psychology, computer science) and manage successfully to combine insights from sciences and humanities.

The book is positivist in its goal to identify theoretical foundations of MmT that have been relatively stable for the past few decades or centuries and are likely to remain so in coming decades or centuries. Feeling some solid ground beneath our feet makes the time and energy we put into research and study seem worthwhile. The book is also relativist in regarding all music theory (including its own) as historically and culturally contingent. Music theory can only be understood in its social context, which includes cultural, economic, historical, linguistic, religious, and ethical aspects.

In positivist psychology, quantitative research methods are considered intrinsically superior to qualitative ones. If the aim is to find a true answer to a question, a statistically significant empirical finding is required. That can explain the widespread use of rating scales with relatively many points in psychological experiments. In a balanced multivariate design, significant effects can then be found by analysis of variance. But empirical research in psychology can also be qualitative: people with different expertise or backgrounds can be asked to introspect about important issues, and their statements can be analyzed. Given the important role that such studies can play, we cannot claim that quantitative methods are inherently superior. Instead, the best method for a given question depends on the question. Qualitative approaches are more important in exploratory work, laying the foundations for later quantitative studies. Qualitative studies play little role in the present book, but for a different reason: the book is about musical pitch, seen as a quantifiable variable in both cognitive and psychoacoustic approaches.

In recent decades, academic schools and departments of psychology have prioritized quantitative methods, and the idea that quantitative methods are inherently superior has been maintained and promoted by leading psychological journals, conferences, societies, and departments. The idea is problematic if it reflects a reluctance to engage with complementary or contrasting approaches from the humanities. To understand the structure of music, music psychologists need to work with music theorists and vice versa and acknowledge that the thinking of today's music theorists has a valuable and indispensable qualitative element. To get to the bottom of questions about musical structure, we need to understand and combine quantitative and qualitative approaches.

Qualitative methods can allow competing theories to be compared. This book presents different theoretical approaches to MmT that can account quantitatively for empirical data on its perception and cognition. What if two different quantitative approaches account equally well for a given data set? What if the correlation coefficient between predictions and empirical data for two competing theories are not significantly different? In such cases, qualitative considerations become important. We must then think carefully and systematically about the underlying ideas.

Academic Foundations

Academic disciplines in humanities and sciences have foundations that undergraduate students are expected to learn—aspects upon which most representatives of a discipline agree or upon which there is a reasonable consensus. The modern foundations of physics include basics such as classical mechanics and associated mathematics. They also include more slippery issues such as reconciling classical gravity with Einstein's four-dimensional concept of space-time in general relativity. The modern foundations of theology might involve not only close readings of biblical texts and related literature, but also discussions about the advantages and disadvantages of different approaches to their interpretation. Leading representatives of a given discipline may differ in their selection of foundations for undergraduate teaching—materials that are unlikely to change much in the next few decades, saving a major and unexpected paradigm shift.

What are the foundations of Western music theory? Students need to know rudiments of music notation and diatonic structures, and they need to be able to hear those structures (ear training). They should be able to read music in different clefs, notate diatonic triads and standard cadences in all major and minor keys, play these fragments on a keyboard, and transcribe simple melodies and chord progressions by ear. Knowledge of non-Western, historical, or instrument- or style-specific notations is a bonus.

But skills of that kind are surely not the *foundations* of the academic discipline of music theory. Nor is pitch-class set theory, which is primarily a classification system. Schenkerian analysis has foundational character, but there is little consensus about its central tenets; for example, few psychologists regard the *Ursatz* as psychologically real. The history of music theory contains an enormous wealth of ideas, but it would seem overly positivist to ask which of those many theories have foundational character.

The foundations of a theory of MmT should perhaps answer (or try to answer) questions of the following kind (cf. Eberlein 1994): Why are the pitch-time patterns of (tonal) Western music like they are and not quite different? Where do they come from, and why do certain patterns happen more often than others? Music theorists have struggled with these questions for millennia. In the nineteenth century, there was a feeling that convincing answers were getting closer, inspired by the glory of the European tonal canon. Confidence collapsed in the early twentieth century with the rise of post-tonal modernism, coupled with promising developments in comparative musicology and ethnomusicology. (At around the same time but for different reasons, physicists also realized that a theory of everything was out of reach; the search continues.) Clearly, the world of musical structures was much more diverse than an initial study of Western MmT would suggest. MmT could no longer be regarded as a kind of universal reference or evolutionary climax, just as the most successful pale-skinned male composers could no longer be regarded as geniuses. The shock of Nazism and the Holocaust and their implications for German music and *Musikwissenschaft*—previously seen as central and fundamental for the musicological project in all cultural contexts—hammered the final nails into musicological dreams of cultural superiority.

In that context, fundamental answers to music-theoretical questions may seem farther away than ever, and relativism more necessary than ever. The relativism of today's music theory is part of its identity; the same applies to the humanities as a whole. But extreme relativism can also hinder the development of a canon of fundamentals that might plausibly answer basic questions about musical structure. Students will then be presented with diverse ideas that lack a solid foundation.

Another reason for a perceived lack of music-theoretical foundations might be interdisciplinary barriers. Despite progress in some quarters, there is still relatively little constructive communication between music theorists and music psychologists. If music theory has foundations, those two academic groups, representing relevant humanities and sciences, respectively, should more or less agree about them. But as chemist/novelist Charles Percy Snow (1959) wryly observed, humanities scholars and scientists seldom agree about anything. He may have been exaggerating, but there is indeed a persistent lack of deep communication between humanities and sciences, and it is evident from the programs of conferences and the contents pages of journals in each area. It is as if we inhabited two different worlds. This practical separation means that many humanities scholars in the area of musical structure have missed the chance to take on board the most interesting ideas of the music psychologists and consider their implications within a humanities context. Similarly, many music psychologists are unaware of issues that humanities scholars consider important and discuss in great detail—issues that could inspire future scientific investigations.

One aim of this book is to bridge these epistemological gaps within research on musical structure in an attempt to formulate something approaching music-theoretical or music-psychological foundations. But bringing together humanities and sciences can be a risky enterprise. If experts in humanities and sciences often disagree about fundamental issues, they will surely disagree about this book. Music psychologists will regard some of the ideas as dated, although some of the musical applications are new. Music theorists may suspect that my brush is too broad—my claims are too general and do not sufficiently acknowledge musical diversity. My aim is indeed to explore what can be said about MmT in a broad definition, which necessarily involves skimming over details of specific repertoires and exploring new versions of the big picture.

Within this complex historical and intellectual context, the book will summarize leading psychological research from recent decades on the psychoacoustics of MmT in a critical approach that acknowledges different perspectives. Psychoacoustics will be regarded as a scientific discipline in the sense of empirical research about relationships between physical properties and subjective experience of sound. At the same time, the pitch, loudness, and timbre of musical tones will be regarded as subjective phenomena that depend in complex ways on physical parameters. The incorporation of subjectivity as an intrinsic and inherent component of the theory will establish a link between empirical scientific research and artistic subjectivity.

Domains of Interest

Different music theories reflect different domains of interest. A domain of interest could be a stylistic category, a cultural group, or a historical period.

It could be specific, such as the mature works of composer X, or general, such as all music ever. Often, the domain is implied rather than stated.

The domain of MmT is unclear. On the one hand, MmT is a culture-specific phenomenon. On the other hand, it is globally popular, when defined loosely as music based on major and minor triads and/or keys. To take one of countless examples: Japanese children who are exposed to both Western and Japanese music acquire the tonal schemata of Western tonality, the politically dominant style, more quickly (Matsunaga et al. 2020). The broader historical context is one of colonialism and the bloody conquest of exotic cultures by nineteenth-century Europe (Agawu 2009). Given this tragic and in many ways deeply embarrassing history, it is important to reject universalism. As I will attempt to show, it is possible to do that while at the same time acknowledging both psychological and historical aspects of the problem.

With these points in mind, the theory presented in this book addresses four domains of interest, from general to specific:

(1) The physical level

The most general level is the physical universe and the physical principles that can explain it. In the nineteenth century, Newtonian classical mechanics seemed like the key to such universal understanding. It also explained musical acoustics, including the physics of periodic vibrations and the mathematics of spectral analysis. But it became clear in the early twentieth century that classical mechanics did not apply to objects that are very small (quantum mechanics) or travel very fast (relativity). Classical mechanics nevertheless remained an indispensable foundation of physics and engineering, as well as musical acoustics.

(2) The biological level

The second level is planet Earth and the diverse life that it supports. This level includes universals of human biology and physiology, which in turn depend on how humans and other animals interact with their physical and social environments. Of musical relevance is the anatomy and physiology of the vocal tract (relevant for singing and speech), the physiology of body movements (relevant for musical gesture and dance), the separation of high and low frequencies on the basilar membrane in the inner ear, and the recognition of perceptual patterns by neural networks in the brain. Humans share these musically crucial features with other primates and, beyond that, with other mammals and other animals; a relevant difference is the lower larynx in humans (Fitch and Reby 2001).

(3) The Western level

The third level is the European (or Western) culture of the past millennium. Vocal polyphony emerged in the Notre-Dame school in twelfth-century Paris and developed in different directions over the centuries, driven by music's religious and secular functions in combination with culture-specific principles such as the contrast between consonance and dissonance. The music that we are addressing in this book cannot be separated from its specific cultural and historic context.

(4) MmT

The fourth level is that of specific periods or styles within tonal "common practice" or MmT. At this level, we can ask: What principles (acoustical, psychological, or cultural) might co-determine those styles? For example, what is the difference between the harmonic language of jazz and mainstream classical music from the perspective of the psychoacoustics of pitch and consonance/dissonance (C/D)? What broader social or psychological principles can (at least partly) explain such distinctions?

The music whose structure we want to understand in this book exists on all four of these levels: physical, biological, Western, and tonal.

(1) Physics

Sounds in the everyday human physical and social environment are highly complex and constantly changing. Environmental sound sources produce a mixture of harmonic, nonharmonic, and almost-harmonic spectra. The sounds that reach the ear from vibrating objects are generally superpositions of direct and indirect sound. These simple observations have music-theoretical consequences: they can explain why animal ears (including human ears) are constantly carrying out a kind of running spectral analysis, and why animal auditory systems are so sensitive to pitch-time relationships (chapter 6). If things were not like that, music as we know it would be totally different or impossible.

(2) Biology

The biological aspect includes the physiology of human language. Understanding speech involves tracking the fundamental frequency of voiced sounds and the relative amplitudes of partials and non-voiced sounds (e.g., in fricatives). That happens quickly, as speakers articulate one phoneme after the other. These linguistic abilities are presumably evolutionary adaptations that helped pre-sapiens hominids survive and reproduce. They also belong to the foundations from which musical melody and polyphony emerged. The perception of musical timbre is similarly related to the perception of speech phonemes.

From an evolutionary viewpoint, complex music-cognitive processes may be byproducts, taking a piggyback ride on neural mechanisms that presumably evolved for processing speech. But the reverse is also possible: aspects of speech may have piggybacked on aspects of music (Altenmüller et al. 2013; Fitch 2006). The remarkable musical sophistication of infants (Trehub 2003) is of interest, both in itself and insofar as it helps adults understand their musical experience. Infant behavior, like all behavior, depends on a mixture of nature (genetics) and nurture (learning), which includes twenty weeks of prenatal auditory and motor learning (Ullal et al. 2013).

(3) The West

This book focuses on Western polyphony—not African (Arom 2004) or Asian (Wang 2015); the author lacks the necessary expertise to be more inclusive. Over centuries of creative experimentation, Western polyphony morphed into MmT. Exactly how that happened is a fascinating story. A polyphonic passage can often be reduced to a chord progression, suggesting that both harmony (vertical aspects, relative to the score) and counterpoint (linear or horizontal) play important roles.

(4) MmT

Since the seventeenth century, music theorists have agreed that the most commonly heard chords in MmT have roots, relative to which the other tones can be perceived. From a modern psychoacoustic viewpoint, the perception of musical root progressions is comparable with fundamental frequency tracking in speech. Sonorities in MmT also differ in C/D, more dissonant sounds resolving onto consonant sounds.

By combining approaches from these four domains, we can ask how MmT works—why, in a given tonal style, some pitch-time patterns are preferred over others. In attempt to answer this broad question, I will do my best to accommodate diverse epistemologies, while also admitting that my approach could be biased by my background as pianist, physicist, psychologist, psychoacoustician, and music psychologist. My academic background in science and psychology motivates me to ask what is "going on" in the "mind" of the average music listener (the inverted commas reflecting the difficulty of operationally defining such concepts), whether consciously or unconsciously. This question cannot be adequately answered without also considering music from historical, theoretical, analytical, and practical perspectives.

Description versus Prescription

The approach to understanding musical structure in this book is primarily descriptive—not prescriptive or suggestive (D. Temperley 1999). Descriptive approaches tend to be empirical, exploring what music is like by observing it. Leonard Meyer (1973) explained: "Because music fascinates, excites, and moves us, we want to explain, if only imperfectly, in what ways the events within a particular composition are related to one another and how such relationships shape musical experience" (17; as cited in D. Temperley 1999, 68).

Prescriptive approaches are more intuitive, based on the intuitions of performers, composers, or improvisers. They indicate or imply what music *should* be like—how we should hear, analyze, perform, or compose it. Colleagues in music theory and analysis often adopt a prescriptive or suggestive approach because they want to explore different ways of hearing a piece, comparing those ways of hearing with the (possible) intention of the composer. They want to evaluate aesthetically different ways of hearing musical structure and are writing for an elite audience of highly trained colleagues in the field of music theory and analysis. John Rahn (1980) clarified the aim of prescriptive analysis under the evocative heading "The great adventure": "To analyze music is to find a good way to hear it and to communicate that way of hearing it to other people. Probably there is no single best way of hearing any given piece, but some ways are generally recognizable as being better than others. Each person makes his own decision as to the degree and quality of insight communicated to him by a particular analysis of a particular piece. If these individual judgments did not tend to converge into clumps, music theory would be even more anarchical than it is. An essential ingredient for this clumping process is willingness to pursue actively someone else's truly alien insight" (1). The argument is convincing, but I will nevertheless avoid looking for a good way to hear music. Instead, I will try in this book to describe what music sounds like in the perception of most people, with or without musical training.

The descriptive approach of music psychologists can lead to misunderstandings. A music theorist, reading a psychologist's description of how music is perceived, may get the impression that the psychologist is saying music *should* be perceived that way. But the psychologist may be simply describing how the music *is*, without any value judgment—and trying to be objective about it, to the extent that that is possible.

The focus in this book on MmT is not a value judgment. MmT is not better or worse than other ways of structuring music. From a psychological perspective, MmT is interesting because so many people listen to and perform music of that kind. There may be reasons for that, of course, and they too are interesting, but an academic interest in MmT is initially independent of the reasons for its popularity. I will therefore tend toward a descriptive approach that focuses on quantifiable subjective experience rather than more suggestive aspects such as aesthetic evaluation. I will assume an important role for perception and experience, including categorical perception, seen as an aspect of the conversion of musical sound into cognitive representations that happens when we listen to and understand music. But by adopting a descriptive approach, I am not rejecting prescriptive or suggestive approaches. I agree with David Temperley (1999) when he wrote, "My aim is *not* to recommend descriptive theory over suggestive theory, or to denigrate suggestive theory in any way" (79, italics in original).

A descriptive approach does not imply passivity on the part of either the researcher or the music listener. On the contrary, I will be assuming a performative listening attitude. Listeners do not passively respond to the tonal relations in music. Instead, they constantly try to make sense of and interpret what they hear in the context in which they hear it. They actively focus attention on different parts of the musical structure. From one moment to the next, they expect certain continuations to happen based on a mixture of nature (here, the pitch content and structure of music just heard) and nurture (their experience of similar musical patterns in similar musical styles).

Listeners also actively construct music's psychological and social meanings—political, philosophical, religious. According to Andrew Chung (2019), "through the performative theory of musical utterance as efficacious action, [I propose] a unified theory of musical meaning that bridges gaps between musical reference on the one hand, and musical effects on the other. It offers a way to understand musical meaning in ways that are deeply contextual (both socially and structurally): imbricated with the human practices that not only produce music but are produced by it in the face of its communicative capacities" (abstract). Along similar lines, Chris McRae (2020) explained: "Performative listening is a way of theorizing and practicing listening as an embodied and sensory act that is constituted by and constitutive of relational, cultural, and historical configurations. As a critical and reflexive way of engaging in interpretive research and encountering others, performative listening works towards four commitments: listening with curiosity, listening with and to the body, listening for context and location, and listening with accountability" (399).

As an example of the descriptive-prescriptive dichotomy, consider the use of pitch-class set theory (e.g., Forte 1973) to analyze and understand MmT. The original intention of pitch-class set theory was to label pitch combinations in atonal and pantonal music in a consistent and systematic way that accounted for octave equivalence and was independent of enharmonic spelling. The theory emerged from a humanities-based tradition of partly suggestive music theory and analysis: when authors of musical analyses invoked pitch-class sets, they were often implying that the listener *should* parse the music into those structures. The claim (often tacit) could even be that the music can only be understood appropriately if the listener hears it in a certain way. In this book, pitch-class sets will instead be used descriptively as a useful way of systematically enumerating and labeling all possible pitch patterns within given constraints. Within that universe of possibilities, pitch-class sets will be used to identify patterns that typical listeners *might* normally hear.

Related to the descriptive-prescriptive dichotomy, the ways in which music is perceived, as described in this book, are assumed to be typical among listeners who are musically enculturated but not necessarily musically trained, keeping in mind that those people may perceive the structures in question neither consciously nor deliberately. We ask: Is the pitch, melody, or chord in question aurally relevant? If so, to what extent, and how likely is it that a music listener will attend to it consciously or that it will contribute significantly to musical experience? (In this book, the word "audible" is often a shorthand for "aurally relevant.") Another central question is whether the described general principles of music perception are valid, at least for the average listener. If the described perceptions are indeed typical, we musicians and music theorists should be able, if we wish, to listen consciously and intentionally in the described ways.

In the descriptive tradition of music psychology, the book focuses on empirical studies. It then compares their results with the musical intuitions of music theorists (including the author). Each approach (intuition and empirical studies) has its pros and cons, and each can enrich the other. Both music psychologists and music theorists, working alone, can miss important points about musical structure or perception, or misleadingly emphasize some aspects at the expense of others. Psychologists investigating melodic expectation have investigated differences between leaps and steps, or between continuations in the same direction and reversals (Narmour 1990), whereas historical theorists of melodic structure such as Heinrich Koch (1782) were more interested in the hierarchical structure of phrases, such as the difference between antecendents and consequents. When Huron (1991) showed that J. S. Bach tended to avoid simultaneous perfect 5th and octave intervals in his polyphonic works such as fugues but not in homophonic works such as chorales, the idea may seem obvious in retrospect, but without a computer-database analysis, music theorists may never have formulated such a statement. Without Huron's descriptive approach, the idea may never have become explicit.

Ecology

Consistent with music's cultural embeddedness, empirical studies of music cognition should as far as possible be ecologically valid. They should investigate how people behave in real, familiar musical situations. They should avoid creating strange new artificial situations, or asking participants to do something that lies outside their domain of (musical or other) expertise. David Butler (1992) highlighted the mistakes that empirical psychologists and psychoacousticians can make:

> Thus, we find ourselves facing what is probably the most troublesome methodological dilemma in music perception research. The more control we gain over the auditory stimulus, over the listening environment, and over the response task, the more music we must take out of both stimulus and response. Musicians in their native habitat seldom encounter sine tones, and musical sounds are anything but steady-state. Musicians tend not to perform their music in anechoic chambers. Musical tones are rarely played or heard—whether sequentially of

> simultaneously—out of the context of other tones (or the mental images of tonal contexts supplied inferentially by the listener). Music and musical behavior are extremely complex. (6)

An ecological approach to understanding musical structure focuses on musical interactions between human actors and their physical and social environments. Musical structure depends on both psychology (auditory perception, experience, and memory) and physics (the physical structure of musical sounds). Experience is crucial: music is art, and art depends on subjectivity.

Only perceptible sounds are relevant. If you cannot hear some aspect of a musical structure—whether it be a single tone, a melodic or rhythmic pattern, or some more abstract structure—or if the structure has no aesthetically interesting effect on your musical experience, then (at least from a phenomenological perspective) that structure can hardly be relevant for questions of how the music works. Fred Lerdahl (1992) criticized pitch-class set analyses of Boulez's *Le Marteau sans maître*, explaining that if you cannot hear a pitch structure such as a mathematical manipulation of a tone row, or if it doesn't affect your musical experience in an interesting way, it is hardly musically relevant, except insofar as it may have inspired the composer. Terhardt et al. (1982b) made a related point. If a partial within a complex tone cannot be heard at all due to masking by other sounds, it is aurally irrelevant. It cannot, therefore, contribute to the sound's perception or the listener's auditory experience. If a listener is presented with two successive sounds that are almost identical, the only difference being the physical presence or absence of a specific pure tone, and if that listener cannot hear any difference, then the pure tone is inaudible and hence aurally irrelevant. That being the case, a scientific theory of musical structure should first establish what structural features are aurally relevant before proceeding.

I will be assuming that subjective auditory experience depends on interactions with real environments and their affordances. An affordance is a concept from ecological psychology: how I can use something, or what use it is for me. An environment's affordances are the possibilities it makes available to the perceiver. An affordance is a "resource or support provided by the environment to an organism, furnishing or affording the organism with an opportunity to act in a particular way" (oxfordreference.com). A melody or piece of music is an affordance if I enjoy it (hedonic affordance; Golembiewski 2013) or it reinforces my cultural identity (cultural affordance;

Einarsson and Ziemke 2017). Seen in that way, affordances can explain why people are motivated to engage with music.

In an evolutionary approach to psychology, affordances promote survival and reproduction, directly or indirectly. Put simply, the (neuro-)physiology of music can explain important aspects of *how* music works, whereas social and environmental interactions are more likely to explain *why* music is like it is and not different. Affordances can explain the origins of musical phenomena by clarifying their functions.

Subjective experience often depends on affordances. The things that we perceive consciously in everyday life tend to be the things that are important for our survival and reproduction. Pitch is useful if it carries important messages as part of speech prosody (Frick 1985), whose function is interpersonal communication. The link is illustrated by autism, in which difficulties or atypicalities in emotional social communication correlate with difficulties in vocal pitch perception (Schelinski and Kriegstein 2019). In that way, affordances can explain the importance of pitch in music. In speech, prosody informs the listener about the emotion, intention, and attitude of the speaker. Given the link between emotional communication in speech and emotional communication in music (Juslin and Laukka 2003), the importance of pitch for speech is reflected by the importance of pitch for music.

In this book, I will tend to avoid or sidestep neurophysiological explanations for the perception of pitch and musical structure, on the assumption that they are not well enough understood for that purpose. Instead, I will assume that subjective experience (including musical experience) is determined mainly by environmental affordances. Subjective experience is a highly complex emergent property of neurophysiology (see chapter 4 on metaphysics); the psychophysical relationship (the relationship between acoustical measurements and subjective experience) is simpler and easier to apply in music theory than relationships between neurophysiology and acoustics, or between neurophysiology and experience.

Categorical Perception

Like other aspects of everyday environments, musical pitch and time are perceived categorically (Burns and Ward 1978), and the categories are hierarchically structured (cf. Krumhansl 2000). Colors are perceived as red, green, or yellow; within those categories, there are shades of red, shades of green, and shades of yellow. Walking down the street, we perceive objects in categories: cars, people, trees. Within cars, we distinguish small from large; within people, women from men; within trees, bushy from tall. Upon meeting people, we perceive facial expressions categorically as expressing different basic emotions (Etcoff and Magee 1992).

When musical pitches are perceived in semitone categories in the chromatic scale, they are also spontaneously perceived in other categories (e.g., relatively high tones, or tones played by the trumpet), and those categories are hierarchically related to other categories, such as the scale of E♭ major or the sound of brass instruments.

In speech, categorical perception distinguishes phonemes; the ability to discriminate sounds that are almost identical can improve near categorical boundaries (Liberman et al. 1957). In music, tones are perceived as belonging to scale steps (even if the listener cannot name those steps): "Intervals are more like islands than single ratios, and melody generation is about navigating between these islands to maintain the distinguishability of pitches over the course of a melody" (Pfordresher and Brown 2016, 17). Tones are also heard as either in tune or out of tune. In this way, categorical perception differentiates perception from sensation, sensation being a more primitive and less differentiated feeling for the surface of a stimulus.

Across cultures, intervals and scale steps are psychological categories that are learned by exposure to music from that culture and develop gradually in the context of multigenerational oral-aural traditions. Often, those scale steps are represented in music notation. The pitch-time categories of Western music correspond to notational symbols, such as those for C♯ or for quarter notes. But singers are usually not aware of the category labels as they sing; nor do listeners hear the labels. An exception is the musician with good absolute or relative pitch who is listening analytically.

There is an interesting analogy with vision. The color red corresponds to a range of wavelengths (approx. 620–750 nm), just as the tone C♯ (in a given context) corresponds to a range of frequencies (roughly the standard frequency of C♯ plus or minus a quarter tone). But color categories are largely determined by the physiology of the retina (including cone cells that are sensitive to long, medium, or short wavelengths), whereas pitch categories are entirely learned. There are no musical intervals inherent in the physiology of the basilar membrane; musical scales are cultural products.

A theory of MmT that emphasizes categorical pitch perception is consistent with an empirical psychological or ecological psychological approach. It is also one that questions the role of Pythagorean ratio theory (see chapter 10).

Nature versus Culture

Nature is the physical and biological world as it existed before humans started changing it. Culture is not so easy to define. Wikipedia, consulted on August 26, 2022, defined culture as "an umbrella term which encompasses the social behavior, institutions, and norms found in human societies, as well as the knowledge, beliefs, arts, laws, customs, capabilities, and habits of the individuals in these groups."

Explaining the foundations of MmT inevitably means taking a stand on nature-versus-culture issues. Consistent with the Wikipedia definition, I will assume that musical structure is primarily cultural in nature. Music is not somehow essentially natural, as many music theorists of the past have implied, although some aspects are based on natural phenomena. Rameau argued that the aesthetic qualities of musical chords are grounded in their similarity with the harmonic series as it occurs in nature (Christensen 1993)—an idea that resonated with Schenker (Cook 2007). The idea is not incorrect, but it can be misleading. Although the sounds of nature can be heard as music, what we normally call "music" is usually entirely artificial or anthropogenic.

We can also regard the harmonic series as embodied. In our everyday experience, the harmonic series is mainly produced by the human vocal folds—part of the natural human body—as they vibrate within voiced speech sounds. That implies a strong and perhaps primordial link between musical harmony and speech. The importance of the harmonic series for MmT (and for the theory presented in this book) is consistent with an embodied approach to music perception, in which the human actor is seen as both a biological and a social entity with whom one interacts socially, just as one interacts with music.

Despite these links with nature, it can be misleading or counterproductive to claim that one musical structure is more natural than another. A major triad or tonality is hardly more natural than its minor equivalents, given that all such structures are ultimately cultural (Athanasopoulos et al. 2021; Lahdelma et al. 2021). Western music is often about producing beautiful sounds—whether to praise a god or to please audiences, and whether those audiences are aristocratic or popular—and the reasons why musical sounds are perceived as beautiful typically involve a mixture of nature and nurture.

Parsimony

Some ideas and theories in this book will be criticized for being too simplistic. But I have a good reason for favoring simpler theories. The principle of parsimony or economy of explanation suggests that given a choice between a simpler and a more complex theory, we should prefer the simpler one, insofar as the two work equally well.

The principle is attributed to the fourteenth-century empiricist philosopher and theologian William of Ockham, hence the term "Ockham's razor." He argued that "Plurality is not to be assumed without necessity" and "What can be done with fewer assumptions is done in vain with more" (Edwards 1967, 307). If two explanations exist for the same phenomenon, the simpler explanation is preferable because it is easier to falsify; if it cannot be falsified, we can have more confidence in its validity. If we have a choice between two models or algorithms for modeling a cognitive phenomenon in psychology, we should prefer the simpler model, other things being equal: "Choose the model that provides a sufficiently good fit to the data in the least complex manner, thus ensuring good generalizability" (Pitt et al. 2002, 487; see also Honing 2006). Simpler models are easier to evaluate, improve, and apply across disciplines. Even Schenker (1935/1979) seemed to apply the principle: "My concepts show that the art of music is much simpler than present-day teachings would have it appear. However, the fact that the simplicity does not lie on the surface makes it no less simple. Every surface, seen for itself alone, is of necessity confusing and always complex" (xxiii).

Parsimony is particularly important when trying to answer apparently simple questions. Take this one: Why is Western music usually based on a seven-tone diatonic scale? A parsimonious answer might run like this. If we stack seven 5ths and transpose higher tones down into the same octave, starting from F, we get F, C, G, D, A, E, and B—the tones of a C-major scale. This simple procedure is sufficient to reduce all intervals between adjacent scale steps to whole tones or semitones, avoiding adjacent half steps such as C-C♯-D (Pressing 1978). If we stop at six 5ths, a minor 3rd remains (if B is missing, there is a minor 3rd between A and C). If we stop at eight 5ths, we get adjacent half steps: starting from F and stacking 5ths, the seventh tone is B (completing the diatonic) and the eighth is F♯ (midway between F and G). The F♯ is then perceived as a chromatic neighbor—not as a member of the diatonic set—so the process of creating new scale degrees stops.

Simple explanations of that kind can render more complex mathematical treatments unnecessary. Norman Carey and David Clampitt (1989) demonstrated that pentatonic, diatonic, and chromatic scales are well formed in the sense that single intervals can generate all the elements. That may be true, but it is not the simplest, most direct answer to the question of why the diatonic scale is like it is.

Countless examples of overly complex theories could be offered, but for the moment, one more will suffice. Possibly inspired by David Huron and Joshua Veltman (2006), Daniel Harasim and colleagues (2021) set out to derive the optimal number of modes (church modes or tonalities) in different historical periods. They used geometric modeling on the assumption that certain cognitive pitch structures have psychological reality, and a large data set of 13,000 scores in MIDI format. It is certainly interesting and indeed necessary to use statistical analysis of real musical scores as a foundation for investigation and theory. But the failure of their method to separate major and minor modes in the nineteenth century was surprising, and their method considered only the tones in the score—not how they were perceived (including variations in pitch salience and implied tones at audible partials and missing fundamentals). Musical structure is based on what music listeners hear, which differs from what musicians see in scores. Nor did they consider the conceptually simple hypothesis that major and minor scales arose from Schenkerian prolongation of major and minor triads.

Interdisciplinarity

This book is inspired by the contrasting contributions of musical thinkers from the humanities and musical practice on the one hand (Heinrich Schenker, Allen Forte, Fred Lerdahl) and the sciences on the other (Carol Krumhansl, Ernst Terhardt, Al Bregman). To my knowledge, approaches of that kind have never been combined in a deep way. Humanities scholars in music theory and analysis may have a thorough understanding of Schenker's hierarchical analyses and Forte's pitch-class set theory but only superficial acquaintance with Krumhansl's cognitive structures and Terhardt's theory of pitch perception; the opposite may be the case for scientists in the area of music psychology.

Any broadly interdisciplinary enterprise that crosses the boundary between humanities and sciences has its hazards. In different disciplines, the same terms may have different meanings. For example, the term "dualism" is used in two ways in this book, one (psychophysical dualism) referring to the mind-body problem in philosophy, the other (harmonic dualism) to the relationship between major and minor triads in music theory. Another example is the apparently innocent word "resolve" (or resolution). In music theory, dissonances resolve onto consonances. In scientific research on pitch perception, a partial of a complex sound is resolved if it can be separately heard. Again, the word is the same, but the two meanings have nothing to do with each other. The problem can be addressed by careful definitions—but that is just the start. There are also different priorities, values, and underlying assumptions in different academic disciplines, which can lead to misunderstandings and conflict.

Eric Clarke (1989a) addressed differences between music theorists and cognitive psychologists as they talk about the same thing: musical structure. Exposing a certain naivety among some psychologists about the effect of culture on music perception and cognition (including "the almost complete rejection of a historical perspective in cognitive psychology" and "attempts . . . to account for socio-historical conventions in terms of innate perceptual characteristics," 11), Clarke noticed a "somewhat unstructured conceptual leakage that can . . . be observed between the two [disciplines]" (1) and warned that "an indiscriminate and uncontrolled borrowing of bits and pieces sometimes seems to be little more than an attempt to grasp at anything that will bolster up inadequate theories, with scant concern for the compatibility or otherwise of their theoretical origins" (2).

My approach to interdisciplinarity in this book is similar in some ways to Clarke's and different in others. Despite his warning, I tend to include approaches to similar questions from different academic disciplines in the same text, even if they initially seem incompatible. I then creatively explore how those contrasting ideas might complement and enrich one other.

Many ideas in this book were inspired by interdisciplinary explorations and could prove to be controversial. They include the following. Musical intervals are culturally learned, approximate pitch distances—not ratios of small integers. In musical practice, major-3rd intervals are neither pure/Just (386 cents) nor Pythagorean (408 cents), but they do vary across and even outside that range. In unaccompanied diatonic chant, some tones are more consonant or stable than others, depending on the relationship between their audible partials and the current scale, which can explain the origin of the leading tone. Major and minor triads existed in musical practice from the start of tonal polyphony in the twelfth century as compositional possibilities, although the musicians of the time had no vocabulary to describe them (or chords in general) and no concept of root or inversion; in ensuing centuries, major and minor triads emerged as the most consonant trichords and became the most common trichords in Renaissance music for that reason. Most theories that attempt to explain why the minor triad is important, despite the relatively large numbers in its Just ratio (10:12:15), and beyond that most music theories of any kind that are based on number ratios, are problematic and misleading. Arnold Schoenberg's (1911) theory of "strong" chord progressions, and the drama of dominant prolongation in Western tonal music of the eighteenth and nineteenth centuries, can be explained by an implication-realization theory that involves the perception of missing fundamentals in major and minor triads.

Often, academic creativity arises from controversy and conflict. With that in mind, this book deliberately presents controversial ideas if they seem to be both challenging and promising—ideas that may be understood differently by expert colleagues in contrasting relevant disciplines in humanities, sciences, and musical practice. Resolving differences of that kind (to use the word "resolve" in yet another way) could lead to new insights. The ideas will be received differently by different colleagues, depending on their academic background—whether music theory, music cognition, or some other discipline—and what they believe about reality and metaphysics (see chapter 4). That is surely an exciting prospect! Let the interdisciplinary discussion begin.

3 Rudiments

Readers of this book who are qualified in a relevant nonmusical discipline such as psychology, neuroscience, acoustics, computer science, sociology, or philosophy may lack a solid grounding in music notation and pitch-class set theory. With such readers in mind, this chapter introduces some musical rudiments that recur repeatedly in later chapters, including diatonic and chromatic labels for musical intervals, pitch-class set labels for chords, and theories of octave equivalence, scales, keys, and harmonics.

Intervals

Musical structures such as scales and chords may be specified using either diatonic labels (relative to diatonic scales, with seven tones per octave) or chromatic labels (relative to the twelve-tone chromatic scale). Diatonic intervals are counted inclusively, whereas chromatic intervals are counted exclusively. A familiar example of inclusive counting is the French expression *quinze jours*, fifteen days, for which the English translation is two weeks or a fortnight. In this case, the French count the days inclusively, whereas the English count them exclusively.

Inclusively counted diatonic scale-step labels start with one. The interval between two identical tones is a unison. The diatonic interval between C and E is a 3rd—named after the three diatonic tones C, D, and E. In this book, I abbreviate "third" to "3rd" for diatonic interval labels and for partials of complex tones (e.g., "the 3rd harmonic"). Exclusively counted chromatic interval labels start with zero, since the pitch difference between two identical tones is zero. The chromatic interval between C and E is [4] because E is four semitones higher than C (C–D is two semitones and D–E is two semitones). I use

Table 3.1

Relationships among chromatic interval labels, diatonic interval labels, and harmonic numbers.

Example of Interval	C-C	C-C♯/D♭	C-D	C-D♯/E♭	C-E	C-F	C-F♯/G♭	C-G	C- G♯/A♭	C-A	C-A♯/B♭	C-B
Chromatic label	0	1	2	3	4	5	6	7	8	9	10 (T)	11 (L)
Diatonic label	P1	m2	M2	m3	M3	P4	A4, d5	P5	m6	M6	m7	M7
Alternative diatonic labels		A1	d3	A2	d4	A3		d6	A5	d7	A6	d8
Harmonic numbers	1, 2, 4, 8		9		5, 10			3, 6			7	

The first row, "Example of interval," shows each interval with C as the lower tone (e.g., C-C♯/D♭). In the second row, chromatic labels are interval sizes in semitones, counted exclusively. In the inclusively counted diatonic labels in the third row, P = perfect, M = major, m = minor, d = diminished, and A = augmented. The harmonic numbers in the last row are octave generalized (transposed down by octaves so they are all in the same octave register; see section on harmonics).

square brackets [ ] for chromatic interval labels. The relationship between chromatic and diatonic labels is shown in table 3.1.

Diatonic intervals are classified as either perfect or imperfect. In ancient theory, perfect intervals were consonant, and imperfect intervals were dissonant. Perfect intervals are the (perfect) octave, 5th, and 4th; imperfect intervals are the 2nds, 3rds, 6ths, and 7ths. A perfect interval can be augmented (A = one semitone wider) or diminished (d = one semitone narrower). An imperfect interval can be major (M) or minor (m), the difference between the two being a semitone ("major" means larger and "minor" means smaller). For example, a major 3rd (M3) is a semitone wider than a minor 3rd (m3). An augmented imperfect interval is a semitone wider than the corresponding major interval, and a diminished imperfect interval is a semitone narrower than the corresponding minor interval. For example, an augmented 3rd (A3) is a semitone wider than a major 3rd (M3), and a diminished 3rd (d3) is a semitone narrower than a minor 3rd (m3). In this way, intervals can be labeled in two different ways, called enharmonic equivalents. For example, a m3 is equivalent to an A2, since both span [3] semitones.

Octave Equivalence

Géza Révész (1913) proposed a two-component theory of musical pitch that acknowledged the equivalence of tones at octave intervals. According to him, tones first have pitch (*Tonhöhe*), which is their approximate height on a continuous scale from low to high—similar to today's register (octave register, pitch register). Second, musical tones have *Tonigkeit*—a series of twelve psychological categories in each octave. Besides being high or low, a tone can be C, C♯, D, and so on. The corresponding modern terms for Tonigkeit are chroma in music psychology and pitch class in music theory; I will use pitch class because of the important role that pitch-class set theory plays in this book.

The two-component idea is reflected by standard pitch labels. The tones of the diatonic scale are labeled from A to G in each octave. If you add a number after the letter, it denotes register. In American Standard Pitch Notation, C4 is middle C, which is the C in register 4 (about 262 Hz). Pitch class C♯ could be in any octave register, but pitch C♯4 is in register 4, a semitone higher than C4. Each new register begins on C (not A), so the notes a semitone below and above middle C are B3 and C♯4/D♭4, respectively.

Sometimes, when thinking about musical structure, we only consider pitch classes and ignore octave register, assuming it to be secondary. A lot can be learned from a music theory that initially regards octave-related tones as equivalent, just as a lot can be learned in physics from a mathematical equation that has been simplified by removing terms that approach zero under certain conditions.

In an octave-generalized theory of MmT, the tonic tone, the tonic triad, the diatonic triads, and the major and minor scales are octave generalized. In reality, things are more complex. Scales can start and end in different registers, and chords can be voiced differently, with different chordal inversions, spacings, and doublings. A perfect authentic cadence sounds especially final because the bass and the soprano voice simultaneously come to rest on the tonic (Caplin 1998). Whether a tonic triad functions as a tonic depends on which tone is in the bass: the triad's second inversion (6/4 chord) has dominant function if the 4th and 6th above the bass resolve to a 3rd and a 5th (double suspension). In these examples, octave register does matter, which contradicts the principle of octave equivalence.

Scales

A scale is a set of tones (also called scale degrees) that are used for making music. A standard diatonic scale corresponds to the white keys on the piano, in any transposition up or down by a constant number of semitones. In this case, no tone is assumed more important than any other (no tonic or reference pitch is defined). A major scale is a standard diatonic scale with a specific defined tonic or reference pitch (e.g., C in the case of the set C, D, E, F, G, A, B). A harmonic minor scale (e.g., C, D, E♭, F, G, A♭, B, where C is the tonic) is more complex; although often regarded as diatonic, it is not standard diatonic.

A standard diatonic scale can be created by adding perfect 5ths and subtracting octaves until a seven-note scale is reached. If we start with F (any F in any octave register), the next tone is C, then G, D, A, E, and B. Putting the tones in order of rising pitch, the scale is C, D, E, F, G, A, B. If we start with C, we get C, D, E, F♯, G, A, B, which is the same as C, D, E, F, G, A, B but transposed up a perfect 5th or down a perfect 4th. If we start with G, we get C♯, D, E, F♯, G, A, B—and so on.

The process can be visualized on a structure called the circle of 5ths: the series C, G, D, A, E, B, F♯, C♯, G♯/A♭, E♭, B♭, F, and back to C—often illustrated on a clockface, with C at 12 o'clock, G at 1 o'clock, D at 2 o'clock, and so on. A standard diatonic scale is a set of seven adjacent points on this clock. No matter where you start on the clock, if you put together seven adjacent points, you will always get a standard diatonic scale.

Diatonic scale steps are labeled $\hat{1}$ to $\hat{7}$, in both major and minor keys, the first scale step ($\hat{1}$) being the tonic and the seventh ($\hat{7}$) being the leading tone. Relative to the chromatic scale, the pitch of these scale steps relative to the tonic ranges from [0] to [11] semitones. Scale degree $\hat{1}$ is [0] semitones above the tonic, $\hat{2}$ is [2] semitones, $\hat{3}$ is [4] semitones in a major key and [3] semitones in a minor key, and so on. In this way, we can represent a major scale as [024579L] (where L = 11; the commas can now be omitted), a harmonic minor scale as [023578L], a rising melodic minor scale as [023579L], and a falling melodic minor scale as [023578T] (where T = 10). Those are the most common scale types in the theory of MmT.

Keys

In Schenkerian theory (Forte and Gilbert 1982), a musical key or tonality can be regarded as a prolongation of its tonic triad (I or i). At a different hierarchical level, any scale step (not only $\hat{1}$) can be prolonged, so that a whole passage is perceived relative to it. For example, a passage of music might be perceived as a prolongation of ii, the supertonic triad. In that case, non-Schenkerians might refer to a passing modulation to ii, especially if the dominant of ii (V/ii, pronounced "five of two") is present. In this case, the sharpened tonic scale degree ♯1 is often introduced—also called the lower or rising leading tone of $\hat{2}$ because it lies a semitone below $\hat{2}$. A rising leading tone is said to tonicize the tone toward which it moves (here, $\hat{2}$).

In chord progressions in a given key or tonality (such as C major or B♭ minor), chords are perceived relative to the tonic triad. From a psychological perspective, the tonic triad acts as a psychological reference. It is therefore appropriate to use the same notation for triads and keys: "I" can mean either tonic triad or tonic key, "ii" can mean either supertonic triad or supertonic key, and so on.

But the key is generally ambiguous, making it possible to hear the same chord in different ways, even in the same context (Lerdahl 2001). As an example, imagine a G-major triad (i.e., a chord comprising the tones G, B, and D, each in any register) in a C-major context (i.e., in the key of C major). Its normal harmonic function is dominant or V. More precisely, it is the dominant of key I, also called V/I ("five of one"). But the same chord can be perceived as the tonic triad of G major (I/V) or the mediant III of the mediant key iii, E minor (III/iii, "three of three").

Harmonics

The ear analyzes all incoming complex sounds into partials or bands of noise in a continuously running spectral analysis. Partials (also called pure-tone components) are pure tones, which have sinusoidal waveforms. The partials of sounds produced by environmental objects can in principle have any frequencies at all.

Most musical tones, such as those of bowed violins and trumpets (but also voiced speech sounds such as /i/ and /m/), have a periodic waveform. That means the graph of air pressure against time (measured using a microphone) has a basic form that repeats many times per second (the fundamental frequency). According to the theory of spectral analysis (also called Fourier analysis), periodic tones have a harmonic spectrum: the frequencies of the partials correspond to a harmonic series. If the partials are generated independently as pure tones (e.g., with one loudspeaker per tone), the resultant sound is indistinguishable from the original, provided the amplitude and phase relations between the partials are also reproduced. Such tones are called harmonic complex tones, and their partials are called harmonics (or harmonic partials).

The diatonic intervals between successive harmonics, or elements of the harmonic series, are the octave (between harmonics 1 and 2), perfect 5th (between 2 and 3), perfect 4th, major 3rd, minor 3rd, and so on. Relative to the chromatic scale, the intervals between adjacent tones in the harmonic series, in semitones, are (approximately) twelve, seven, five, four, three, and so on (cf. table 3.1).

To give each harmonic a musical pitch label, we need to notate both its pitch class (C, C♯/D♭, D, D♯/E♭, and so on) and its octave register. A harmonic series on C2—two octaves below middle C, and two leger lines below the bass clef—is C2, C3, G3, C4, E4, G4, B♭4, C5, D5, E5. Here, C2 is the fundamental or 1st harmonic, and C4 (the 4th harmonic) corresponds to middle C. The series has been arbitrarily limited to ten harmonics because usually no more than ten are aurally relevant. Some harmonics are out of tune relative to twelve equal divisions of the octave (12-EDO, the usual tuning of the modern piano; see chapter 10), the mistuning being most noticeable for the 7th (near B♭4).

The harmonic series plays a role whenever speech is perceived in any cultural context. Every voiced speech sound is analyzed by the inner ear into its audible partials. The auditory system needs somehow to recognize either the periodicity of the waveform or the harmonic pattern of the partials if it is to track fundamental frequency in a noisy environment (see chapter 5). The harmonic series also plays a role in Western music theory, which has traditionally emphasized consonance. One way to achieve consonance is to favor musical intervals from the harmonic series, which are familiar to the human auditory system, being present and audible in every voiced speech sound.

Another way to achieve consonance is to avoid psychoacoustic roughness. When musical chords are created by superposing harmonic complex tones, roughness is reduced by favoring harmonic intervals between the fundamental frequencies. The roughness is caused when upper partials almost coincide. Roughness is a sensation caused by a fluctuating amplitude envelope or fast beating (faster than twenty beats per second), and it is one of the main psychological foundations of dissonance in Western music (see chapter 14).

Triads

Diatonic triads are labeled I, ii, iii, IV, V, vi, vii° in major keys and (depending on interpretation) I, ii°, III, iv, V, VI, vii° in minor keys. Uppercase symbols denote major triads ([047]; transpositions of CEG), and lowercase symbols denote minor triads ([037]; transpositions of ACE). A superscript circle (°) denotes a diminished triad ([036]; transpositions of BDF). Sometimes, these distinctions are dropped and uppercase Roman is used regardless of the quality of the triad. Some common jazz/pop chord symbols are shown in table 3.2.

In the key of C major, chord I (the chord built on scale-degree $\hat{1}$) is CEG or C major. Chord ii is DFA or D minor, iii is EGB or E minor, IV is FAC or F major, V is GBD or G major, vi is ACE or A minor, and vii° is BDF or B diminished. In C minor, chord III is usually E♭ major (E♭GB♭); the leading tone $\hat{7}$ (B♭) is not raised to B♮, given the dissonance of the augmented triad.

Table 3.2
Selected jazz/pop chord symbols (left column) and their meanings.

Chord symbol	Chord name	Trichord label (semitones above root; T = 10, L = 11)	Diatonic intervals above root					
			2nd (9th)	3rd	4th	5th	6th	7th
-	major triad	[047]	-	M	-	P	-	-
m	minor triad	[037]	-	m	-	P	-	-
sus	suspended triad	[057]	-	-	P	P	-	-
°	diminished triad	[036]	-	m	-	d	-	-
+	augmented triad	[048]	-	M	-	A	-	-
7	(major-minor) 7th	[047T]	-	M	-	P	-	m
m7	minor 7th	[037T]	-	m	-	P	-	m
M7	major 7th	[047L]	-	M	-	P	-	M
°7	diminished 7th	[0369]	-	m	-	d	-	d
ø	half-dim. 7th	[036T]	-	m	-	d	-	m
11	dominant 11th	[0257T]	M	-	P	P	-	m

For diatonic intervals, P = perfect, M = major, m = minor, d = diminished, A = augmented.

Pitch-Class Sets

The identity of a musical chord or scale depends on the intervals between the tones in semitones (Vincent 1894). If we assume octave equivalence, that means applying modulo-12 (clockface) arithmetic, in which $11+1 = 0$, $11+2 = 1$, and so on. Pitch-class set theory (Forte 1973; Hanson 1960; Rahn 1980) allows all pitch patterns in the chromatic scale, including all chords and scales, to be labeled in a simple, unambiguous manner.

Pitch-class set theory was originally developed to analyze and understand the atonal (post-tonal) repertoire. Later, it was used to investigate more complex tonalities (e.g., Walker 1989; Yih 2000). But pitch-class set analysis can help us understand the structure of any music composed in the chromatic scale. From a scientific perspective, it has the advantage of neutrality: all twelve tones are assumed equivalent, and no prior assumptions are made about their relationships. The musical relationships that we perceive are assumed to depend on our auditory experience of similar relationships—not on relationships that are visible or salient in musical scores.

A major triad in root position (such as CEG, with C in the bass) has a major 3rd (E) that is [4] semitones above the root C, and a perfect 5th (G) that is [7] semitones above the root C. So, its chromatic label is [047], where [0] represents the lowest or bass note. Relative to E, the chord is [038] because G is [3] semitones above E, and C is [8] semitones above E. Relative to G, the same chord is [059]. A minor triad in root position is [037], a diminished triad is [036], an augmented triad is [048], and a suspended triad is [057].

In pitch-class set theory, all major and minor triads belong to one pitch-class set, here called {037}. That is the set's prime form, found by considering all possible transpositions and intervallic inversions of the set and minimizing the sum of the numbers. The label {037} is preferred to {047} because the sum of the three numbers is 10 for {037} and 11 for {047}. Of all the chordal inversions of major and minor triads, the sum is lowest for [037].

Howard Hanson (1960) counted all possible subsets of twelve chromatic scale tones (collections of pitch classes), grouping together transpositions and intervallic inversions. The result was a list of six intervals (dichords), twelve trichords, twenty-nine tetrachords, thirty-eight pentachords, fifty hexachords, thirty-eight heptachords, twenty-nine octachords, twelve nonachords, and six decachords. The symmetry of this list (e.g., the equal number of dichords and decachords) was due to complementarity. For example, for every trichord there is a complementary nonachord (3 + 9 = 12). The sets [047] (major triad) and [1235689TL] are complementary because together they include all twelve pitch classes.

The word "inversion" is used in two different ways in music theory, which I will call intervallic inversion and chordal inversion. The intervallic inversion of an interval of [1] semitone is [12] – [1] = [11] semitones, and of [2] semitones is [12] – [2] = [10] semitones. That usage is common in pitch-class set theory. In traditional tonal theory, inversion has a different meaning: it means taking one of the higher tones in a chord and transposing it into the bass. Take the major triad CEG: written this way, we understand that C is the lowest tone and G the highest. If we transpose the 3rd (E) down into the bass, we get ECG, a first-inversion C-major triad. If we transpose the 5th (G) into the bass, we get the second-inversion triad GCE (or GEC).

Note that the expressions "first inversion" and "second inversion" always imply chordal inversion. In the terminology of pitch-class sets, {037} includes both [047] and [037], which are intervallic inversions of each other. The set {037} also includes the chordal inversions of major and minor triads that one gets by transposing one of the upper voices into the bass: [038] for the first-inversion major triad, [059] for the second-inversion major, [049] for the first-inversion minor, and [058] for the second-inversion minor. Similarly, {025} is a pitch-class set that includes not only [025] and [035], but also [03T], [079], [029], and [07T].

A pitch-class set is a set of pitch classes that is invariant under both transposition and intervallic inversion. For example, the set {037} includes all major and minor triads (which are intervallic inversions of each other), regardless of the root (for which there are twelve options—in other words, the set is invariant under transposition).

Pitch-class theory distinguishes pitch-class sets from Tn-types. Here, T stands for transposition (not T = 10) and reminds us that each chord can be transposed to twelve different scale steps; n is the size of the transposition in semitones. To simplify the language in this book, I will refer to trichords and tetrachords instead of Tn-types, with a trichord comprising three pitch classes and a tetrachord comprising four (cf. Parncutt, 2009). For example, trichords [047] and [037] belong to pitch-class set {037}. My treatment of Tn-types is similar to that of John Rahn (1980), except that I use single symbols for 10 (T) and 11 (L). In modulo-12 arithmetic in mathematics, 10 = A and 11 = B; to avoid confusion, I prefer to reserve the letters A and B for musical note names (i.e., for absolute rather than relative pitches).

In pitch-class set theory, there is no difference between enharmonically equivalent pitches or intervals. Pitch classes C♯ and D♭ are the same; we might call both pc1, both being one semitone above C. The size of a minor-7th or augmented-6th interval is the same: [T] semitones. Both Mm7 chords (such as CEGB♭; called C7 in jazz and pop scores) and augmented-6th chords (such as CEGA♯) are labeled [047T] because in both cases the tones are [4], [7], and [T] semitones above the root. The prime form of this tetrachord in pitch-class set theory corresponds to its first chordal inversion {0368} because the sum of the four numbers is smallest in that case. Similarly, a major-7th chord in root position is [047L] (where L = 11), but the prime form is {0378}.

The intervals corresponding to a prime form are given by the interval vector: a list of how often intervals of [1] to [6] semitones occur between its notes.

Figure 3.1
All nineteen possible trichords (Tn-types) in music notation in close position on C4. The labels above the figure are familiar from music theory; those below the figure are numbers of semitones relative to the bass. From Parncutt, Sattmann, et al. (2019) by permission of University of California Press.

For major and minor triads, the interval vector is<001110>. The first two 0s mean the chord contains no intervals of [1] or [2] semitones. That is, there are no minor or major 2nds. The three 1s in the middle of the vector mean the chord contains intervals of [3], [4], and [5] semitones (minor 3rd, major 3rd, perfect 4th)—once each. The final 0 means there is no interval of [6] semitones (tritone). This vector tells us that the intervals between the tones of major and minor triads are always minor 3rds, major 3rds, and perfect 4ths, and their intervallic inversions: major 6ths, minor 6ths, and perfect 5ths. That property of major and minor triads distinguishes them from other chords.

Trichords

Figure 3.1 shows all nineteen possible trichords in music notation, in close position relative to C4. There are twelve pitch-class sets of cardinality 3 (i.e., comprising three pitch classes). Of those, seven are asymmetrical: they change when intervallically inverted. If we create those inversions and add them to the list, we get nineteen trichords altogether. For example, intervallically inverting [013] (here, CD♭E♭) produces [023] (CDE♭), and intervallically inverting a minor triad [037] (CE♭G) produces a major triad [047] (CEG).

The trichords [013] and [023] belong to the same pitch-class set {013} because they have the same interval vector<111000>: both contain interval classes of [1], [2], and [3] semitones, and neither contains interval classes of [4], [5], or [6] semitones. Similarly, major and minor triads belong to pitch-class set {037}. They include only interval classes [3], [4], and [5], so their interval vector is<001110>.

The pitch-class set theory used in this book is mainly confined to Hanson's (1960) original insights, but the presentation is closer to that of Forte

(1973). The approach is somewhat simplified, with a focus on the main elements of MmT. Trichords and tetrachords are identified by counting semitones relative to a reference pitch that is usually the conventional root, for example [047] for a major triad, [047T] for a Mm7 chord, and [047L] for a major 7th chord. If these representations are regarded as root-position chords, chordal inversions can be notated similarly, for example [038] for a first-inversion major triad and [0368] for a first-inversion Mm7.

II Extramusical Foundations

4 Metaphysics

An interdisciplinary theory of MmT is grounded in several nonmusical disciplines, including philosophy. What is the ontology and relevance of music's physical *reality (the vibrations produced),* subjective *reality (the private or shared experience of listeners), and* abstract *reality (the information content of scores)? Are physical tones, perceived pitches, and musical notes different things, or are they different representations or perspectives of the same thing? How might such realities or perspectives interact? Research on the perception and cognition of musical structure may skip over such issues as if they were trivial, but in a detailed multidisciplinary approach, they are intrinsic, and in psychoacoustics, they are the focus of empirical research. Which philosophy is correct:* physical monism *or* psychophysical dualism*? Each ontology has its own theoretical advantages and disadvantages, so perhaps a middle path is appropriate. Rather than reifying the "mind," we should consider how sensations, emotions, thoughts, and the self depend on physical and social interaction and environmental affordances. If subjectivity is an epiphenomenon of physicality, subjective causality (free will), although central for subjectivity, is an illusion in a physical monist approach.*

Objectivity

In the nineteenth and twentieth centuries, the academic discipline of physics explained a lot about the physical world. The problems that could not be completely solved concerned things that were too tiny (quantum mechanics) or too fast (relativity) for us to imagine. Advances in physics had an enormous impact on everyday lives and standards of living, enabling revolutionary developments in engineering, both positive (light bulbs, telephones, cars, washing machines) and negative (military technology, climate change). We can now confidently assert that the universe, which includes all matter and energy, obeys the laws of physics, insofar as those laws are understood. That makes us confident about the existence of the physical universe.

In music, a physical measure of the frequency of a tone is inherently and entirely objective. But the word "objective" has different meanings that need to be distinguished. It does not necessarily mean "exact": physicists make mistakes, and physical measures have inherent uncertainty. Here, the word "objective" refers to the inherently physical nature of either the measurement process (e.g., counting the number of periodic repetitions per second) or the thing being measured (sound as physical changes in air pressure).

Those who believe that only the physical world exists adopt various arguments to demonstrate the nonexistence of subjectivity. In the following, I will not distinguish between physicalism, materialism, and monism, although the distinction may be important in other contexts. There are also various kinds of monism, including physical monism, property monism, substance monism, or existence monism (Schaffer 2018). From a physical viewpoint, the monists are right that the physical world is closed and complete, the mystery of what happened before the big bang being a possible exception. Clearly, the laws of physics apply regardless of our subjective experience or our illusion of free will.

Subjectivity

Does a subjective world exist, somehow quite different from the physical world? Even within a physically dominated worldview, it is logically possible that subjective experiences exist, although they are not themselves physical in nature and hence not subject to the laws of physics. If so, they must lie outside of physicality—not outside in the sense of space-time coordinates, but in another sense, for which we have no words. Physicalists are free to find such a proposal ridiculous, and perhaps it is. But that does not make our (or their) subjective experience any less real or less important.

If physicalists acknowledge the existence of subjective experience, they do so on the understanding that it depends directly and entirely on the physical world. But we know from psychophysics that any attempt to explain subjective (e.g., musical) experience in terms of the physical world, reducing it entirely to physical processes, would be endlessly complex. Very roughly, the brain contains 100 billion neurons that are connected by 100 trillion synapses. That is a lot of information to keep track of. Further, the relationship between physiology and experience is different in every person's brain, for at least two reasons. First, every person has a different genetic foundation for the development of neural networks (nature). Second, every person has different lifetime experiences, which affect the Hebbian connections in their neural networks (nurture).

Like the word "objective," the word "subjective" has various meanings. It can imply that people disagree about something (a subjective opinion). It can suggest that the observer and the object of observation are not separate, such that the observation interferes with the thing being observed. It can refer to the inherent subjectivity of qualitative research in psychology or of the humanities as a whole.

Here, I wish to use the word "subjective" in yet another way. Music, and musical phenomena such as MmT, are inherently subjective, regardless of how differently different people experience music and regardless of how investigations of music may affect the music being investigated. Roger Scruton (1983) explained:

> If the description of music is so dependent upon metaphor, we might go on to conclude that music is not, strictly speaking, a part of the material world. By that I mean that any scientific description of the world of sound should not mention—as an independent fact of the matter—the phenomenon of music. For there is no explanatory function to be filled by the concept of music that will not equally be filled by the concept of organized sound: no scientific method need discriminate between these two (the extension of each concept in the material world being identical). Hence, by the axiom of simplicity, the concept which describes the material essence of what is heard (the concept of sound) is the only one that we need employ. If there is an additional *fact* of the matter, it is that we (beings of a certain kind) hear music. Music belongs uniquely to the intentional sphere, and not to the material realm. Any analysis of music must be an exercise in intentional rather than scientific understanding. (86, italics in original)

Music as subjective experience is an everyday thing. It happens when we listen to music in a concert or using an electronic device, or during our own individual or group performance. It happens when artistic patterns of sound evoke mysterious emotions or other subjective responses that we can barely describe. But it also happens throughout our waking lives. Whenever we are aware of being awake and alive, we are experiencing subjectivity.

The argument may become clearer if subjectivity is analyzed into different kinds:

- *Interpersonal subjectivity* is uncertainty about an idea or claim because different people disagree about it. For example, experts in different relevant disciplines may disagree, given their different knowledge or approach. Different artists or art lovers may disagree because they have different tastes.
- *Observational subjectivity* is subjectivity that arises from observing a phenomenon from within rather than from outside. It is due to the non-separation of the object and the observer. In cultural studies, it is a virtue; in quantum mechanics, it is inevitable.
- *Metaphysical subjectivity* refers to private experience or the "mind" as opposed to the physical world or knowledge. It is about being a subject with sensations, thoughts, emotions, agency, and desires, but it is not the self or agent. It includes music and other art forms when they are considered as subjective experience.

My focus here is on metaphysical subjectivity. I wish to argue that it exists within certain limits to which we will return. But I disagree with the extreme idealist position of eighteenth-century Irish philosopher George Berkeley, according to which only mental phenomena exist. Most readers of this book will need no convincing that Berkeley's proposal can be rejected. There can surely be no doubt about the existence of the physical world. Whether physicists will one day discover a theory of everything, I don't know. Therefore, I don't know if such a (mathematical) theory of everything exists. But the laws of physics, as currently formulated by physicists, are certainly referring to something that exists, however obscure those theories may be in some of their details.

To explain subjective experiences scientifically, we need a scientific theory of subjectivity. Many readers will immediately doubt that such a thing exists. Isn't science about the physical world? Isn't the physical world objective? Often, science is indeed only about the physical world. In such cases, it is appropriate to regard reality as entirely physical and ignore metaphysical questions. But if we are trying to understand something that is obviously and intrinsically subjective, such as musical experience, we need to define carefully what we mean by subjective. We need to analyze different aspects of subjectivity and consider the problem from different disciplinary viewpoints. Like anthropogenic global warming, the mind-body problem doesn't disappear if you ignore it, and denying its importance doesn't solve it.

The purpose of this chapter is not to contribute to current nonmusical discourse on the philosophy of reality. Instead, my aim is to integrate relevant academic disciplines in humanities and sciences into a discussion about musical structure and, in so doing, to expose and explain some of my theoretical bias. My explanations will, I hope, allow readers to clarify their thoughts and come to their own philosophical conclusions.

The word psychoacoustics can be understood in different ways. A psychoacoustic approach to music theory might ask how subjective experiences (such as pitch) depend on physical measurements (such as the frequencies and amplitudes of the partials of complex musical sounds). The question is not trivial, and it overlaps with issues in the philosophy of mind. There are important open questions here that continue to be discussed in an interdisciplinary exchange that brings together philosophy and neuroscience. But the first question we need to answer is whether the two worlds in question—the physical world and the world of subjective experience—actually exist.

The physical aspect should be straightforward. Einstein apparently remarked that "Reality is merely an illusion, albeit a very persistent one," but in that case, he was referring to problems about the nature of physical reality that arise in the physics of very high speeds (relativity) and very small distances (quantum mechanics). None of that is musically relevant. To understand music, we need to understand the Newtonian world of classical mechanics, about which there can be little metaphysical doubt. With few exceptions, readers of this book will agree that the pages in this book and the ink on those pages exist as part of the physical world. Even if there were no humans to read them, the pages and the ink would still exist. Even if no one witnessed a tree falling in a forest, the tree would still have fallen. The physical vibrations that we perceive as music exist similarly, and so on.

There is less agreement about the ontological status of the world of subjective experience, at least among scientists. One thing that we can say is that the discipline of music theory has always depended on, and been motivated by, that subjective world—also called imagination, imagery, mental representation, fantasy, or subjectivity. Hugo Riemann (1914–1915) appropriately distinguished subjective experience from music notation and music's physical realization in performance as follows:

> The alpha and omega of music as art lies not in music as it sounds but rather in the mental representation [*Vorstellung*] of tone relations, which lives in the musical imagination [*Tonphantasie*] of the creative artist prior to its inscription in

> musical notation and subsequent re-emergence in the imagination [*Tonphantasie*] of the hearer. Both the registration [*Festlegung*] of artistic creation in musical notation and the actual realization of the work in performance are but means of transplanting a musical experience from the imagination [*Phantasie*] of the composer into that of the musical listener. (Riemann 1914–1915, 20; transl. Hyer 2011, 139; italics in original)

Descriptions of subjective experience differ fundamentally from descriptions of physical reality. Therefore, empirical data about subjective experience are fundamentally different in nature from physical measurements or objective observations. A comprehensive scientific examination of music must therefore include a systematic and transparent approach to these issues.

Subjectivity cannot be measured in the way that physical things can be measured (using combinations of physical units: kilograms, meters, seconds)—at least not directly. In that sense, we may claim that subjectivity is nonphysical in nature—a kind of non-substance. But psychologists and psychophysicists often measure quantitative aspects of subjectivity. An example is the brightness of light, a subjective variable that depends on various physical parameters: the light's physical power per unit area, the distribution of that power across the spectrum, the context in which the light is perceived, and individual differences. If we measure the experience of light intensity by asking an experimental participant to compare different intensities with each other, we are not performing a physical measurement. But we are indeed measuring something, and that something is subjective experience, which is mediated by consciousness. The pitch of a musical tone is a similar case. Pitch is measured when a listener hears a tone and estimates its height, either absolutely or relative to another tone. That is a purely subjective measure—no matter how accurate or reliable—and it implies that every pitch that we hear in music is purely subjective in the same sense.

Despite these arguments, many readers will still be wondering whether the subjective world actually exists. One argument for its existence is reliability: subjective states can be described in different ways that often agree with each other. We may also compare an introspective or first-person description with a third-person description in which others describe their experience. To believe some introspections or reports but not others, as the eliminative materialists do, would be inconsistent. "Eliminative materialism (or eliminativism) is the radical claim that our ordinary, common-sense understanding of the mind is deeply wrong and that some or all of the mental states posited by common-sense do not actually exist and have no role to play in a mature science of the mind" (Ramsey 2002).

For the purpose of this book, I will avoid hair splitting of that kind and simply accept the existence of subjectivity, if only because the book would make little sense without it. There is nevertheless an interesting argument against the existence of subjectivity, and that is the illusory nature of subjective intentionality. Our decisions do not cause things to happen because the neurological correlates of the decisions happen long before we become aware of them. In that sense, free will is an illusion; our minds are not causing things to happen in the physical world. Whenever we have the impression that we have decided something, we are actually observing a decision that has already happened—our bodies and brains have in some sense already decided (Banks and Pockett 2007; Heisenberg 2009; Soon et al. 2008). David Oakley and Peter Halligan (2017) explained: "Despite the compelling subjective experience of executive self-control, we argue that 'consciousness' contains no top-down control processes and that 'consciousness' involves no executive, causal, or controlling relationship with any of the familiar psychological processes conventionally attributed to it."

That may well be true, but it does not imply that subjectivity does not exist. Instead, it implies that the physical world is totally unaffected by subjectivity. In physical reality, our bodies and everything that they are made of are reacting physically with the physical environment according to the rules of physics, however formulated—and remembering that today's physicists have not yet understood everything and probably never will. The mathematical laws of physics say nothing about subjectivity.

The point can be explained using the zombie argument (Chalmers 1995). I can be forgiven for thinking that all other people besides myself are philosophical zombies with no consciousness because I can only directly experience my own consciousness (although the idea of experiencing one's own experience is admittedly circular and hence paradoxical). We all assume that other people have consciousness because it is normal to do so. Besides, other people are otherwise similar to me, and human survival might depend on everyone agreeing that everyone has consciousness. So, we had better believe it, and a person who did not believe that would be regarded as crazy and possibly dangerous.

Given this background, I will tentatively embrace the philosophy of epiphenomenalism, according to which the physics and physiology of the body, including the brain, cause mental events but not vice versa (Robinson 2010). The relationship between the mind and the body is asymmetrical: from the perspective of the physical world, the subjective world does not exist, whereas from the perspective of the subjective world, the physical world is distorted by subjectivity. Our subjectivity is an epiphenomenon of the physical world—corresponding to it but not causing anything in it.

Berkeley was not entirely wrong. One might cogently argue that subjectivity (in the sense of consciousness or mind) is as important as the physical world, for without it, there would be nothing—at least not from a subjective human viewpoint, which is the only viewpoint available to us. When we die, the show will be over, for each one of us. One might then argue that music exists only in the subjective world of our sensations and emotions, because without those sensations and emotions, there would be no music—at least not in the everyday sense of the term. Of course, musical sensations and emotions correspond in complex ways to events in the brain, the body, and the physical world, but that observation does not make the purely subjective phenomena that we call sensations and emotions any less real—certainly not from an artistic or aesthetic perspective.

One might argue that ordinary everyday mental states (experiences, thoughts, self) exist merely because they are usually held to exist. We have no other way of knowing that they exist than our own experience of them. Therefore, there is no reason to believe they do not exist. Besides, without subjectivity, there would be no music, and there would be no point writing this book. We would not care about music if we did not experience its subjective qualities. Without musical experience, music would be meaningless, and without any kind of experience, everything about our lives would be meaningless (assuming that "meaning" is an inherently subjective thing).

Despite these arguments, we have to admit that subjectivity is less well understood than the physical world is. It is harder to describe or measure and more variable or imprecise. How, then, does subjectivity work? The empirical and philosophical literature on this question is diverse, and while there may be consensus on some points, the diversity of opinions remains striking.

Panpsychism is the idea that consciousness (or soul or spirit) is inherent in all matter. From a scientific perspective, panpsychism can be rejected because it assumes that consciousness can be located in physical space. We may think that our consciousness is in our head—specifically, between and just behind the eyes (Bertossa et al. 2008)—perhaps because that is where we see the world from. But that, surely, is an illusion. We can also ask whether nonhumans have subjectivity (consciousness), and if so, where it is located. Again, the question seems impossible to answer, given the zombie argument.

What we *can* say is that human consciousness developed, both phylogenetically and ontogenetically, in parallel with language. Both are part of behavioral modernity, a set of behavioral and cognitive traits that distinguished early humans from their closest evolutionary relatives. Archeological evidence for the emergence of behavioral modernity includes "carvings, personal ornaments, musical instruments, depictions on cave walls, and new stone and bone technology" (d'Errico 2003), which lead to technological dominance, the invention of agriculture, and population growth (Hill et al. 2009). Since then, the difference between humans and related nonhuman animals has accelerated. The most linguistically competent nonhuman animals may be able to learn a small vocabulary of quasi words, but they can hardly combine words into simple sentences, let alone complex ones (Conway and Christiansen 2001). If they have consciousness of the subjective kind being discussed here, we guess on this empirical basis that it is much less developed than human consciousness.

Popper's Worlds

Karl Popper and John Eccles (1977) distinguished three worlds of reality, which I will call the physical, the experiential (or subjective), and the abstract. It is important to clarify from the outset that these worlds are not necessarily separate. They can also be regarded as different ways of observing, perceiving, or documenting a single reality. In practice, the difference between different realities, on the one hand, and different perspectives on the one reality, on the other hand, can be small. In both interpretations, we are challenged to clarify the differences among three categories and to use correspondingly clear terminology.

World 1 is the physical world. It includes musical tones, defined as physical vibrations or oscillations in air pressure. World 2 is subjective, made up of experiential phenomena. They include emotions and sensations in different sensory modes, but also intrinsically music-structural phenomena such as the shape, energy, or character of a musical phrase. Popper used the term "mental processes" to refer to this world, but both terms ("mental" and "processes") are problematic in an ecological approach. World 3 is about information or knowledge. It includes musical scores and the (musical) information that is processed by the brain or a computer in a cognitive approach (Popper referred to "abstractions").

There may be a fourth world that is fundamentally different in nature from the other three. The self, which includes such concepts as agency, intention, will, action, soul, and spirit, is not included in Popper's three worlds, although we might associate it with World 2. One could argue for its separate existence as a kind of World 0.

Worlds 1 and 2 are the familiar two sides of psychophysical dualism. World 3 can be compared with Plato's theory of ideas or forms—the reality behind the physical world, according to him. Plato argued that only ideas or forms really exist because only they encapsulate the true nature of things. Mathematical statements such as $2+2=4$ are absolutely correct in all times and places, but corresponding physical claims are not because the world is constantly changing. When describing World 3, Popper focused on knowledge and culture, which are created by humans and hence constantly changing, whereas Plato's ideas were more absolute. In a general approach, both can be considered part of World 3 because both are abstract forms of information or knowledge.

The three-worlds idea, as a foundation for psychophysics, has often been implied or anticipated. Hermann von Helmholtz (1877) called World 1 *körperliches Ohr* and World 2 *geistiges Ohr*. He also presented a World-3 theory of perception based on symbols. In his words: "Our representations of things cannot be anything other than symbols, naturally given signs for things, which we have learned to use in order to control our motions and actions. When we have learned to read those signs in the proper manner, we are in a condition to use them to orient our actions such that they achieve their intended effect" (Steege 2012, 77).

Tone perception involves all three worlds. A harmonic complex tone can first be described physically (with frequencies and amplitudes). Second, it evokes subjective qualities (pitch, loudness, timbre). Third, we can identify the sound source that produced the tone and give it a label. The source's detected presence can be regarded as a sign or piece of information: "the sufficient sensory sign for the presence of a specific sounding body" (Helmholtz 1856, 526; Steege 2012, 78). Helmholtz distinguished perceptual information of this kind from parallel subjective experience or cognitive representation

(*Vorstellung*), which belongs to World 2, and from the physical stimulus, which belongs to World 1.

Music (and hence MmT) is by nature a subjective, experiential, cultural phenomenon. As such, it belongs to World 2. But music would be nothing without its correspondences to Worlds 1 and 3. Starting with World 1, there is a complex correspondence between musical experience and associated brain states and processes. We should therefore avoid jumping to physicalist (materialist) conclusions (assuming that everything is ultimately physical) or making unwarranted assumptions about the mind–brain relationship. The psychoacoustic approach is instead to separate Worlds 1 and 2 clearly, and then to perform some experiments to understand how they are related. Regarding World 3, music is often notated in musical scores, which can be regarded as instructions to performers. In the Western tradition, the academic discipline of music theory regards the musical score as a foundation for musical analysis. In a psychological approach, in which we want to know how music works psychologically, World 2 is a more appropriate foundation. Clearly, we need to understand the relationship between Worlds 2 and 3 if there is to be productive collaboration between the academic disciplines of music theory and music psychology.

Along similar lines, Milton Babbitt (1965) mentioned "three usual representations of the musical event: the graphemic (what Michael Kassler calls the 'written musical experience'), the acoustic (Kassler's 'produced musical experience'), and the auditory (Kassler's 'received musical experience'). The absence of a one-to-one relation between any two of these domains and the lack of sufficient knowledge to translate generally and totally from one to another, particular from the graphemic to the auditory, make it essential that output information be interpreted only in the domain of the input information" (76).

Geraint Wiggins and colleagues (2010) adopted Babbitt's categorization, noting that his representations correspond to Popper's three worlds and clarifying that today's sound files (e.g., MP3) belong in the graphemic or abstract category. Wiggins took "the philosophical stance that the mysterious thing that is Music is actually something abstract and intangible, and which does not have real existence in itself, but which is *described* by all three of these representational domains; in a sense, something like the notion of the Platonic ideal, as expressed, for example, in *The Republic*, but *without actually*

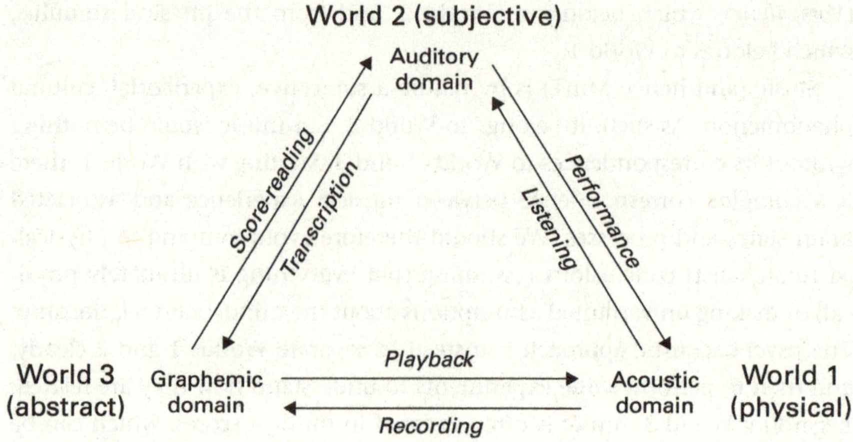

Figure 4.1
Babbitt's three domains and relationships among them according to Wiggins et al. (2010), with added labels for Popper's three worlds. Adapted from Wiggins et al. (2010).

existing in the real world. It is therefore reasonable to think of all of these domain-specific representations as being aspects of Music, but none of them is Music, individually" (232–233, italics in original).

Wiggins et al. (2010) added in a footnote that "the assumption that music is a real-world object—in the sense of Popper's World I—and is therefore amenable to study, in the same way as physical objects, under the standard scientific method is at the root of many misconceptions" (233).

Babbitt's three domains, and the musical relationships between them according to Wiggins et al., are illustrated in figure 4.1. The figure also shows the corresponding worlds of Popper. The arrow labeled "performance" implies that musicians imagine the sounds to be played before playing them; the music exists first in the imagination (auditory domain, World 2) before being created in the acoustic domain (World 1). "Playback" refers to the realization of a digital music file; it can also refer to music performance when reading from a score (also imaginable in the diagram as score reading followed by performance). Wiggins et al. (2010) argued that whereas music exists in all three domains and cannot be identified with just one of them, "the place where Babbitt's three domains come together is really in the human mind/brain" (234), implying that a definition of music primarily involves World 2: "music is *fundamentally* a psychological entity" (234).

The academic discipline of phenomenology attempts, among other things, to describe subjective experience and its structure. The normal approach is to use language; in this book, I will assume that quantitative phenomenology (using numbers) is also possible. Subjective experiences such as pitch or loudness can be measured psychologically and compared with corresponding physical parameters—frequency and sound pressure level (SPL), respectively (S. S. Stevens and Galanter 1957). We can argue that the subjective qualities of a tone exist, even if the traditional psychoacoustic paradigm can never provide a complete description of them (Haverkamp 2019). That is due to the inherent ineffability of experience, including musical experience (Raffman 1993).

Edmund Husserl (1900) introspectively analyzed conscious phenomena, their intentional structure, and how they depend on perception, which is directed toward objects in the world. He considered thoughts and ideas—how they are formed in the conscious mind. In that respect, we might say retrospectively that he was interested in Popper's World 3, the abstract world of thoughts and language—perhaps more so than World 2, the world of sensations and emotions. His main discoveries concerned "transcendental-phenomenological method, the phenomenological structure of time-consciousness, the fundamental role of the notion of intersubjectivity in our conceptual system, the horizon-structure of our singular empirical thought" (Beyer 2020). On the assumption that typical music listeners (with the exception of some musicians and music theorists) are not analyzing what they hear in terms of melodic intervals, harmonic functions, and so on, but rather are being carried on a subjective wave of sensations and emotions, Husserl's approach is not directly relevant for this book.

What is the relationship between World 2 (subjectivity) and World 3 (information)? I am regarding World 2 as subjective-experiential, comprising subjective experiences such as sensations and emotions. Underlying these sensations and emotions is some kind of information processing, but we do not experience that processing itself—we only experience the sensations and emotions. We do, however, experience thinking, and thoughts may be considered information (World 3) to the extent that they could be written down. Thoughts are also accompanied by sensations and emotions (World 2). Our subjective experience may depend on the information processing, but it clarifies things to regard the information itself as a different kind of reality.

Like any other philosophical theory, Popper's idea of worlds is not perfect, and readers of this book are welcome to disagree with it. One might argue that it failed to solve problems inherent in Cartesian mind-body dualism; instead, the goalposts were shifted. It is also unfalsifiable (impossible to disprove) and hence unscientific according to Popper's (1934/1959) own approach to the philosophy of science. A clear separation of the three worlds nevertheless allows us to sharpen our terminology and thinking as music theorists and music psychologists.

Frequency, Pitch, Note

In this book, I will assume that:

1. Frequency (in hertz) is a physical property of a tone and belongs to World 1.
2. Pitch is an experience (how high or low a tone sounds) and belongs to World 2.
3. A note is an instruction to a musician to play a tone—part of World 3. Its name (e.g., C♯5) is a piece of cultural information. The existence of a musical score is independent of how it is stored—whether written down, stored in a computer, or memorized by a musician.

These ideas are deliberately simplistic. The reason for applying Popper's theory is equally simple: to clarify discussion of music-theoretical and psychoacoustic issues with the aim of better understanding MmT. That is interesting because assumptions about the nature and structure of reality can influence how psychological and musical theories are constructed. Epistemological differences in questions about musical structure—differences among music-theoretical, grammatical, cognitive, psychoacoustic, and mathematical approaches—can sometimes be traced to differences—often tacit—in underlying assumptions about reality.

One of the tasks of the academic discipline of psychoacoustics is to explore relationships between frequencies, pitches, and note names empirically. Frequency is not the same as pitch for two reasons. First, the exact pitch of a pure tone depends on its physical intensity and on the presence of other sounds or maskers (pitch shift; see chapter 13). Second, a complex tone includes both many frequencies and many pitches, and the relationships among them are complex. Nor is frequency the same as note name: a note can be played with different tunings, covering a range of frequencies, and still be considered in tune or appropriate in a musical context. Pitch is different from note name for a similar reason. The pitches that we perceive in tonal music do not necessarily correspond to the notes in the score. Even if all tones are played equally loudly, the intervallic relationships mean that some sound more important (salient, stable) than others due to a mixture of masking and harmonic pattern recognition. Some notes are inaudible, and some perceived pitches do not correspond to notes (chapter 13).

In the long tradition of basing music analysis on musical scores, these complex relationships have often been pushed under the carpet. More recently, they have been neglected by cognitive psychologists. One aim of this book is to explore how these complex relationships affect musical structure.

The word "pitch" is used differently in psychoacoustics and mainstream music theory. Whereas music theorists often use the word to refer to the vertical position of a note on a musical staff, or a given scale step relative to a give scale (which I call "note name"), psychoacousticians use the word in the sense of how high a tone is perceived to be, both absolutely and relative to other tones in its context. The latter is how the word is usually used in this book. To clarify this, I will sometimes write "perceived pitch."

Some terms in this book do not correspond clearly to Popper's worlds. Pitch class (chroma in music psychology) is an example. It can refer to all three worlds—a frequency, a pitch, or a note name. A pitch class is a category within which a frequency or pitch falls. The category corresponds to a note name for which the octave register is not defined (i.e., the category can be placed in any octave register). Pitch class can also be defined operationally as the outcome of a psychological experiment. Pitch class C♯ may be defined as any tone, note, or pitch, in any tuning system and in any octave register, that a musician would normally call C♯. This practically oriented definition implies a certain permissible range of tuning (see chapter 7).

Monism versus Dualism

Tacit assumptions about reality and its structure can affect the way scientific theories are constructed. If the assumptions remain unstated and unanalyzed, conflicts between researchers or schools of thought can arise. Such conflicts can be observed in areas as diverse as pitch perception, cognitive psychology, and music theory. Some of those conflicts could be resolved, at least in part, if researchers were more open about their assumptions.

In the psychoacoustic literature, the word "psychoacoustic" is used in at least two different senses, depending—it seems—on the writer's beliefs about reality. Some psychoacousticians are monists, believing that only the physical world exists, whereas others are dualists, believing in either a separate world of consciousness (or subjectivity) or a fundamental difference between physical and subjective measurements. Both monists and dualists may believe that their perspective is so obviously correct that it is not worth explaining.

Monist psychoacousticians tend to focus on relationships between sound as physically measured and the auditory system as physically measured. Dualist psychoacousticians tend to focus on relationships between sound as physically measured and aspects of subjective experience as measured in standard empirical procedures. In both cases, the psychophysical relationships can be expressed mathematically.

This book aims for a middle path. I acknowledge the validity of the monist position, but I argue that the central questions posed in this book are too complex to answer within a monist framework. For practical reasons, we are forced to adopt a dualist perspective. The question of whether the dualist perspective is correct may ultimately be unanswerable.

Psychoacoustic research on pitch perception depends on subjective pitch judgments by conscious listeners. Monists aim to understand such behavior objectively, that is, independent of consciousness or subjectivity. They regard the body, including the brain, as a complex physical system that obeys the same universal laws of physics as the rest of the universe—which it in fact does. However, the psychophysical problem has not yet been solved at this level. We are only beginning to understand the complex relationships that exist between physical properties of sounds impinging upon the ears and biophysical measurements of the auditory system. Physical measurements involve temporal periodicity of neural firing patterns, and temporal patterns are in some respects mathematically equivalent to frequency spectra. The link between the two involves the mathematics of spectral analysis. It is unclear which aspects of this complex system influence conscious experience and which do not. Besides, experienced pitch is probably an emergent property of a complex system. As such, it does not generally correspond to the activity of a specific neural event or unit.

Given this complexity, we must be careful not to jump to conclusions when comparing predictions of psychoacoustic models with psychoacoustic data. Correlation does not imply causality. If a time-domain model (based on the timing of specific neuronal responses) succeeds in accounting for a given data set, that does not necessarily mean that the corresponding physiological processes are causing the corresponding conscious experiences (such as the experience of perceiving the pitch of a tone). First, the mind-body problem is not solved and may be intrinsically unsolvable. Second, auditory neural networks are large and complex, and everything that happens within such a network can be influenced by both time- and frequency-domain phenomena, the latter being related to the position of excitation on the basilar membrane.

Relevant neural networks have mixed temporal and spectral (tonotopic) characteristics (e.g., Erfanian Saeedi et al. 2016). The word "tonotopic" refers to the one-to-one correspondence between the frequency of a pure tone and the place of maximum excitation on the basilar membrane (Pantev et al. 1989). Given the complexity of relevant biological neural networks, pitch does not depend only on tonotopic processing (e.g., M. A. Cohen et al. 1995), nor does it depend only on temporal processing (e.g., Meddis and Hewitt 1991a, 1991b). From an ecological or evolutionary viewpoint, in which the brain takes advantage of whatever useful information is available, pitch probably and generally depends on a mixture of temporal and spectral representations.

Unanswerability

There is something about the relationship between the physical world and subjectivity that is inherently unknowable. David Chalmers (1995) called it the "hard problem of consciousness" (200). Like Chalmers, I assume here that at some level, the problem of consciousness is intrinsically unsolvable: we can know neither why nor how we have consciousness. Attempting to answer such questions invariably leads to confusion, and jumping to conclusions is unscientific. It is better to admit what we do not know.

The difficulty may be related to Gödel's incompleteness theorem in mathematics, according to which no system (in this case, the physical universe) can explain itself without recourse to something outside of itself. The theorem implies that there is something about consciousness that cannot be explained by the physics and physiology of humans and their environments alone (Buechner 2010).

I will therefore reject reductionism in this case, whether strong or weak, and support (or at least not reject) a kind of mysterianism. There is something mysterious about the subjective-objective distinction that we may never understand. Science has solved many difficult problems in the past, including some that previously seemed intractable. Mysterianism is not a spiritual or religious position; whereas religious people believe something, mysterians do *not* believe something. It would be a logical fallacy, and perhaps unrealistic or even arrogant, to conclude that a problem of this kind will one day be solved just because other difficult scientific problems have been solved. Perhaps it will, and perhaps it will not.

As Chalmers (1995) pointed out, there are also "easy problems" of consciousness that can indeed be addressed. They include regular psychophysics and related neuroscience; questions about which parts of the brain are active in what kind of way when we have a certain kind of experience; the dependence of musical pitch on temporal versus spectral processing of sound in the brain; and questions about which parts of the brain are active, and in what way, when musical pitch patterns are being perceived. Those are nevertheless very complex questions, despite being easy relative to the hard problem of consciousness.

A Middle Path

Consider the following middle path. When asking relevant questions about the nature of reality and given the persistent tension between mind-body dualism and physical monism, we may argue that subjectivity does indeed exist, for without it our musical experience would not exist. However, subjectivity exists only as an inseparable epiphenomenon of the physical world. In other words, subjectivity corresponds in some way to the physical world but does not represent the ultimate cause of anything in the physical world.

This seems like a rational response to the unanswerability of questions about human consciousness. Rather than insisting on either physical monism or psychophysical dualism, we can adopt a compromise—a kind of limited dualism. We can agree that "consciousness is ontologically distinct from anything physical" (Weisberg n.d.), but a weaker version of that statement may be preferable, namely: conscious perception is ontologically distinct from physical measurement. Thus, I agree with property dualists that physical properties are distinct from mental properties, but not with substance dualists because the word "substance" is inappropriate to describe subjectivity. I don't believe that the mind and the body can be separated from one another: no "soul" or "spirit" leaves the body after death or during out-of-body experiences.

If in this approach the physical world is too complex for the purpose of understanding musical experience, one practical alternative is to take subjective experiences seriously, as if they really existed, and investigate them carefully and directly. If we decide to follow that route, we first need to be clear about the meaning of subjective experiences. Are we referring to them as things in themselves—Immanuel Kant's noumena? Or are we referring to what we reliably know about them—phenomena?

For practical reasons, and at the risk of oversimplifying Kant's theory, I will focus here on phenomena—observable facts or events—because I have no other choice. I can only talk plausibly about something that I can observe. If we treat subjective experiences as Kant's phenomena, we can express what we know about them in either words (qualitative description) or numbers (quantitative results of psychophysical or psychological experiments). Any subsequent discussion or analysis can then be based on either or both. That does not mean that the information is reliable, however. Our subjective descriptions are generally distorted by the subjectivity of our interpretations, which in turn depend on our specific circumstances (Carruthers 2011). Moreover, not all subjective experience can be accessed by introspection, much of it being ineffable (indescribable).

Consistent with ecological psychology (J. J. Gibson 1979) and 4E cognition (embodied, embedded, extended, enacted; Menary 2010), I will further assume that the contents of our consciousness correspond not only to the physical state of our brain, but also to the physical state of our body and our environmental and social situation. Subjective experience is as an emerging property of the relationship between organism and world, and conscious subjective experience often corresponds to affordances (cf. E. Gibson and Rader 1979).

The "Mind"

According to Oxford World Encyclopedia (2014), the mind is a "hypothetical faculty postulated to account for the ability of conscious beings to think, feel, will, or behave. The mind is considered to control, or consist of, so-called mental processes. Dualist philosophers, such as René Descartes, have distinguished between mind and matter as two totally independent entities."

The operative word here is hypothetical. The brain is observable, but the mind is not. Ecological psychologists complain that cognitive psychologists reify the English word "mind" by considering it to be a thing (Latin: *res*) or talking about it as if it was more specific or clearly defined than it actually is (Notterman 2000).

What is the mind, and does it exist? To answer that, it helps to analyze the mind's contents. Introspection suggests that its contents are of three different kinds:

- *Sensations and emotions* can sometimes be described using language, and sometimes not (they are "ineffable"). The experiences themselves may be considered fundamentally nonlinguistic. For example, I can experience the color red without labeling it.
- *Thoughts* are like imagined conversations. They include knowledge and anything that can be expressed using language. But thoughts cannot be transcribed exactly like an interview, and if we speak more than one language, we are sometimes unsure which language our thoughts are in. Thoughts are nevertheless linguistic in nature (or at least I am assuming them to be so), which makes them fundamentally different in nature from subjective experiences. An imagined scene or melody, in this definition, is quite a different thing from a thought, although imagined scenes and melodies are often linked to, or experienced simultaneously with, thoughts.
- *The self* (soul, spirit, ego) involves feelings of being, agency, intention, and control. As we have seen, neuroscience has questioned the existence of free will. But our impression of having free will is very strong, and if we did not believe in it, society would surely collapse. For one thing, it would be impossible to convict murderers because there would be no difference between accidental and deliberate killing.

In everyday language, we conveniently group this together and call it "mind." When James Taylor sang "In my mind, I'm going to Carolina" in 1968, he was referring to all three aspects: the imagined experience of being in Carolina, his thoughts about Carolina, and his personal identity as belonging to Carolina. But in a research context, the word "mind" can be misleading and problematic:

- The mind contains fundamentally different things. Why put all those eggs into one basket?
- If the mind is defined as everything that is not physical, the definition is negative. A positive definition is lacking.
- The mind, regarded as a container for the mind's contents, is not observable. Only the listed contents are observable.

As we have seen, the relationship between mind and brain involves a potentially unsolvable problem (Chalmers 1995). That being the case, it may be misleading, or at least premature, to talk in a specific way about the mind. The word is hard to define (so it is unclear what it means) and hard to translate (suggesting an arbitrary bias toward the English language). That casts doubt on expressions such as "functional architecture of the mind" or "modularity of the mind" in the tradition of Chomsky and Fodor. When cognitive psychologists compare the mind to computer software and the brain to hardware, they are surely oversimplifying the problem (or perhaps overcomplicating it). The mind does not have some kind of narrative center, and if we compare it to software, then it is like several or many programs running in parallel (Dennett 1991). The contents of the mind and its cognitive modules do not necessarily have clear physiological correlates.

Whereas I will often talk about subjectivity in this book, I will avoid talking about the mind. I will instead focus on interactions between people and the physical and social worlds in which they find themselves, and relationships between physical events in the real world (outside the brain and body) and subjective experiences, in the case where both the physical events and the corresponding subjective experiences are reliably measurable.

The Ecology of Speech and Music

In the history of psychology, the 1960s saw a cognitive turn led by Noam Chomsky (1959)—a reaction against B. F. Skinner's (1957) behaviorist approach to understanding language. In a nutshell, Skinner had treated humans like nonhuman animals, ignoring their conscious awareness. Chomsky focused on what was uniquely human: complex language. He tried to explain the uniquely human ability to generate coherent sentences in terms of abstract cognitive structures by means of cognitive processes (also called "mental processes").

In the 1960s, many psychologists were convinced by Chomsky and rejected Skinner. But Chomsky's solution was also problematic. Today, the ontological status of Chomsky's mental structures and processes remains unclear. Do they exist, and if so, in what sense? Is reification a problem, as it is with the mind? The discussion is ongoing.

Later, in the 1970s, there was (or could have been) an ecological turn in psychology (Withagen 2018), based on the work of James Gibson (1966) and Roger Barker (1968), in which behavior was understood to depend primarily on environmental interactions, and perception was considered direct and not mediated by internal cognitive structures. In a cross-cultural approach that spans philosophy and psychology, we can recognize ecological approaches to psychology as a rejection of Kant's (1781) idea that knowledge of perceived objects depends on cognition (reflection, prior knowledge, abstract thinking) such that direct, intuitive perception is impossible. According to Kant, concepts need percepts to exist, and vice versa. The ecological turn was less successful than the cognitive turn that preceded it, and it may have been suppressed by the political power of the cognitivists. Perhaps it was too easy to continue comparing the brain to a computer?

Within their academic niche, the ecological psychologists compared human with nonhuman animal behavior, but differently from Skinner. Instead of focusing on conditioning, habituation, reward, and punishment, they focused on environmental interaction. The ecological idea had interesting repercussions for hearing and music. The research area called sound studies investigates cultural implications of sound reproduction technologies, while soundscape ecology studies acoustic communication among living organisms (including humans) and their environments.

Chomsky proposed the existence of universal grammar: structural rules of language that are innate to all humans everywhere, independent of sensory experience (hence, independent of cultural and geographical context). Today, many psycholinguists are instead examining what actually happens in the empirically observable real world when language is being acquired, including the role of statistical learning in interactions between infants and caregivers (Saffran 2003). In such an approach, language becomes a side effect of other physical and social phenomena, and the question of universal grammar does not arise. The grammatical rules that cognitive linguists focus on in their analyses emerge gradually from countless imperfect examples of language. Relative to those examples, grammatical rules are abstract; like the mind, they risk being reified.

The same can be said for cognitive representations in general, including the information that is contained in a musical score or MIDI file, insofar as a musician or music listener has somehow stored it in memory. But there are interesting differences between musical scores and corresponding cognitive representations. Most interesting for this book, some tones sound more stable or more salient than others, and some tones are implied but not notated. In addition, musical scores evoke cultural associations that are addressed in the discipline of musical semiotics.

A statistical learning approach to cognition is ecological if it focuses on environmental interaction (Thelen and Smith 1994). It is consistent with the idea that subjective experience (Popper's World 2) arises from environmental interaction and affordances (aspects of World 1 that are relevant for survival and reproduction) as the organism picks up relevant information (World 3) from the environment (World 1). Organisms are constantly exploring the environment by perceiving and interacting with it; multimodal perception, action, learning, and self-organization are inseparable. The brain is part of the physiological, physical, and social worlds with which it interacts.

Regarding World 3, memory (in the general sense of information storage) is located not only in the brain, but also in the body and in the world—hence the idea of extended mind (Clark and Chalmers 1998). Today, we complement our memory with our mobile phones, and they in turn complement their memory with the Internet. Previously, we used paper notebooks for the same purpose, and each book referred to other books, creating a kind of memory network. Beyond that, every person or physical object with which we interact extends our memory capacity. For example, I can hardly prepare a meal without food, utensils, and some kind of stove, and the design of these objects must be familiar enough for me to be able to use them. It also helps if other family members are around to answer my questions. Therefore, my memory for cooking is partly encoded in those foods, utensils, stoves, and family members.

In such an approach, elements of grammar—nouns, verbs, adjectives, adverbs, prepositions—emerge from human interaction with the physical world (Roy and Reiter 2005). We perceive and recognize things, their actions, their features, the way they do things, their interactions. If there is anything innate about that, it is the general ability to form psychological categories (calling different kinds of dog "dog") and to attach arbitrary sonic labels to them (dog, chien, hund).

On that basis, we can argue that the sound of language, as perceived and experienced, has an ecological origin. We produce sounds with the vocal tract, but we also imagine them before producing them (audiation, auditory imagery). In everyday silent reading, we imagine what the text would sound like if it was read aloud (Abramson and Goldinger 1997), and even how our vocal tract would feel while doing that (J. D. Smith et al. 1995), which helps us understand the text. In music, performers imagine sounds before playing them, whether the music is improvised, sight-read, or played from memory.

We can similarly argue that linguistic grammar has an ecological origin. When we label groups of things (e.g., different kinds of dog), putting similar things together in categories, language naturally becomes hierarchical, implying that there is nothing special about hierarchy that needs to be genetically encoded in advance. A kind of natural grammar is present in the observable world and our interaction with it. Theoretical grammar (the rules for correctly speaking and writing language such as those found in grammar books) is an idealized approximation to real speech, which is full of grammatical errors such as incomplete sentences. Music is similarly full of errors, but they tend be less clearly defined, and musical modernism may approach what might be called an "anything goes" aesthetic. In music, tones, intervals, and rhythms are also perceived categorically, and musicians apply labels to the categories, whereas in language, phonemes, syllables, words, and phrases are categorized. We should therefore not be surprised to find strong analogies between musical grammar and the grammar of language (Lerdahl and Jackendoff 1983).

Children acquire intuitive knowledge of linguistic syntax by interacting with caregivers and the physical world. Similarly, musicians acquire intuitive knowledge of musical syntax by interacting with music and the people who produce it, starting early in life with motherese. The analogy is not watertight because musical functions are less clearly defined and more ambiguous than parts of speech. Eric Clarke (1989b) argued that "certain theoretical distinctions that may be useful in linguistics, such as the distinction between syntax and semantics, maybe inappropriate for music. Instead, a more general framework based on semiotics and communication theory may offer a more flexible and ultimately more powerful context within which to assess both the general relations between language and music as well as some of the more specific correspondences" (9).

Whereas syntactic ambiguities in language are unusual and warrant special mention (e.g., "Helen saw Jill on the hill with the telescope"), musical ambiguities are commonplace. In Riemann's harmonic theory, for example, an E-minor chord in the key of C major is related both to a C-major chord (*Tonika-Leittonwechselakkord*) and a G-major chord (*Dominant-Parallel*). Many other music-syntactic ambiguities can be found in examples provided by Lerdahl and Jackendoff (1983); see also Jackendoff and Lerdahl (2006).

In both music and language, children have an innate, socially driven tendency to acquire the rules in personal interaction, despite the ambiguities. As they do that, they produce increasingly complex snippets of language and music that their caregivers consider to be increasingly correct. In both cases, the process includes attaching meaning to quasi-arbitrary sounds and storing the information in neural networks. The act of accompanying an infant on its journey toward language and music—constantly playing with sounds created by the vocal tract, repeating fragments, learning to articulate in seemingly endless repetitions, in constant embodied interaction with people and the world—shows us what is important about language and music acquisition. The important elements are the child's innate sociality, its constant motivation to interact, and the multimodal complexity of the social interactions within which language and music are acquired. The rules of linguistic grammar and music theory are secondary.

Music and language are similar to everyday sounds in some ways and different in others. One difference involves the locus of attention. Everyday auditory perception is distal, focusing on sound sources and their affordances—the physical world, or Popper's World 1. Music perception, like the perception of visual art, tends to be proximal, focusing on the sound's quality or acoustic surface (e.g., pitch and timbre), or World 2. In language, the focus is on something else again: the meaning of the language, in World 3. Despite these differences, we can learn interesting things about music perception from research on the perception of everyday sound and language. An important link is the theory of auditory scene analysis, which will be discussed next.

5 Auditory Scene Analysis

The auditory system constantly and effortlessly recognizes sound sources in the environment: people talking, dogs barking, wind in the trees. The same system recognizes pitch-time patterns in music: melody, texture, activity, intensification, and so on. The theory of auditory scene analysis clarifies how the brain processes incoming audible sounds to construct a model of the outside world. Bregman's auditory scene *is equivalent to Terhardt's* spectral-pitch pattern*—a graph of (subjective) spectral pitch against perceived time, a third dimension being perceptual salience. The auditory scene comprises only audible features; its contents are limited by masking among partials and noises. A sound's or partial's audibility depends on parameter settings such as window length and spectral acuity, which in turn depend on evolution. Auditory scene analysis allows familiar patterns to be recognized by a combination of primitive (innate) and schema-based (learned) processes. In an ecological approach, environmental sound sources and their affordances are perceived directly, and auditory scene analysis happens instantaneously.*

Background

Albert Bregman's (1990) theory of auditory scene analysis was not a music theory, nor was it expressed in musical terms, but the implications for music theory were profound. The scientifically oriented music theorists who took advantage of the connection included David Huron (2001), Stephen McAdams (McAdams and Bregman 1979), and James Wright (Wright and Bregman 1987).

Bregman's approach was compatible with Ernst Terhardt's (although the two researchers worked completely independently). Terhardt (1972) focused on one aspect of what Bregman would later call auditory scene analysis: the grouping of simultaneous partials by harmonicity, an aspect that Carl Stumpf (1883, 1890) had earlier called fusion (*Verschmelzung*; Schneider 1996). Like Krumhansl (1990), both Terhardt and Bregman used computational methods for various purposes, including the design of experiments, the analysis of data, and the construction of models. Both also acknowledged the importance of subjectivity and phenomenology (Bregman 2005; Terhardt 1998).

Bregman's background was in philosophy and psychology. For his PhD, he studied memory in an empirical psychological approach. Along the way, he discovered auditory grouping effects (also called auditory stream segregation or streaming). Later, he coined the term "auditory scene analysis," which became a new field of research. Although based in cognitive psychology, the theory of auditory scene analysis is influenced by ecological psychology: how an organism interacts with the environment, depending on affordances (Windsor and De Bézenac 2012).

Spectral Analysis

The input to auditory scene analysis is a running spectrum. To understand that, we need to understand how the ear separates high from low frequencies.

Sound is an aurally perceptible vibration. It is air-pressure variations in the frequency range of human hearing. Sound is perceived when the peripheral auditory system (in particular, the basilar membrane within the cochlea) processes incoming sounds. The basilar membrane resonates to low frequencies at one end and high frequencies at the other, and there are additional neural tuning processes.

Spectral analysis is a mathematical procedure. The input is a waveform (not necessarily periodic—it can be almost any mathematical function) that is confined to a time window. The output is a spectrum: a graph of amplitude against frequency. Since the shape and duration of the analysis window is variable, not all spectral analyses are the same. It can therefore be misleading to refer to *the* spectral analysis or *the* running spectrum of a sound.

The basilar membrane performs a kind of running spectral analysis. The membrane can be imagined as a series of overlapping band-pass filters (e.g., the hair cells themselves) that only admit acoustic frequencies in a given range. The effective bandwidth of these filters is called critical bandwidth (H. Fletcher 1940), equivalent rectangular bandwidth (B. C. J. Moore and Glasberg 1983), or *Frequenzgruppenbreite* (Zwicker 1954, 1961), and it corresponds approximately to a constant distance of 1 mm along the basilar membrane (Greenwood 1961). Relative to musical pitch, critical bandwidth is two to three semitones at higher frequencies and 50–100 Hz at lower frequencies, depending on how it is measured. It is approximately proportional to the logarithm of frequency at high frequencies and approaches proportionality with frequency at low frequencies.

The effective parameter settings of spectral analysis in human hearing are a consequence of evolution. Hearing has evolved to promote social interaction and survival. So, spectral-analysis parameters are optimized for understanding speech and detecting environmental sounds in ancestral settings that could spell mortal danger.

That optimization process involves the relationship between temporal and spectral analysis. At any moment in time, incoming sound is analyzed within a temporal window that lasts for a fraction of a second. The window's effective duration (i.e., the uncertainly in the ear's temporal estimates) changes with frequency. It is also inversely related to the effective width of the corresponding spectral window (critical bandwidth, or the uncertainty in the ear's frequency estimates). This time-frequency uncertainty principle (Oppenheim and Magnasco 2013) is a hard physical constraint: it is physically impossible to improve both the temporal and the spectral discrimination of the ear (or any other frequency analyzer) simultaneously because they depend inversely on each other. The longer the effective window, the smaller the critical bandwidth: in that case, frequency is perceived more accurately (simultaneous pure tones of almost the same frequency are better distinguished), but time and rhythm are perceived less accurately. Conversely, the shorter the effective window, the wider the critical bandwidth: we perceive frequency less accurately, but we perceive time and rhythm more accurately (our ability to discriminate rhythmic patterns improves). There may be individual differences in these parameters; for example, musicians may have superior pitch discrimination (Nikjeh et al. 2008).

The effective window length of human hearing is short enough to separate different sounds in fast speech, where the rate of syllables is typically four per second. It is also long enough (25 or 200 ms according to Luo and Poeppel 2012) to allow the frequencies within a single harmonic complex tone to be separated from each other—up to about the 10th harmonic in the central musical range—so that fundamental frequency can be tracked at the same time as voiced speech sounds such as /i/ and /u/ are separated in noisy environments. This evolutionary compromise between temporal and spectral accuracy affects how we experience all sound including music.

In a first approximation, both temporal and spectral windows can be imagined as square, with sharp edges. But reality is more complex: auditory spectral and temporal windows have different shapes that are related to each other and can be investigated empirically (Glasberg and Moore 1990; Moore et al. 1988).

Masking

Not all sounds are relevant for hearing. Sound must be audible to be relevant. Sounds that are masked such that they become completely inaudible, contributing nothing to sound as experienced or sound as it affects behavior, are aurally irrelevant. Psychoacoustic masking is part of spectral analysis: the longer the effective temporal window, the more clearly simultaneous partials can be separated, and the less they mask each other.

To illustrate that, imagine two people talking while walking down the street. A large, noisy bus goes past. For a moment, they cannot hear each other and have to guess what was said or repeat themselves. Masking of that kind depends on interactions between individual partials in complex sounds (H. Fletcher 1940). The degree to which two simultaneous pure tones can be discriminated by the basilar membrane depends on the the frequency difference between the tones (or the distance between the two corresponding maxima on the basilar membrane) and the effective filter bandwidth at that frequency (critical bandwidth). The effect is complicated by beats and difference tones; a clearer picture can be obtained if, instead of a pure tone, a narrow band of noise is used as the masker (Egan and Hake 1950; Terhardt 1998).

Masking is an important aspect of musical ensemble performance. A piano accompanist can play loudly and still not drown out a violinist or flautist who is playing very high pitches because in that situation, the main sounds from each instrument are relatively distant from each other in frequency. A piano accompanist has to be more careful if the soloist is a cellist or bassoonist playing relatively low pitches.

The Auditory Scene

The input to auditory spectral analysis is a waveform, visualized as a graph of air pressure against time. The output is a graph of (spectral) pitch against

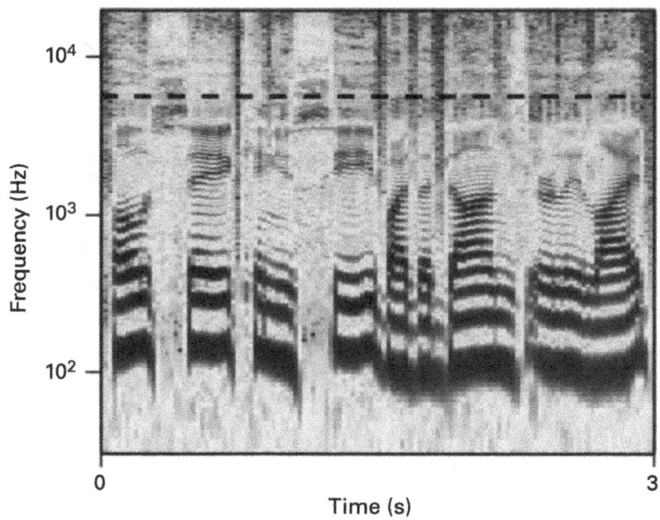

Figure 5.1
Example of a running spectrum. The horizontal bands represent the harmonics of a speaking voice. Where the horizontal bands stop, the sound is no longer voiced. Sounds with more energy at higher frequencies are brighter, whether voiced (e.g., /i/ rather than /o/) or unvoiced (e.g., fricatives). From B. B. Monson et al. (2014).

perceived time, illustrated in the case of running speech in figure 5.1. This is Bregman's (1990, 1993) *auditory scene,* also called auditory running spectrum or *spectral-pitch pattern* (Stoll 1982). It depicts the changing audible tone and noise components of a complex sound. I will assume that it represents everything that is audible (or aurally relevant) in a sound. It is also the input to further cognitive sound processing. Whereas the input to each ear is two-dimensional (the dimensions being time and sound pressure), the auditory scene is three-dimensional (time, spectral pitch, and salience), although it is usually shown graphically in two dimensions (time and spectral pitch).

Before the industrial revolution, an audible pure tone of constant frequency was seldom heard (except perhaps as tinnitus). Today, such tones are often heard, generated by loudspeakers and machines that (often unintentionally) produce hums and whistles of constant frequency. In a graph representing the auditory scene, a pure tone is a horizontal line. If the frequency of the tone gradually falls, the line falls. If the tone is complex and the frequency ratio between pairs of partials remains constant, the partials are represented by parallel horizontal lines. Bands of noise are also represented.

Perception is a process whereby the organism reconstructs the outside world from the incomplete information received from the senses, focusing on affordances: aspects of the environment that are important for survival and reproduction. The perceptual reconstruction of sound sources is an example of a more general ability to construct reality and meaning from a combination of sensory input, past sensory experience, and knowledge (Raskin 2001). Our brain may construct a model of a person talking based on the sound of speech; or it may construct a model of a person playing a bassoon from the sound of the music. McAdams (1984) called this construct an *auditory image*—a metaphor that is useful for understanding the structure of musical compositions such as pitch-time patterns and orchestration. Whereas he defined an auditory image as a "psychological representation of a sound entity exhibiting a coherence in its acoustic behavior" (291), his idea is perhaps best conveyed by this passage: "Imagine that you are walking blindfolded through the streets of a city. What do you hear? A combination of chugging and whirring metal and the popping of rubber on coble stones is heard as a passing car. A rhythmic clicking of toe nails and jangling of small metal medallions is heard as a dog trotting by. A small herd of children goes giggling and screaming by on bicycles. You walk past a jack-hammer pounding the street with metal and your ears with painful pressure waves" (298).

In each of these examples, the auditory system constructs the sound source based on incomplete auditory information, combining that with information retrieved from memory about the kinds of objects that normally produce these kinds of sounds, how they are normally perceived in all senses, how they normally behave, how they are experienced emotionally, and so on. The aspects of these sound sources that are most important for us are their affordances: the danger that the car might run into us (or stop and pick us up, if we want a lift), the possibility that the dog might be friendly (or angry, biting us on the ankle), the delight that so many children are having so much fun (or the annoyance that they are disturbing the peace), and the threat that the jackhammer might damage our hearing (or the satisfaction that urgently needed roadworks are at last happening).

McAdams (1984) also emphasized the active nature of sound perception, both in everyday life and in music: "What is most compelling as a result of all of the research on auditory organization is the fact that the will and focus of the listener play an extraordinarily important role in determining the final perceptual results. Music listening (as well as viewing visual arts or reading a poem) is and must be considered seriously by any artist as a creative act on the part of the participant" (319). The listener can actively choose what aspects of the auditory scene, including what implied patterns in the scene, to focus on (attend to). In that way, music listening can be *performative*—an interesting concept for understanding post-tonal music (Utz 2022) that is consistent with an ecological approach to perception. How we hear music changes as our auditory system interacts with music and learns from the interactions. Adopting this perspective, Helga de la Motte-Haber and Reinhard Kopiez (1995) considered representations of music in memory and the aesthetics of new music reception, including electroacoustic music.

Spectral-pitch-time patterns can be used to recognize objects of all kinds in the physical environment by hearing alone (Bregman 1990). But that does not mean we first track the pitch and then recognize the object. On the contrary, the brain may first recognize an object and only then extract its pitch, timbre, loudness, and so on. Neil McLachlan and Sarah Wilson (2010) explained: "From an ecological and evolutionary perspective, when the auditory system detects a new sound it should first decide on its identity (to decide whether it is a threat or opportunity), then its location, and finally what other information may be contained within the sound. This primacy of sound source identification is likely to be reflected in neurobiological mechanisms of audition, where identification mechanisms would initiate quickly and interact with a broad range of other mechanisms" (175).

If that is true, our subjective experience of pitch may depend more on environmental interaction than on neurophysiology. Moreover, to the extent that pitch depends on neurophysiological states and processes, those states and processes can be seen as an end result of multigenerational, evolutionary (phylogenetic) developments as earlier generations tuned themselves to their typical physical and social environments. Whereas the brain is constantly changing, being plastic in both phylogeny and ontogeny, our experience of musical pitch, and subjective experience in general, depends primarily on environmental interaction.

It is not easy for a listener to analyze the contents of the auditory scene consciously because perception normally focuses on affordances, which in this case are sound sources in the environment. To help us monitor sound sources, our brain automatically groups together parts of the auditory scene and separates other parts according to Gestalt principles. These processes mostly happen without conscious awareness; we are only aware of the output.

How, then, can we know the contents of the auditory scene? Psychoacousticians answer this question either empirically or theoretically. Empirically, they investigate the audibility of specific sounds in the auditory scene, which depends on masking; theoretically, they model corresponding perceptual processes.

Perceptual Grouping

Auditory scene analysis involves assigning complex patterns of sound to sources in the physical and social environment, so that the perceiving animal can better interact with the environment and hence better survive and reproduce. It involves pattern recognition by biological neural networks, in which the input is the auditory scene and the output is the object(s) that are recognized and their perceptible features. Any sound source with a familiar frequency–time pattern can be recognized from its characteristic Gestalten, from rain falling on a tin roof to a cat saying "meow."

If we hear the same kind of sound many times, each time with small variations (e.g., the experience of seeing and/or hearing a dog barking), we learn what these not-quite-identical repetitions have in common. In ecological psychology, the commonalities are called *perceptual invariances*. Our brains store information about these invariances and "look for them" subsequently in incoming sounds. If a new sound is similar to a familiar sound, we link the new sound to the familiar sound and in that way perceive the sound source. When we hear the sound of a dog barking in everyday life, our attention is focused not on the sound itself (proximal perception) but on the source of sound (the dog; distal perception). That is especially true if we are in danger of being attacked (i.e., if the dog affords danger), but also if it affords companionship or protection.

Aural recognition of sound sources can be divided into two processes. The first is the grouping together of elements within a complex sound. A truck going past makes several sounds simultaneously: the roar of the engine, the swish of the tires on the ground, the rattle of the loose exhaust pipe. If we are familiar with the sound of trucks, our auditory system guesses that these sounds belong together, and we experience just one truck, despite the many sounds. The second process, which happens in parallel with the first, involves separating the foreground from the background. If our attention is focused on the truck (because we are afraid of being run over), this sound is in the foreground. Recognition of the truck involves not only joining together the elements of the truck sound, but also separating them from other sounds in the background such as other vehicles, people talking, the rain falling, or music playing.

Similarly, music is often heard as foreground melody with background accompaniment. The origin of familiar foreground–background contrasts in music may be traced to more general nonmusical perceptual principles. That is an example of how auditory scene analysis in music is based on auditory scene analysis in everyday interaction with the world. Auditory scene analysis can also explain how tones are grouped in musical structures. Several different perceptual cues are involved, and each one has interesting musical implications.

The first and simplest cue is the onset time of each partial. If several pure tones begin at exactly the same time, they are probably partials of a complex tone produced by a single sound source. We are therefore more likely to hear the partials as belonging together (perceptual fusion). If the tones in a musical passage tend to begin simultaneously, the listener is more likely to experience them as fused. If they do not, the listener is more likely to hear separate melodies. If a composer wants a listener to hear a progression of fused chords, the rhythm will be almost the same in all voices, whereas if the listener should hear the voices separately, each voice will have a different rhythm (Huron 2001).

A second cue that causes partials to be grouped, and one that plays an important role in this book, is harmonicity. If we hear several spectral pitches (pitches of pure tones or partials), we are more likely to hear them as fused (as if coming from a single source) if they correspond to part of a harmonic series. That is because many environmental sounds have harmonic spectra, as do human voices and many musical sounds (wind and string instruments), whereas others do not (bells and drums). The principle of harmonicity can explain aspects of the harmonic vocabulary of MmT. Major triads and Mm7 chords tend to fuse perceptually because their spectra are similar to a harmonic series.

A third cue that causes partials to fuse into one sound experience is coherent modulation of partials. If partials change simultaneously and in the same way, the brain thinks they are from the same source. Partials may get louder and softer, or higher and lower in frequency, at the same time and in parallel. Again, the idea can explain aspects of the classical theory of harmony and counterpoint. Parallel perfect intervals (octaves and perfect 5ths) are often considered undesirable or forbidden, as is similar motion in general (contrary motion tends to be preferred), if the intention is to make the separate parts audible. If, however, the intention is for the parts to blend, they may move exactly, or almost exactly, in parallel. Examples of the latter can be found in *La cathédrale engloutie* by Claude Debussy and Maurice Ravel's *Bolero*; in both cases, at some point the main melody is played in parallel chords.

These ideas also have implications for music performance. If you are playing multivoiced music and want to achieve perceptual fusion, you need to play both in time and in tune—within certain tolerance limits. You should also avoid vibrato because vibrato in one voice will be out of time with vibrato in another voice, contradicting the principle of coherent modulation and undermining perceptual fusion.

The foundations of auditory scene analysis first began to be understood during the scientific revolution when mathematicians developed the idea of spectral analysis. That inspired Rameau to explain the foundations of musical harmony in terms of the harmonic series, a phenomenon that had only recently been understood by Sauveur and d'Alembert as they investigated the physics of vibrating strings (Christensen 1987). That it took such a long time for Western music theorists to realize they could hear out harmonics from complex tones is surprising if traditions of overtone singing already existed in Central Asia, North America, and South Africa (Pegg 1992). Even Aristotle had suggested hearing out the 2nd harmonic when he wrote: "Why is it that in the octave the concord of the upper tone exists in the lower, but not vice versa?" (*Musical Problems*, cited by W. Thomson 1958, 44).

The grouping criteria that are applied in auditory scene analysis may at first seem universal, applying to all (musical) cultures. But they are also susceptible to learning, and so there are also intercultural differences. Gamelan musicians in Indonesia may respond differently to the harmonic series as it occurs in musical sounds due to their extensive experience of, and active interaction with, the nonharmonic complex tones of gamelan bells (McLachlan, Marco, and Wilson 2013).

6 Auditory Sensitivity

The ear is remarkably sensitive to small frequency differences. Small intervals and complex musical scales enable large numbers of pitch structures to be differentiated. Similarly, tiny timbral differences can be heard between real and synthesized musical instrument sounds. The ear is also sensitive to subtle rhythmic nuances. That can explain why pitch and time are the most important parameters in standard Western music notation, whereas loudness and timbre are notated only qualitatively ("ff," "crescendo," "flute," "pizzicato"), and monaural phase relationships are musically irrelevant. The ultimate origin of these sensitivities is ecological and evolutionary. They arise from the physics, physiology, and psychology of auditory perception, in the everyday settings in which our ancestors evolved. Direct and reflected sounds combine before reaching the ear, and reliable information about sound sources is extracted from that combination. Our sensitivity to pitch means we are also sensitive to pitch shifts that arise from physiological nonlinearities or distortions in auditory spectral analysis as the ear struggles to cope with the enormous range of frequencies and the even more enormous range of amplitudes in everyday sounds. In typical harmonic complex tones in speech and music, about ten harmonics are aurally relevant; that is, they can affect conscious perception of pitch, loudness, timbre, or perceived duration.

Musical Just-Noticeable Differences

Why does musical structure mainly involve pitch-time patterns—patterns of pitch and onset time? Why not patterns of other perceptible parameters such as timbre, loudness, and perceived duration? Why is melody, considered as a pitch-time pattern, so important in music? Why is (Western) music notated like a graph of pitch against time, with pitch on the vertical axis (the musical staff) and time on the horizontal axis (note values and measures separated by barlines)? How is it possible, after many centuries of musical creativity and countless compositions in diverse musical styles, that musicians are still composing tonal melodies that are perceived to be original—even when the same old harmonic clichés are repeated over and over again?

These questions can be answered in two main ways. The first involves the recognition of patterns in pitch-time space, which as we saw in chapter 5 is a central process in auditory perception. The second involves the auditory system's sensitivity to pitch and onset time by comparison to other parameters.

The ear's sensitivity to pitch can be evaluated by counting just-noticeable differences (JNDs) of frequency across the range of hearing. Josh McDermott, Michael Keebler, and colleagues (2010) commented: "The typical pitch JND of about a fifth of a semitone is very small compared to the dynamic range of pitch (roughly 7 octaves, or 84 semitones); intensity and brightness JNDs are a much larger proportion of the range over which those dimensions can be comfortably and audibly varied" (1949). Moreover, "the widespread use of pitch as an expressive medium may not be due to an advantage in supporting complex structures involving intervals and contours, but rather in the ability to resolve small pitch changes between notes" (2010).

Consider the following order-of-magnitude estimates. If the JND or difference limen for the frequency of two successive tones is 10 cents or 1/10 semitone across a ten-octave range, then we can distinguish some 1,000 pitches. Note here that the JND for harmonic complex tones (such as musical tones) is smaller than that for pure tones in the central pitch range (Henning and Grosberg 1968), and the JND tends to be smaller in the middle pitch range and larger at the extremes (for a review, see Parncutt and Cohen 1995). If, by comparison, the JND of SPL is 1 dB across a 100-dB range, we can distinguish 100 loudnesses. Very roughly, the ear can distinguish ten times more pitches than loudnesses.

Similarly, music notation can represent many more pitches than loudnesses. Each of the eighty-eight keys on the modern piano can be represented as a pitch in music notation, and the ear can hear smaller intervals and distinguish different tunings of the same note. That allows composers and performers to generate a very large number of distinguishable pitches, an even larger number of scales, and an even larger number of possible pitch patterns within those scales. By comparison, the number of possible dynamic (loudness, intensity) indications is very small: *pianissimo*, *piano*, *mezzopiano*, *mezzoforte, forte, fortissimo,* and a few others. Again, we distinguish roughly ten times more pitches than loudnesses.

The ear's extraordinary sensitivity to spectral frequency is illustrated by experiments with musical sound synthesis. Julius Smith (1996) explained: "A musical instrument should be 'alive' in the hands of the performer. A performance is naturally an interaction between the player and the instrument. While the main attributes of each note are predictable from a score, for example, many subtle qualities are unpredictable, contributing to the delight of the player and audience" (44).

Artificial musical sounds, when created by analog synthesizers, seldom sound natural. The sound of a single tone on an instrument such as a guitar or a trumpet is very complex and hard to imitate. Analog oscillators simply cannot do it, and even complex computer programs can fail. If a tiny sonic detail is not quite right, we recognize the sound as fake. Computer musicians and researchers in music acoustics and psychoacoustics have solved this problem by recording natural sounds and storing them in databases, with different recordings for different pitches and loudness levels (given the complex physical dependencies). They have also created physical models of musical instruments that simulate their complexity (e.g., Chaigne and Askenfelt 1994). In both cases, the importance of high fidelity in the imitation of natural sounds is evidence of the ear's extraordinary sensitivity to tiny changes in frequency.

The auditory system is also very sensitive to timing and rhythm, but for a different reason. The ability of the vocal tract and brain to produce and perceive rapidly changing sound signals is a physiological prerequisite for the emergence of language in evolutionary biology (Lieberman 1973). As prehumans evolved, the ability of the vocal tract and brain to deal with fast changes and timing differences improved as linguistic ability became increasingly important for survival and reproduction.

Monaural Phase

By comparison to its extraordinary pitch-time sensitivity, the ear is remarkably insensitive to monaural phase relationships (i.e., phase relations between partials heard with one ear only). Whereas the auditory system is very sensitive to phase differences between the two ears (which are important and well-known cues for sound-source localization), each of the two ears, when considered alone, is almost phase deaf—at least when it comes to perceiving complex, everyday sounds.

Consider a complex everyday sound pattern that is heard in one ear only. It could be a natural-music or speech signal, such as a short excerpt from a pop song or podcast. The phase deafness of hearing means that the excerpt will sound almost identical after phase relations between partials have been shuffled—even when the results are reproduced with the best high-fidelity audio equipment. That is true, even if the shuffling radically changes the waveform.

The effect can be demonstrated by performing a running spectral analysis with parameters corresponding to human hearing and resynthesizing with changed phase relationships (Heinbach 1988; Mummert 1997). When the sound is converted to a running spectrum, all information about phase relations is lost. When the sound is resynthesized, the new phase relations can be determined as follows: "To minimize rapid phase changes at the boundaries of two part-tone patterns, the sinusoid's initial phase depends on whether or not a nearby part tone (within ±0.05 Bark) exists in the previous part-tone pattern . . . The quality of the resynthesized signals is very good and in many cases perceptually indistinguishable from the sampled original signal" (Heinbach 1988: 118; one Bark corresponds to one critical bandwidth). If the listener hears no difference between the original and resynthesized sound, that is evidence that monaural phase relations are perceptually irrelevant.

Differences in sensitivity to physical parameters can be explained in an ecological approach, in which perception is regarded as active interaction with the environment. Based on research on acoustic communication systems of insects, Heiner Römer (1998) explained:

> Behavioral ecologists, for example, focus their interest on the interactions between an animal and its conspecifics, its predators and prey, food resources, territories, etc. Like ethologists, they are primarily concerned with the adaptive value and/or the evolution of behavior rather than the mechanisms being used. By contrast, sensory ecologists are more concerned with the mechanisms that enable an animal to produce or utilize signals and how the information about identity or location of the sender is transmitted to the receiver(s) . . . Thus, in order to estimate the auditory performance of an insect in the real world, physiologists face a dilemma; they have to leave their properly controlled anechoic chambers and deal with the tremendous variability of the signal in the transmission channel resulting from many different factors acting on it and that are almost impossible to quantify simultaneously. For

> this reason, the sensory ecology of hearing in insects is still in its infancy, but there is increasing acknowledgement of the importance of ecological factors for the evolution of acoustic communication systems. (63)

Consider how frequencies, amplitudes, and phase relations of the partials of a complex sound are affected by acoustic reflections from the ground and objects in everyday human environments, and by superpositions of direct and indirect sound:

- The *frequencies* of the partials are completely unaffected, provided all senders, receivers, and reflectors are stationary (otherwise there could be a Doppler effect). Environmental sound sources such as people speaking, dogs barking, or cars driving past produce tones with continuously changing frequencies. Because those frequencies are unchanged by reflection, frequency contours provide the auditory system with reliable information about sound sources. In the long evolutionary prehistory of humans, primates, and other animals, auditory systems have become highly sensitive to spectral frequency because that sensitivity enables reliable information about sound sources to be detected, which in turn promotes survival and reproduction.
- By contrast, *phase* relationships between direct and indirect sound are quasi random: they depend on the distance of both source and perceiver from relevant acoustic reflectors, including the ground. Therefore, the superposition of direct and indirect sound, which happens before sound reaches the ear in everyday environments, effectively shuffles the phase relations among the partials. The shuffling effect of reflections on phase relations is especially strong in concert halls, where music is often heard or recorded (Griesinger 2015). Phase relationships therefore do not provide the ear with reliable information that could help a source of sound to be recognized, except in the first few tens of milliseconds before the arrival of the first reflection.
- The *amplitudes* of the partials represent a less reliable source of information about sound sources than frequencies, but more reliable than monaural phase relations. Amplitudes change relative to each other when reflectors absorb some of the sound energy, the proportion depending on material and frequency. For example, thick curtains absorb most high-frequency energy. For that reason, the auditory system in monaural perception is less sensitive to amplitude than to frequency.

The ear cannot rely on phase relationships among partials for information about sound sources in the environment. The frequency-time patterns of the auditory scene are much more reliable and informative. That observation, which depends on both environmental interaction and evolution, can explain why the auditory system is so insensitive to phase relations among the partials of a complex tone (as heard with one ear, or the same in both ears).

In fact, monaural phase relations (between partials heard simultaneously in one ear) do perceptually affect timbre—to a small but measurable extent (Plomp and Steeneken 1969). The research has tended to document effects of phase differences within artificial tones in controlled experiments, but not in real musical tones or speech. The effects may be confined to sounds that are relatively unimportant in everyday life, such as piano tones in extreme registers (Galembo et al. 2001). Changing phase relations can also affect pitch ambiguity (e.g., the probability that the 1st as opposed to the 2nd harmonic will be pitch matched in an experiment), but not exact pitch (Meddis and O'Mard 1997; Shackleton and Carlyon 1994); Brian Moore (1977) commented that "results clearly show an effect of relative phase of the components on the pitch of complex tones, although the effect is generally a change in the distribution of matches among preferred values, rather than a change in pitch values" (355–356). A partial that is out of phase with other partials is more likely to be heard separately (Kubovy and Jordan 1979).

Ray Meddis and Michael Hewitt (1991b) commented that "it has been widely believed for some time that the human ear is phase insensitive. However, evidence has accumulated in recent years . . . that the phase relationships of tone components can produce clearly perceptible effects under certain circumstances. Nevertheless, there are many situations in which the manipulation of phase leads to no noticeable effect. It is an important test of any model that it should be phase sensitive for certain stimuli but phase insensitive to others" (2883).

From a neuroscientific viewpoint, the ear's tendency toward monaural phase deafness is surprising, given that the physiology of pitch perception in the brain is largely a time-domain process (Patterson et al. 2002). Changing phase relations can drastically change the waveform, and physical representations of spectral frequency in the central range along auditory pathways are often temporally phase locked to the incoming stimulus (Heil and Peterson 2017). But we tend to experience the sounds as if there were no phase locking or as if the phase locking were unimportant. There is a striking mismatch here between what we experience when we perceive sound (World 2) and what seems to be going on in the brain at the same time (World 1).

The empirical findings on phase (in-)sensitivity are consistent with an ecological approach, in which perception is regarded as an active process of exploration and interaction with the environment, and affordances are crucial. It is also an example of differences among Popper's three worlds, reminding us to separate Worlds 1 and 2 when investigating the perception of pitch-time patterns in music. Our experience of sound is quite different from the simplest physical representation of sound, the waveform; subjective experience is sometimes unaffected by radical changes in the waveform. Our experience depends more directly on the auditory scene. If the auditory scene contains only audible (not completely masked) partials or noises, any change in the auditory scene will change the experience of sound.

Psychophysical Scaling

A basic issue in psychoacoustics is the quantitative relationship between the pitch and the frequency of a pure tone (regarding both as one-dimensional parameters). One approach is to scale pitch such that it doubles when the impression of pitch height doubles. The result is called the mel scale (S. S. Stevens et al. 1937; S. S. Stevens and Volkmann 1940).

The shape of a graph of pitch against frequency then appears to depend on two main factors. The first is the structure of the basilar membrane in the inner ear. Pitch in mel appears to correspond to distance along the basilar membrane (Zwicker and Feldtkeller 1967/1999). The second factor is familiarity with music, in which the tones on the piano keyboard are perceived to be equally spaced (at least in the middle registers). In fact, they are equally spaced in the logarithm of frequency. Listeners who are familiar with this logarithmic relationship in music tend to hear a logarithmic relationship between the pitch of any tone, pure or complex, and (fundamental) frequency over a wide range of several octaves.

The result of mixing these two relationships is that the relationship between frequency and pitch approaches a logarithmic relationship (pitch depending on log frequency) except in low registers, where it approaches linearity. Much the same applies to the relationship between position of maximum excitation on the basilar membrane (distance from the apex) and the frequency of a pure-tone stimulus, consistent with the dependence of spectral pitch on position on the membrane. These general conclusions apply for cats, guinea pigs, chinchillas, monkeys, and gerbils as well as humans (Greenwood 1990). However, research on the mel scale is intrinsically approximate, the results depending on method and the musical background of the participants. These complications have little effect on the sense of pitch height in music, which is based primarily on recognition of specific intervals (octave, major 3rd, etc.) rather than pitch distances on a continuous scale, and uses complex rather than pure tones, for which perceived pitch is scaled closer to log frequency.

The dependence of critical bandwidth on frequency (related to the mel scale) can explain why polyphonic musical patterns, defined by their musical intervals and rhythms, are harder to hear at lower frequencies due to masking. It can also explain why musical patterns sound more dissonant when transposed into the bass register: there is more sensory roughness. These constraints affect the way polyphonic music is structured (Huron 2001).

Aural Relevance

A complex tone produces an excitation pattern on the basilar membrane in the inner ear. The pattern has local peaks corresponding to resolved partials. If there is no peak, as is typical for harmonics of complex tones above about the 10th, the corresponding partial is inaudible.

In their pitch algorithm, Terhardt et al. (1982b) explained that "the extent to which a tonal component is 'aurally relevant' is described by the so-called sound-pressure-level excess (SPL excess) LX. The SPL excess has been defined as the difference between the SPL of an individual component and that SPL which represents the masking power of the rest of the spectrum, taking account of the ear's frequency selectivity" (682). There is an unstated assumption here: a partial that is not aurally relevant is assumed not to influence auditory perception in any way.

If we accept that definition of aural relevance, the next task is to operationalize it in an experiment. In each trial, the listener might hear a complex sound, followed by a short silence, and then the same sound again, but with a small difference: in the second sound, the partial in question may or may not have been removed. The listener would then be asked whether the sounds are the same or different, or to rate their similarity. In different trials, the order of the two sounds would be changed and the same question asked.

In the same series of trials, other such pairs would be presented, in a random order that was different for every listener. The duration of the sounds would be typical of music or speech (say, 300–500 ms). If a group of trained listeners consistently reported that two such sounds were different when they were in fact physically different, and the same when they were in fact the same, then the partial in question would be aurally relevant.

Other researchers have asked a different question. Can a listener focus attention on a specific partial? If so, the partial is *resolved*. According to Christopher Plack and Andrew Oxenham (2005), "for a harmonic to be 'resolved', a trained listener must be able to hear out the harmonic as a pure tone with a distinct pitch" (22). The experiment reported by Reinier Plomp (1964) is an example. Listeners could hear three sounds, one at a time, by turning a switch. The first sound was a harmonic complex tone with twelve harmonics of equal amplitude; the second was a pure tone corresponding to one of the partials; and the third was a pure tone lying midway between the partials. The task was to decide which of the two pure tones corresponded to a partial in the complex tone. Plomp found that listeners could resolve five to eight harmonics of a harmonic complex tone for fundamental frequencies over a wide range, from 64 to 2,000 Hz. Along similar lines, Brian Moore and Kengo Ohgushi (1993) found that listeners can hear whether a successively presented pure probe tone is higher or lower than a partial within an inharmonic complex tone when the spacing between the components exceeds about 1.25 times the equivalent rectangular bandwidth or critical bandwidth. In those cases, the partial was resolved.

In this book, the *audibility* of partials in musical sounds usually refers to relevance—not resolution. That is first because a test of aural relevance is more sensitive; weaker or more heavily masked partials can be shown in this way to be audible. All aurally relevant partials should be included in music-theoretical considerations—not only those that listeners can focus attention on. Second, the listener's ability to focus attention on a partial (analytic listening) is not the issue here. We do not normally need that ability in everyday life, nor do musicians normally need it. In psychological methodology, it is problematic to ask experimental participants to demonstrate an ability that they may not have or do not usually need. The task of deciding whether two almost identical sounds are the same or different is more natural: in everyday life, we often attend to the difference between two sounds, and in music, we are sensitive to tiny differences between synthesized and real tones. An experiment about aural relevance is relatively ecological and holistic, whereas an experiment about the resolution of a partial is more cognitive and analytic.

The theory of virtual pitch assumes that virtual pitches depend only on spectral pitches, and spectral pitches are evoked only by aurally relevant partials. An aurally relevant partial may be unresolved in the sense that a listener cannot focus attention on it. We might therefore predict that unresolved pitches sometimes evoke or affect virtual pitches. That has indeed been demonstrated for both humans (e.g., Kaernbach and Bering 2001) and for nonhuman primates. In the latter case, Xindong Song and colleagues (2016) showed that "the pitch strength of a harmonic tone is dominated by resolved harmonics; pitch of resolved harmonics is sensitive to the quality of spectral harmonicity; and pitch of unresolved harmonics is sensitive to the salience of temporal envelope cues" (781).

Brian Moore and Stuart Rosen (1979) demonstrated that harmonic complex tones (such as those in speech and music) typically have high, unresolved harmonics that can contribute to virtual pitch, in agreement with Jan Schouten (1940) and Jan Schouten and colleagues (1962). Brian Moore (1989) argued that "Terhardt's model cannot explain the pitch produced by high, unresolved harmonics, whereas in our model they arise as part of the same process as is used for lower harmonics" (186). In fact, Terhardt had argued that high harmonics are aurally relevant if they contribute to pitch, as they did in the experiment by Moore and Rosen. If the SPL excess of a partial is positive, the partial is assumed to contribute to virtual pitch and other perceptible sound parameters, including timbre, regardless of whether a listener (even the most expert) can focus attention on it. Seen that way, the experiment by Moore and Rosen provided data against which the masking algorithm of Terhardt et al. (1982b) could be tested.

Moore was also referring to physiological processes that enable pitch perception—specifically, the relative importance of temporal and spectral information. Temporal information, including phase locking between auditory neurons and time intervals between peaks in the fine structure of the waveform, tends to be more important for unresolved partials, in which activation patterns on the basilar membrane overlap and maxima are close (B. C. J. Moore 1989). Conversely, spectral information is more important for resolved partials (Dai 2010). But Terhardt's algorithm does not address these physiological foundations, given that *all* spectral and virtual pitches may depend in complex, unspecified ways on *both* spectral and temporal information.

It may have been more appropriate to criticize Terhardt's masking algorithm or to suggest an improved version. The mathematical formulation for SPL excess in Terhardt et al. (1982b) was based on earlier work by Eberhard Zwicker and Richard Feldtkeller (1967/1999) and adjusted to model the aural relevance of individual partials (spectral pitches), but the formulation was not perfect and had not been extensively tested. The simple frequency-domain approach was problematic, given the known, complex role of time-domain processes (Terhardt 1998; Biberger and Ewert 2016), including comodulation masking release (Verhey et al. 1999).

An experiment to investigate aural relevance is more sensitive than an experiment to investigate the resolution of partials. Listeners can often hear the effect of a partial on a sound's pitch or timbre in a holistic manner, even if they are unable to focus attention on the partial. Michael Hall and Richard Pastore (1992) presented major and minor triads in which the 3rd of the chord was barely perceptible. They found that "accurate chord perception was maintained at distinguishing tone intensities nearly 20 dB below the threshold for separate detection" (752). In other words, listeners could hear whether the chord was major or minor when the tone that enabled that distinction was aurally relevant but unresolved.

As an alternative way to understand these distinctions, consider sound files in the coding format MP3. They are created by compressing regular digital sound-files, such as those in .WAV format. The compressed files are intended to sound exactly the same as the original: the average person should not hear any difference. Compression is achieved by removing parts of the running spectrum that are not aurally relevant according to psychoacoustic models and resynthesizing (Heinbach 1988).

Audible Harmonics

How many harmonics are relevant for the construction of musical chords? Is the 7th of a chord based on the 7th harmonic of a complex tone? What about the 9th or the 11th?

Those are old questions in the history of music theory. In a psychoacoustic approach, we can find out empirically how many partials are audible in a typical isolated harmonic complex tone, such as a voiced speech sound or musical instrument tone. The answer depends on two main factors: the physical characteristics of the tone (fundamental frequency and amplitude envelope) and the effective filter widths or window lengths of the auditory system as it generates a running spectral analysis (critical bandwidth).

The expression "number of audible harmonics" can be understood either physiologically or experientially. Physiologically, it refers to harmonics for which information is separately coded in the auditory nerve. It is difficult to investigate that empirically, so it is appropriate to take a more experiential psychoacoustic approach. That is especially true when the ultimate aim is to understand musical structure as subjectively perceived.

When aural relevance is operationalized as described above, the 7th harmonic of a typical harmonic complex tone in the middle of the musical range is usually aurally relevant but not necessarily resolved. That is because the interval between the 7th and 8th harmonics is close to a critical bandwidth. In the central pitch range, partials mask each other if their amplitude is approximately equal and they lie well within one critical bandwidth (B. C. J. Moore and Ohgushi 1993; Shackleton and Carlyon 1994).

For music-theoretical purposes, we may assume that about seven harmonics are resolved in typical harmonic complex tones (J. G. Bernstein and Oxenham 2003; Hartmann and Goupell 2006; Plomp 1964). But roughly ten are aurally relevant—on average, and regardless of musical expertise—in the sense that information about their separate spectral pitches is sent via the auditory nerve to the auditory cortex, and a listener can hear a difference if the partial in question is turned on and off. Consistent with that assumption, Tammo Houtgast (1976) found that the tendency to hear pitch at a subharmonic of a pure tone decreased as (sub-)harmonic number increased from seven to eleven. That being the case, the analogy between the root of a chord and the pitch of a harmonic complex tone depends on harmonics up to about the 10th. Higher harmonics are probably neither aurally nor music-theoretically relevant (see chapter 15).

III Musical Foundations

7 Matthew Shirlaw

A plausible theory of musical structure should combine older insights with recent research, and balance contributions from humanities and sciences. Which music theories of the past could serve as foundations? Many nineteenth- and twentieth-century scholars attempted to explain harmony scientifically. This chapter focuses on a Scottish contribution: that of Matthew Shirlaw from the early twentieth century. He addressed many pertinent issues, including the origin of harmony in harmonic complex tones, the role of the 7th harmonic, the origin of the minor triad (despite its complex frequency ratio), the origin of the major and minor scales, and the relationship between scales and harmony. From a modern perspective, Shirlaw's discourse was often problematic. Today, we can construct a more realistic theory by asking similar questions and taking advantage of late-twentieth-century developments in psychology and psychoacoustics.

Context

An impressive example of positivist theorizing about MmT was that of Scottish organist and musicologist Matthew Shirlaw (1873–1961). His book (Shirlaw 1917) was a comprehensive critical survey of theories of harmony, starting with Zarlino, Rameau, and Tartini and proceeding to Kirnberger, Hauptmann, Helmholtz, Oettingen, and Riemann before closing with some lesser-known English theorists. The book was mysteriously reprinted in 1955, when Shirlaw was more than eighty years old, as if nothing in musicology and music theory had changed in the meantime.

Soon after the reprint of his book, Shirlaw published a more concise statement of his position in *Music Review* with the confident title "The Science of Harmony" (Shirlaw 1957). In retrospect, it is surprising that the article was published at all. By the 1950s, music-theoretical speculation had become unfashionable, especially when it addressed questions about MmT. Music theory and musicology had moved on. Leonard Meyer (1957) was wondering about the relationship between information theory and musical meaning. Helen Mull (1957) was interested in the psychological effect of musical repetition. Heinz Kohut (1957) was asking about music's psychological functions. Kazu Nakaseko (1957) was investigating symbolism in ancient Chinese music theory. Lejaren Hiller and Leonard Isaacson (1957) had started to use computers to compose music.

Amid this broadening of musicological and music-theoretical interest, Shirlaw was still wondering about the origin of the minor triad. Was it less natural than the major? What general principle lay behind major and minor triads? What general principle could explain dominant and subdominant harmony? If Shirlaw's manuscript had been subject to anonymous peer review, the reviewers may have considered it too naive for publication, even if they themselves were unsure how to answer Shirlaw's questions.

Many twentieth-century music theorists have complained about positivism of this kind. Norton (1984) commented: "The message to the modern music theorist and musicologist is clear: we cannot do music theory—past or present—in a positivistic vacuum. Until the political, philosophical, social, economic, physical, and cognitive forces of any age are brought properly to bear upon the nature of polyphony, harmony, and tonal consciousness, there can be no adequate description of that music's true historical dimension" (137). Along similar lines, Richard Goldman (1965) asserted: "One of the central facts about our harmonic system or language is, as Tovey points out, that it is invented. Although certain basic elements are taken from the natural overtone series, the developed relationships that form the language are, in a sense, arbitrary. The basic or fundamental tonal relationship, that of dominant and tonic, on which all further development rests, corresponds to no acoustical law, and has no 'scientific' basis. The musical relationship is wholly invented, or 'created by art'" (4).

Norton and Goldman were on the right track, but their rejection of positivism was a little too brazen. Within Western culture, we can hardly imagine an alternative version of MmT, in which the subdominant has the function of the dominant and vice versa—although it has been argued that chords such as IV, ii, or ♭VII can in certain contexts have dominant function (e.g., in rock; Nobile 2016). Nor can we imagine an alternative MmT in which major and minor are symmetrically interchanged. We can hardly conceive of some kind of historical tipping point, when the language of music could have developed along one of two equally likely but totally contrasting paths. Instead of claiming that the dominant–tonic relationship is arbitrary, Goldman might more realistically have claimed that a convincing scientific explanation for this phenomenon, if one exists at all, had not yet been formulated (cf. chapter 20).

Consider empirical findings about the perception of Western music by Western and non-Western listeners (Bai et al. 2016; Fritz et al. 2009). From an ethnomusicological viewpoint, studies of that kind are problematic, focusing on responses of non-Western listeners to Western music from a Western perspective, but not vice versa. For that reason alone, we might reject them out of hand. But despite their ethnocentricity, these studies do demonstrate that aspects of Western music (and by implication of any music) may be universally perceptible, including the basic emotions that the music expresses.

Looking back on Shirlaw (1957) from today's perspective, his article was not only a source of misleading statements. On the contrary, it systematically listed some perennially interesting questions about MmT, for which it should now be possible to formulate appropriate answers. In the following, I will attempt to do that. While my answers will often be problematic due to the unpredictability of cultural contents of any kind, my tentative answers may also approach the truth (in a hermeneutic sense) more closely than corresponding previous attempts in music theory and music psychology. As such, they may serve as an introduction to the rest of this book. They also demonstrate the extent to which scientific research can (or cannot) help us understand how tonality works.

Harmony

Shirlaw (1957) was not alone in claiming that "major harmony arises in harmonic resonance, in the primary set of overtones" (266) and "harmony is generated upward, not downward" (267). Rameau (1737) had introduced the idea of *corps sonore* and explained that a chord is generated by its root, considered as a *son générateur*, an idea that many later repeated. But the term "generation" is misleading. Riemann (1905) noted that a harmonic series of pitches only belong together insofar as a vibrating string or air column can produce them simultaneously (I. Bent 2011).

But even that is not how musical chords originated in musical practice. Chords originated when musicians played different tones simultaneously in musical contexts. They then aurally evaluated the resultant sound, usually while respecting existing compositional rules or conventions. Later, they repeated simultaneities that they liked, or those that worked well in musical contexts. There is nothing particularly generative about that process. If spectral analysis is relevant to this question, we need to consider the spectrum of the whole chord—not that of a single tone.

The Natural 7th

One can hardly deny that a relationship of some kind exists between musical harmony and the harmonic series, even in a modernist context (Väisälä 2002). In that limited respect, Rameau was surely right, despite his positivist, scientifically inspired approach. But Rameau's theory raised an interesting question that has bothered theorists ever since: How many elements of the harmonic series are musically relevant? How many partials of a harmonic complex tone should we consider? As Shirlaw (1957) wrote, "Why stop at number six, why should the higher harmonics be excluded? Why not at least include the natural 7th?" (267).

The question is again misleading because it suggests that intervals are ratios and that Just intonation is superior. Modern empirical psychoacoustics has offered alternative answers that mainstream music theory has hardly considered, perhaps because of their complexity:

- The 7th harmonic is usually audible (aurally relevant) in typical harmonic complex tones such as voiced speech sounds (see chapter 6).
- A partial can be perceived as part of a harmonic complex tone, contributing to the salience of the virtual pitch at the fundamental (Terhardt et al. 1982b), even if it is mistuned relative to harmonic series by as much as semitone (B. C. J. Moore et al. 1986). The 7th harmonic is about 1/3 semitone flat relative to an equally tempered minor-7th interval above the 4th harmonic. It is mistuned enough to sound out of tune in music performance (when realized as a complex tone within a chord), but not enough to be perceived as not belonging to a harmonic series (when realized as a pure tone within a complex tone). To reinforce the root, the minor-7th interval above it need not be tuned exactly to the natural ratio 7:4 but can deviate by a quarter tone or more.

- Similarly, musical pitch is perceived categorically. Tones in the chromatic scale can be mistuned by up to a quarter tone and still be perceived as belonging to the scale step in question.
- To sound in tune, today's performers normally intone close to 12-EDO because it is familiar. Whereas intonation depends on many factors, and good intonation is generally a compromise among different criteria, empirical studies suggest that proximity to familiar interval sizes is important.

These observations are consistent with the root-supporting function of the minor-7th interval above the root in regular 7th chords. They can explain the clarity (non-ambiguity) of the root of the Mm7 (e.g., GBDF) chord by comparison to other 7th chords (Parncutt 1988; Parncutt and Radovanovic, 2023), and why several other tetrads feature a minor 7th above the root. But the exact size of the "natural 7th" interval does not affect the tuning of the minor-7th interval in music performance. That tuning is primarily determined by familiarity with 12-EDO.

The Minor Triad

Like many other theorists, Shirlaw (1957) complained that "the primary overtone series provides us with a major harmony, but no minor" (275). But the relationship between musical chords and the harmonic series is not that simple, and in any case, the overtone series does not "provide" anything. Historically, musical chords arose when musicians listened to and evaluated tone combinations in their music-structural and historic-cultural contexts. New tone simultaneities that passed this informal test, often over a long period of decades or centuries, were eventually accepted as chords.

Chords tend to be more consonant than non-chords. But the perception of consonance also changed historically. Today, in retrospect, we can infer that three main psychological factors contributed (and continue to contribute) to the consonance and dissonance of tone simultaneities in Western culture: smoothness, harmonicity, and familiarity (see chapter 14). Harmonicity is the degree to which a given spectrum perceptibly resembles a harmonic spectrum; roughness is due to fast beating. The major triad is closer to the harmonic series than the minor triad and therefore has greater harmonicity. The Mm7 chord is even closer, but the roughness of its tritone (due to a clash between the 3rd harmonic of the lower tone and the 2nd of the higher) makes it less consonant. The major and minor triads are the most consonant and hence most prevalent trichords in MmT because they are the only trichords in the chromatic scale that include perfect 5ths (making them relatively harmonic) and lack 2nds (making them relatively smooth; Parncutt 1988).

Like many theorists before him, Shirlaw (1957) also claimed that the "minor is simply a clouded or camouflaged major harmony" (275). That is tantamount to saying, "I can't explain why the minor triad is so important, so let me invent something." The "clouded" idea is hardly convincing, given that any chord can be created by shifting one tone of a more familiar chord by a semitone. An augmented or suspended triad is not a "clouded or camouflaged" major triad.

It is more convincing to argue that major and minor triads are the most consonant three-tone subsets of the twelve-tone chromatic set (i.e., the most consonant trichords), their consonance depending primarily on smoothness and harmonicity. The third trichord in order of consonance is the suspended-4th triad (or suspended triad), which is less suitable as a tonic triad, for two reasons. First, it is considerably more dissonant than the major or minor triad due to the roughness of the major-2nd interval. Second, the triad's dissonance can be reduced by moving one tone (the suspended 4th) by a semitone (resolution to a major triad) or a whole tone (resolution to a minor triad)—all the while remaining within the same diatonic scale. That makes the suspended triad unstable; in Schenkerian theory, it cannot be prolonged.

The minor triad is clearly less consonant than the major due to its perceptible departure from harmonicity. But the difference in consonance is smaller, and the minor 3rd cannot be resolved to a major 3rd without introducing a non-diatonic tone. Thus, both major and minor triads make suitably consonant and stable tonic triads, whereas other possible trichords do not. This explanation is surely the best available for the predominance of major and minor triads, being both logically parsimonious and based on widely accepted general theory that is in turn based on replicable empirical data. An additional explanation is not necessary.

Another feature of major and minor triads is the almost equal spacing of their chord tones around the pitch-class circle (e.g., on a clock face, a major triad corresponds to twelve, four, and seven). Music theorists have argued that almost-equal spacing facilitates voice leading from one chord to another (Cohn 1997; Tymoczko 2010). But an argument based on vertical consonance as a combination of psychoacoustic smoothness and harmonicity may be more powerful and more appropriate for two reasons. First, there is a strong empirical scientific literature on the perception of smoothness, harmonicity, and the musical consonance of Western sonorities. Second, a theory based on vertical consonance explains the importance of major and minor triads for MmT more parsimoniously because it ignores horizontal context.

Ignoring context may at first glance appear to be a weakness, but it can also be seen as a strength. A theory based on voice leading should attempt to enumerate and account for all voice-leading situations in which a given chord can appear. To my knowledge, no existing theory does that. Whereas horizontal context certainly affects consonance, the enormous number of voice-leading possibilities in the chromatic scale means that differences between voice-leading contexts for different chords may cancel each other out. If so, such a theory would not make clear predictions.

The claim that "minor is a chromatically altered major mode" (Shirlaw 1957, 267) is as misleading as the idea that the minor triad is a "clouded" version of the major. The minor scales have their own character and identity, to some extent independent of the major. A more promising approach is to consider passages of music in major and minor keys as prolongations of their tonic triads. The major mode arises when the major tonic triad is prolonged, and the minor mode arises when the minor tonic triad is prolonged.

The Tonic

Shirlaw (1957) wondered—again, like many before him—why the tonic of the C-major or C-minor scale is C rather than F. "The three fundamental sounds of C major then, were *f-c-g*, and the three triads . . . from which the sounds of the C major diatonic scale were derived. *C* is tonic. It generates its 5th, *g*. What, then, of *f*?" (266).

Here, Shirlaw again sided with Rameau, thinking that musical elements are "generated" by natural sounds. In that approach, the tone C "generates" G because G corresponds to its 3rd harmonic. But that is not how new musical elements emerge historically. Instead, different possible sounds are tried out in different contexts. Today, taking advantage of the possibilities created by three independent theories—Schenker's hierarchical approach to music analysis, Terhardt's theory of virtual pitch, and Leonard Meyer's and Eugene Narmour's theory of implication and realization—we can answer his question in a new way.

- A passage of music in MmT is a prolongation of its tonic triad. A C-major triad implies chroma F more strongly than F♯, because two chord tones (C and G) correspond to harmonics of F, whereas only one (E) corresponds to a harmonic of F♯ (see chapter 15). Therefore, the C-major scale includes F rather than F♯ (although F♯ does occur frequently, as the leading note of $\hat{5}$).

- A passage of music whose tonic is C generally includes the tone B because rising leading tones tonicize more strongly than falling (see chapter 11).

Taking those two points together, it becomes clear why the tonic of a C-major scale is C and not F.

Scales and Harmony

Inspired by theorists such as Rameau, Fétis, and Riemann—possibly also by Tartini, Hauptmann, and Helmholtz—Shirlaw (1957) claimed that "our scales owe their existence to harmony, which arises independent of any scale" (267). In other words, harmony comes before melody. This old music-theoretical idea seems to imply either that melody is derived from harmony or that melody is generally harmonized; hence, melody is inseparable from its harmonization. But Shirlaw was also thinking of the origin of diatonic scales in musical practice, which involved octave, 5th, and 4th intervals. Their emergence must have involved human sensitivity to those intervals, whether the tones were simultaneous (in which case they could be tuned by minimizing roughness) or successive (tuned by maximizing pitch commonality).

The ancient Greek word *ἁρμονία* or *harmonía* (to join, fit, or fix together) referred, in music, to octave, 5th, and 4th intervals, both in practice and in theory. It also referred to mathematical ideas about interval ratios. In the everyday modern sense, "harmony" does not necessarily refer to ratios; instead, it implies MmT. Harmony is "agreement or accord. A pleasing combination of elements, or arrangement of sounds. The academic study of chords" (Wiktionary "Harmony," accessed March 9, 2022).

In a different text, and based on a detailed study of ancient Greek theory, Shirlaw (1951) had written: "One may observe how, in the course of musical history, a multiplex order of scales tends to give way to some simpler system of a few scales. Such a development is brought about by tonal relationship, concord, harmony. While inharmonic scales may exist literally in their hundreds, those determined by harmony, concord are brought within well-defined limits. It was thus that the Greeks made use of the concords of the fifth and fourth. Nevertheless, they never entirely lost their love of 'free' melody. Their tetrachords were harmonic. But within the tetrachord the sounds were 'free,' might vary indefinitely, were indeed inharmonic" (134).

Today's standard diatonic scale (white keys on the piano in transposition) owes its existence to harmony in the ancient Greek sense but not in the modern sense. The earliest modern harmony, one might argue, was the polyphony that emerged in the twelfth century in the Saint Martial and Notre-Dame schools, eventually including 3rd/6th intervals and triads. Over a period of a century or more, that led to a new feeling of sonority, which according to Sarah Fuller (1986) was "a significant structural resource in the advanced polyphony of fourteenth-century France and, in particular, in the music of Guillaume de Machaut" (38). That resource had previously been lacking, although scales had not. In that sense, scales came before harmony.

In the musics of the world, if there is harmony, melody is older. Even within Western culture and from a historical perspective, we can hardly speak of harmony before polyphony got started in the twelfth century. Even if the word "harmony" is confined to an earlier definition, such as men and boys singing in octaves, or even if it is about pitch relationships regardless of whether simultaneous or successive (e.g., in ancient Greece), we may still safely assume that melody came historically before harmony, contrary to Shirlaw's claim.

8 Hugo Riemann

In the late nineteenth century, Hugo Riemann profoundly influenced thinking about harmony and tonality. Today, German-speaking students of music and musicology learn his theory of harmonic function (or something similar), while in North America, his influence has grown with the neo-Riemannian school. His contribution was grounded in an encyclopedic knowledge of musicology and music theory, but many of his ideas are problematic from a modern psychological viewpoint. They include his reliance on music as notation, his idea of musical intervals as simple number ratios, and his theories of harmonic function, harmonic dualism, undertones, the Tonnetz, *and chord-scale relationships. Today, we can continue where Riemann left off, constructing a more scientifically plausible theory of MmT.*

Music as Notation

Like most music theory, that of Hugo Riemann was largely based on composers' scores. A century later, historical musicologists were criticized for devoting too much attention to notated scores and not enough to performance and perception (Cook 1998). A similar criticism can be leveled at music theorists. Many of us have fallen into the following traps:

- We assume tacitly that the notes in the score are equally important. After all, they look equally important on paper. In fact, the outer voices of a polyphonic texture tend to be perceptually more important (more salient) because they are masked less than inner voices. Tones that are accented are more salient in our perception. Accents can be either immanent in the score (e.g., harmonic or melodic) or performed (dynamics, articulation; Parncutt 2003). While this point may be intuitively clear, it helps to state it explicitly and incorporate a psychoacoustic theory of masking.

- We read meaning into enharmonic distinctions (C♯/D♭). In fact, there is no evidence beyond music-theoretical intuition that enharmonic distinctions consistently reflect different intonations (in performance) or meanings (in perception). The meaning and intonation of musical tones depends primarily on the context in which they appear in the chromatic scale; enharmonic spelling is primarily a matter of conforming to notational convention or making scores easier for performers to read (chapter 12).
- We assume that the pitches we hear correspond to the notes in the score and no others. In fact, those (spectral and virtual) pitches often differ from notes. For example, a C-major triad CEG typically evokes virtual pitches (missing fundamentals) at F and A (see chaper 15) and spectral pitches (audible partials) at B and D (chapter 16).

Like other nineteenth-century music theorists, Riemann can be criticized for failing to address these issues adequately, but such criticism would be unreasonable. Today, with the benefit of hindsight and drawing on the findings of modern psychoacoustics, we can today move toward a more realistic theory of MmT.

Intervals as Ratios

Like the Pythagoreans—and most music theorists since then—Riemann assumed unquestioningly that musical intervals are ratios of small integers. They can at least be defined mathematically as ratios. But as Riemann well knew, Aristoxenus had already criticized that idea way back in the fourth century BC: "Music is something which we hear. Height and depth of pitch are perceived qualities of sound, and need to be investigated as such. They are not rates of vibration, or of any other kind of physical movement, and they are not numerical ratios" (A. Barker 1978, 10).

Aristoxenus did not apply those ideas consistently, and his surviving writings, like those of other Greek musical thinkers, abound with ratio-based treatments of musical intervals. But he was at least aware that such approaches are problematic, and for the next two-and-a-half millennia, he was not alone. In an approach now considered old-fashioned, Richard Crocker (1962) recognized that "it is false to believe that the Middle Ages relied solely on mathematics and excluded the judgment of the ear in determining the nature of consonance. These authors [John of Garland, Anonymous I & II, Lambert,

Philippe de Vitry, Jacob of Liege, Johannes Tinctoris] say, in sum, that the ear takes pleasure in consonance, and the greater the consonance the greater the pleasure; and that for this reason one should use chiefly consonances in composing discant" (4).

These things were unfortunately not clear to Riemann. In his book *Die Natur der Harmonik* (1882), he explained: "The oldest principles of exact harmonic theory are the mathematical definitions of the consonant intervals, adduced through the investigation of the nature of sounding bodies, which can be derived from the teaching of Pythagoras but are doubtless much older than him. . . . The ancients did not yet recognize the consonance of the third 4:5; they theoretically defined the tone in their scale corresponding to our third scale-degree as an octave displacement of the fourth fifth (C-G-D-A-E) and considered it to be a dissonance by virtue of its complex mathematical ratio (64:81)" (Steege 2011, 66).

Like so many, Riemann failed to engage with a logical problem that is inherent to all such thinking. If musical intervals are simple ratios, then a major 3rd must be either 4:5 or 64:81. It cannot be both at once because there is only one major 3rd. The composers that Riemann so admired never recognized two different major 3rds! The interval can be tuned differently of course, but tuning varies on a continuous scale and is not characterized in practice by specific tunings that correspond to simple ratios. Moreover, it cannot be the case that one group of people (e.g., the ancient Greeks) perceived the major 3rd as one ratio and another group (e.g., composers of Renaissance counterpoint) perceived it as another. Nor can the same interval be perceived as one ratio in one context and another ratio in another context (e.g., as 64:81 in melody but 4:5 in harmony). Our musical intuition tells us that there is only one major 3rd, for if there were two, at least some composers would have distinguished them in their notated compositions. But with a few rare modernist exceptions (Harry Partch springs to mind), no one did that. If an interval was a ratio, we would repeatedly have to choose between Just and Pythagorean versions of the same interval—but there is no clear criterion for such a choice, nor is there evidence for it happening in musical practice.

Nevertheless, "throughout his life Riemann was apparently convinced that the mind conceived of harmonic space in terms of just intonation, even though temperament [non-Just interval tunings] was an ever-present fact of life in the outer world of musical practice" (Wason 1988, 84). Insofar as those things might also have been apparent to Rameau (among many other historic theorists), we may join Allen McHose (1947) in concluding that "the error that Rameau and many theorists following him made was to become involved in mathematical and acoustical explanations" (5).

Harmonic Function

Whereas Riemann's ideas were mostly appropriated from other theorists, his use of the term "harmonic function," evidently inspired by the idea of mathematical function, was more specific and original. A chord's function is how it relates to the prevailing tonic and the roles it plays in harmonic progressions (what chords it follows and precedes). Insofar as a chord may be said to have meaning in a tonal context, that meaning, in Riemann's approach, lies its function.

Riemann enjoyed making grand philosophical claims about harmonic function. Brian Hyer (2011) explained: "it is not what a chord *does* that matters, but what it *is*: a functional designation names a chord's *being*. For Riemann, there can be no chord without being, and being derives from a chord's participation in the *Tonalität*, from the borrowed glory of the tonic—this is what the reduction of all harmonies to three tonal functions sums to" (109, italics in original).

Riemann's (1893) theory of harmonic function harked back to Rameau's concept of tonic, dominant, and subdominant harmony. Riemann proposed that each of these chords is a category that includes several chords. The tonic category includes I and vi. Sometimes, it includes iii—but iii can also belong to the dominant category, along with V and vii°.

The ambiguity of harmonic function had been discussed by Rameau (1737), who pointed out that a ii^7 chord can function as either a subdominant relative to the tonic (IV with added 6th) or a dominant of the dominant (v7/V; *double emploi*). That argument impressed Riemann and is still valid, regardless of what one makes of Rameau's complex and somewhat arbitrary mathematical deliberations based on frequency ratios.

The inherent ambiguity of Riemann's chordal classification system (e.g., iii being regarded either as a kind of I or as a kind of V, depending on context) can be seen as either positive or negative. On the positive side, harmonic functions really are ambiguous. On the negative side, ambiguous harmonies typically have other meanings that are not included in Riemann's system (e.g., the root of iii may function as a leading tone to the root of IV). If the system is intended as a practical way of labeling chords, there should be no ambiguities.

Transformation

Riemann taught that a chord can be transformed to another chord without changing its function. In today's jazz theory, a chord can substitute for another (Riemann: *Klangvertretung*). In such cases, the two chords (the original and its transformation) are similar to each other in the sense that they have tones in common.

Riemann classified relationships between triads into four kinds. For clarity in English, I will use standard American terms and labels: dominant D, relative R, parallel P, and leading-tone exchange/contrast (*Leittonwechsel*) L. If we limit the discussion to triads, there is one common tone in the dominant relationship D (also called *Schritt* by Riemann because it moves one step on the circle of 5ths, e.g., C to G). There are two common tones in each of the following three relationships:

- *Leittonwechsel* L, obtained by shifting the root of the triad down a semitone, so the root effectively rises by a 3rd (e.g., major to mediant, I to iii, C to e, CEG to BEG). This relationship was more simply called *Gegenklang* (counterchord) by Hermann Grabner (1944) and Diether de la Motte (1976).
- *Parallel* P, obtained by shifting the 3rd of a triad up or down by a semitone to switch between major and minor (e.g., I to i, C to c, CEG to CE♭G).
- *Relative* R, obtained by shifting the 5th of a major triad up a whole tone, so the root effectively falls by a 3rd (e.g., major to relative minor, I to vi, CEG to CEA; called *Parallel* by Riemann).

Steven Rings (2011) explained: "The parallel operation, or P, transforms a triad into its parallel major or minor [e.g., C → c]; the relative operation, or R, transforms a triad into its relative major or minor [e.g., C → a]; and the leading-tone change, or L, transforms a triad by maintaining its minor 3rd and moving the remaining element 'outward' by semitone to produce another consonant triad ([e.g., C → e]" (228).

Terhardt's theory of virtual pitch (see chapter 13) can explain why some of these relationships are experienced as stronger or closer than others. In operation D, there is one common tone and two common pitches; the two main pitches that are implied in the first chord are realized in the second, producing a strong effect of progression. The chord GBD implies E and C as missing fundamentals, which are realized in CEG.

In the last three cases (L, P, and R), Riemann also considered the size of melodic intervals and hence the overall pitch distance between the two chords. In R, one tone in the chord traverses a whole tone (e.g., G to A), whereas in L and P, one tone moves through a semitone (e.g., C to B, or E to E♭). A smaller interval implies a closer relationship. If we additionally consider the role of implied pitches and pitch commonality (chapter 18), one of the three progressions (R) has higher pitch commonality than the other two, but that advantage is partly counteracted by higher pitch distance (see table 18.1).

As Riemann acknowledged, these ideas were not entirely new, having been presented in part by others, including Fétis (1844) and Rameau (1737). Rameau had also pointed out that the tonic, dominant, and subdominant chords were not limited to triads, the dominant often having an added 7th and the subdominant an added 6th. Nor was the idea of a quasi-gravitational pull between the chords original, having been proposed by Rameau, inspired by Newton. What Riemann brought to the argument (or should have brought) was his experience of nineteenth-century music, in which any major or minor triad can progress to almost any other, and the idea that almost any chord can be assigned to one the three functions based on the tones it contains and how it is approached and quitted. But he refused to accept 12-EDO (equal temperament) as a foundation for understanding MmT, preferring an approach based on simple integer ratios.

Harmonic Dualism

Robert Wason (1988) described harmonic dualism as "that *cul de sac* of nineteenth-century German theory" (69). A lot of ink was spilled trying to explain the relationship between major and minor triads—much of it by Riemann. His approach to harmonic dualism was an elaboration of previous contributions, in particular that of Moritz Hauptmann (1868), which had been inspired by Hegelian dialectics.

The problem that Hauptmann (and others before and since) had tried to solve is this: The major triad is the most common sonority in tonal music; the minor triad is the second most common. The consonance of the major triad is reflected by its ratio, which is 4:5:6 if one assumes that 3rds are Just (a shaky assumption, as we have seen). So far, so good. The problem is that the numbers in the ratio of the minor triad are too big: the ratio is 10:12:15 if the 3rds are just. If consonance was reflected by the size of the numbers, the ratio 5:6:7 would be more consonant than 10:12:15 and only slightly less consonant than 4:5:6. But to ears accustomed to 12-EDO, 5:6:7 sounds like a mistuned diminished triad. It is certainly more dissonant than a regular minor triad. Why, then, does the minor triad play such an important role in music?

Riemann (1905) was on the right track when he wrote that "a healthy, natural musical instinct cries out for the unconditional recognition of the consonant character of the minor chord, as one of the two important forms of consonance" (I. Bent 2011, 174). However, the following statement, in which he concisely expresses his dualist belief, is misleading: "The major and minor chords are not two arbitrary conglomerates of tones existing *inter pares* alongside however many others; rather, they are the two unique, fundamentally different entities in terms of which all other possible combinations are heard" (180).

From a modern psychological perspective, the reason why major and minor triads act as psychological references, relative to which other chords are perceived, is evidently because major and minor triads are familiar (cf. Peretz et al. 1998)—not because they are unique. They are familiar because they are the most consonant trichords in the chromatic scale in a psychoacoustic sense. Much time and effort could have been saved if nineteenth-century music theorists had realized that.

But the harmonic dualists would not have been satisfied by such a simple explanation. They had a strong intuition that there was much more to it—that major and minor triads were in some fundamental and possibly magical sense the foundation of all music—assuming that Western music was the apotheosis of all music, just as Western culture was assumed to be the apotheosis of all culture. Perhaps they also believed that if we properly understood the nature and origin of major and minor triads, we would at last understand the mystery of musical meaning and emotion.

The harmonic dualists proposed that major and minor triads are mirror images of each other and, in that sense, equivalent opposites (antithetical). One line of argument was purely mathematical in the tradition of Pythagoras. Riemann was impressed by Gioseffo Zarlino's (1558) distinction between harmonic division (dividing a string into 1/2, 1/3, 1/4, and so on of the length) and arithmetic division (multiplying the string length by two, three, four, and so on; Steege 2011) and recognized that harmonic division could generate the minor triad, just as arithmetic division generated the major. But those elegant mathematical relationships did not necessarily correspond to anything in the physical world.

Taken to extremes, dualism meant that everything minor was a mirror image of everything major. Whereas major chord progressions tend to move in a particular direction (T-S-D-T, or I-IV-V-I), minor chord progressions, in a strict interpretation of harmonic dualism, should move backwards (t-d-s-t, or i-v-iv-i)—which they usually do not. Similarly, we might expect the root (or *Hauptton* in Riemann's terminology) of the minor triad to correspond to its 5th, but evidence for this surprising assertion is lacking in real music.

Searching desperately for an idea that would save harmonic dualism—an idea to which he clung throughout his life with quasi-religious fervor—Riemann made some remarkably odd pseudo-scientific proposals. For example: "What distinguishes major from minor comes down to the essence of major consonance being the simplest ratios in the increase in speed of vibration, that of minor consonance, by contrast, being the simplest ratios in the enlargement of the vibrating mass . . . By this means, the principle of major can succinctly be seen to lie in growing intensity, and the principle of minor in accumulating mass" (Riemann 1905, 43; transl. Holtmeier 2011, 7).

If we regarded the theory of harmonic dualism as a paradigm according to Thomas Kuhn's (1968) theory of scientific revolutions—given the string of nineteenth-century theorists that promoted it—we could then point to some persistent anomalies or observations that we can make about major and minor triads that contradict dualist theory. From that viewpoint, it is surprising that Riemann never managed to clear his desk and look at the problem from a fresh, non-dualist perspective. But given the historical context of harmonic dualism, that is an unfair criticism. The idea was too deeply embedded in nineteenth-century thought.

Other theorists found contrasting, but equally arbitrary, explanations for the minor triad. While Schenker (1906/1954) managed to avoid Riemann's dualism, his explanation of the minor triad was equally arbitrary: "The minor mode springs from the originality of the artist, whereas the source, at least, of the major mode flows, so to speak, spontaneously from nature" (52). Paul Hindemith (1937) similarly claimed that only the major triad is "given" to us by "nature," after trying in vain to derive the minor triad coincident overtones (Cazden 1954).

Today, we can address this issue, considering all possible pitch combinations within the chromatic scale using pitch-class set theory. Most of these combinations have been available as prepared sonorities or quasi-accidental tone combinations since polyphony first emerged in the twelfth century. As polyphonic traditions developed, the more consonant—the most harmonic and least rough—tended to be preferred (Parncutt, Reisinger, et al. 2019).

Remarkably, Riemann was well acquainted with the two main foundations of such an approach: Stumpf's (1883, 1890) theory of perceptual fusion and Helmholtz's (1877) theory of sensory roughness. But Riemann's thinking also lacked some central ideas. He failed to realize that both Stumpf and Helmholtz could be right at the same time; they did not necessarily contradict each other. He also failed to realize (understandably, given the historical context) that all tone combinations that can be notated conventionally or played on a 12-EDO keyboard could one day be part of music.

Of all nineteen sonorities of three pitch classes (trichords, or Tn-types according to Rahn 1980) that can be created within the twelve-tone chromatic scale (however tuned; figure 3.1), two are clearly more consonant than the rest: the major and minor triads. They combine high fusion or harmonicity in the sense of Stumpf due to the presence of a perfect-4th or -5th interval, and low roughness in the sense of Helmholtz due to the absence of a major- or minor-2nd interval.

These ideas may seem obvious in retrospect—even trivial. But they were hardly thinkable for music theorists and scientists who believed that musical intervals were simple ratios that had been mercilessly distorted by 12-EDO ("equal temperament")—an idea that is still prevalent today (e.g., Duffin 2007). Even Helmholtz—who, like other nineteenth-century psychophysicists, including Gustav Fechner, Ernst Mach, Wilhelm Wundt, and Carl Stumpf, was a trained musician—explained musical intervals in terms of number ratios and opposed 12-EDO (Hui 2012).

These questions were also politically loaded. The idea of basing music theory on twelve equal tones of the chromatic scale was sometimes seen as politically left, whereas the more hierarchical idea of music being derived from, or generated by, a single "ruling" tonic was politically right. Nicholas Cook (2007) explained:

> At one point in his *Musikalische Syntaxis* (1877) Riemann denounces the theorists Melchior Sachs, Albert Hahn, and Heinrich Vincent, using what Rehding (2003, 63) terms "the bewildering epithet 'Social Democrats.'" Riemann's epithet turns out to be not so bewildering when explained: all three theorists attempted to build a harmonic system on the chromatic scale, so—in Riemann's view, which Schenker would certainly have shared—subverting the natural foundations of music in

> the overtone series. As Rehding continues, "The common conservative reproaches against the political Social Democrats can indeed be easily transferred to the music-theoretical realm: they promoted a society without hierarchy." (8)

But the question was not political at all. Any nineteenth-century musician, regardless of political persuasion, could listen to an orchestra perform a piano concerto from the "timeless" German classical repertoire and see and hear directly how the orchestra tuned itself, seemingly effortlessly, to the piano's version of stretched 12-EDO, as if that were the normal way of tuning for all orchestral instruments. Clearly, the octave interval comprised twelve equal pitches that were approximately equally spaced and equally available for making music (limited only by the mechanical idiosyncrasies of specific instruments). The evidence was plain to see and hear. The interesting question was why certain combinations of those twelve tones and certain pitch-time patterns were being used more often than others.

Undertones

If the harmonic division of a string produces harmonic overtones, which according to Rameau really exist in the sound that we hear (and he was right about that), then Zarlino's arithmetic-division exercise (so admired by Riemann) suggests that analogous undertones might exist. If that were true, the tone C4 would have physically present undertones at C3, F2, C2, A♭1, and so on, which might explain the origin of the minor triad (here, F minor), just as the overtones of C4 (C5, G5, C6, E6, and so on) can explain the origin of the major triad (here, C major).

Both Rameau and Riemann took the idea of undertones seriously, at least for a while. Even after Riemann abandoned the idea that undertones might physically exist, he paradoxically maintained his belief in harmonic dualism based on undertones (I. Bent 2011). And it must be said that the two theorists were not completely wrong. Whereas undertones may not exist physically in World 1, they may be said to exist subjectively in World 2, in the sense that the pitch of a complex tone generally corresponds approximately to a subharmonic of a salient partial (De Boer 1956; Terhardt et al. 1982b). But even that observation is insufficient to save Riemann's theory of harmonic dualism. It does not explain the minor triad—at least not in the way Riemann had imagined—because it does not clearly explain the minor triad's root. To answer that question, we need a model that is based on general (nonmusical) psychoacoustic principles and successfully predicts the root of any chord in the chromatic scale. A successful model should not treat the minor triad as a special case. Instead, the minor triad should be treated in the same way as any other tone simultaneity.

Tonnetz

Riemann's *Tonnetz* was a two-dimensional map of harmonic relationships, similar to that of Leonhard Euler (1739). It featured three axes whose intersections created triangles representing major and minor triads. Adjacent triads in the *Tonnetz* were related, with two tones in common. Along the three axes, tones were related by 5ths (. . . F, C, G, D . . .); major 3rds (. . . A♭, C, E, G♯, B♯ . . .), and minor 3rds (. . . A, C, E♭, G♭, B♭♭ . . .).

Like similar maps proposed by other theorists, the *Tonnetz* demonstrated diverse possibilities for chord progression and key modulation. It also gave a rough idea of the musical or psychological distances between tones, chords, or tonalities, and it has been newly interpreted in the context of neo-Riemannian theory. From the perspective of modern cognitive psychology, the *Tonnetz* can be seen as a cognitive representation; like other cognitive representations, its psychological reality is unclear (see chapter 21).

Enharmonic distinctions are prominent in the *Tonnetz*, which redefines and reifies them. One can stack 4ths or 5ths along one axis of a *Tonnetz*, proceeding linearly and accumulating sharps in one direction and flats in the other (creating a line of 5ths; D. Temperley 2011). Alternatively, one can circumnavigate the circle of 5ths, returning to one's starting point. The accumulation of sharps or flats happens faster if you stack major or minor 3rds than if you stack 4ths or 5ths.

Enharmonic distinctions arise from a notation system that is based on an arbitrary choice: the idea that musical pitch notation should start from a standard diatonic scale and notate deviations from that scale as accidentals (see chapter 12). That is not a criticism of conventional music notation; on the contrary, an alternative system that can notate diverse tonal and atonal styles equally compactly and legibly is scarcely imaginable (Read 1987). If, however, conventional notation had originally been based on the chromatic rather than the diatonic scale (an interesting thought experiment), music theorists would have understood musical structure differently, even if the music that subsequent generations of composers had produced had been relatively unaffected.

The *Tonnetz* can be practically useful if it inspires composers to explore new musical patterns. But its theoretical usefulness is limited if it tends to emphasize some relationships at the expense of others and lacks a solid psychological foundation. As an abstract mathematical structure, the *Tonnetz* does not necessarily reflect human perception and cognition, and complex mathematical treatments such as those of Richard Cohn (1997) do not solve the basic psychological problem. A more promising approach is to consider all audible aspects of the stimulus and all possible ways in which relationships between those aspects can be perceived, weighted according to perceptual salience. The theory of pitch commonality and pitch distance (Parncutt 1989; chapter 18) was an attempt to create such a theory.

9 Diverse Epistemologies

This chapter briefly reviews five additional academic approaches to MmT in their respective social and academic contexts: those of Schenker, Krumhansl, Terhardt, Lerdahl, and Forte. Heinrich Schenker's theory of tonality relied on the concept of prolongation. Several decades later, his idea found psychological and empirical expression in Carol Krumhansl's key profiles. Schenker's approach was based on his musical experience and intuition as a composer/theorist and driven by a certain German cultural conservatism, whereas Krumhansl's findings and conclusions were founded on the quantitative results of controlled psychological experiments. Despite their enormous epistemological and cultural differences, both thinkers achieved breakthroughs in the understanding of pitch structures in tonal contexts. Other influential but highly contrasting recent approaches to the theory of MmT include Ernst Terhardt's theory of pitch, consonance, and harmony; Fred Lerdahl's generative theory of musical structure; and Allen Forte's pitch-class set theory. Each approach was a product of its social context, which includes academic, cultural, and political aspects. Today, a new interdisciplinary synthesis is possible.

Heinrich Schenker

Heinrich Schenker (1868–1935) wanted to understand why the greatest German tonal works were so great. Why did they arouse such wonderfully profound and majestic emotions? A modern music psychologist might have phrased the question like this: Why does music in familiar or preferred styles, or styles with which one identifies socially and psychologically, evoke such strong emotion? Regarding social identification, what might be the role of Émile Durkheim's (1912) concept of collective effervescence (Páez et al. 2015)? Does collective effervescence happen in Western concert halls? Why do people who love a certain style of music also tend to hate another, even if both the loved and hated music is highly familiar (E. Schubert 2010)? Why are "responses to aversive music . . . characterized by embodied experiences, perceived loss of agency, and violation of musical identity, as well as social or moral attitudes and values" (Peltola and Vuoskoski 2022, 159)? These issues are beyond the present scope. The point is that music theorists of the past were ultimately motivated by their personal experience of strong or profound musical emotion while listening to their preferred music. They were probably also motivated by the negative emotions they experienced when listening to music they considered inferior.

Searching for such an underlying principle, Schenker developed his concept of deep structure (or fundamental structure). In its simplest form, the *Ursatz* is a simple melodic fragment ($\hat{3}-\hat{2}-\hat{1}$) over a simple bass line ($\hat{1}-\hat{5}-\hat{1}$) that can be interpreted as a prolongation of the tonic triad ($\hat{1}\hat{3}\hat{5}$). Whereas many music scholars in both humanities and sciences agree that Schenker defined deep structure too specifically (e.g., Cook 2007), it does seem reasonable to assume that music has a psychologically real deep structure of some kind, even if our perception of it is not very clear. Schenker's attempt to describe the *Ursatz* (insofar as it exists) was certainly interesting and promising.

Schenker's approach to music analysis was extensively applied and developed by later music theorists, especially in North America (for a convenient summary of terminology in German and English, see Pankhurst 2008). He regarded a musical score or the musical surface as the foreground (*Vordergrund*) of a complex hierarchical structure; the task of the analyst was to uncover the underlying layers, which was done by a process of analytic reduction. In a sequence of steps, the foreground was reduced to deeper structural levels, through the middleground levels (*Mittelgrund*) to the background (*Hintergrund*) or deep structure (*Ursatz*).

Each step involved removing diminutions such as passing notes (*Durchgangstöne*), neighbor notes (*Nebennoten*), and arpeggiations (*Brechungen*); two-note arpeggiations were called consonant skips by Forte and Gilbert (1982). Removing diminutions often meant deciding which of two notes would be part of the structure at the next level or layer (*Schicht*) and which was subordinate and should be omitted. A simple example is the accented appoggiatura: it is subordinate because it is not a chord tone in the current harmony. At a deeper level, only the chord tone to which the appoggiatura leads is maintained.

Whereas removing diminutions is a way of moving to a deeper level, Schenker liked to see the process in reverse, embellishment being a generative process of moving from a deeper level toward the surface. An example of embellishment is the unfolding (*Ausfaltung*) of a linear progression (*Zug*), which might involve an interval between two tones in the background or middle ground being embellished by a stepwise rising or falling line.

In Schenker's approach, analysis of the musical surface may involve dividing the tones of a compound melody into two or more streams (implied counterpoint). A melody becomes compound if it includes leaps (rather than steps) that cause it to split perceptually into a higher and a lower part (implied voices). An arpeggiated melody such as the well-known right-hand opening of Mozart's piano sonata in C major (K. 545) creates two or more streams; in this same example, the Alberti bass in the left hand also splits into streams. There is an interesting interdisciplinary link here: the separation of these streams is accounted for by Bregman's (1990) theory of auditory scene analysis, in which tones that are close to each other in pitch and time are grouped.

Schenkerian analyses can also imply tones at different hierarchical levels that are not present in the musical surface (e.g., Rothstein 1991). If an implied melodic continuation is replaced by a rest, the expected continuation does not disappear. In a psychological approach, such implied tones depend on statistical properties of similar, familiar music. They should not be confused with tones that are implied as missing fundamentals according to Terhardt's theory of virtual pitch.

The Schenkerian middle ground typically comprises more than one level and features simple two- or three-part note-against-note progressions that Forte and Gilbert (1982) called linear intervallic patterns. These patterns follow traditional rules of voice leading; the musical surface can be regarded as an elaboration of each pattern.

Schenker's deep structure involves bass arpeggiation (*Bassbrechung*). Its simplest form, I-V-I, can be seen as a simple arpeggiation of the 5th interval between $\hat{1}$ and $\hat{5}$. Many pieces of music can be reduced to the progression I-III-V-I, such that the tonic triad is arpeggiated in the bass, moving from one prolonged diatonic scale step supporting functional harmony (*Stufe*) to the next ($\hat{1}-\hat{3}-\hat{5}$), and consistent with the idea that a piece of music in MmT is a prolongation of its tonic triad.

Schenker disapproved of the concept of modulation (key change), insisting that a piece of great music stayed in the same key throughout, as represented by its deep structure. He preferred the term "tonicization" (*Tonikalisierung*). Any chord can be tonicized by a preceding secondary (or applied) dominant. These details of Schenker's reductive approach to analysis were grounded in the previous history of music theory, including Rameau, Sechter, and Riemann (Morgan 1978).

Schenker's political attitudes overlapped with those of Riemann, although their music theories differed greatly, and Schenker's general opinion of Riemann was certainly less than positive (Cook 2007). Both Schenker and Riemann regarded German classical music as somehow perfect and timeless, as part of a broadly pan-German nationalist outlook. Their hierarchical conceptions of musical structure reflected their conservative values; in central Europe around 1900, the idea that musical hierarchies reflected political hierarchies was a common one (it may be relevant that Schenker also studied law). Both theorists were convinced that the foundations of MmT (and hence the foundation of all great music) were somehow natural and rooted in mathematical ratios or the harmonic series. Both were elitist—uninterested in the subjective experiences and intuitions of untrained musicians. Their theories were prescriptive rather than descriptive: they told us how music should be heard correctly—not how it is typically heard. In that sense, and in that manner, both were positivists.

Schenker had a dark side that should not be forgotten. Like many of his circle, he was a racist and, in his later years, an admirer of Hitler. His writings contain references to "less able or more primitive races," "inferior races," "wild and half wild peoples," and "the Slavic half-breed" (Ewell 2020, 8). His belief in the inequality of tones in a musical score reflected his belief in the inequality of humans: "It is therefore a contradiction to maintain, for example, that all scale tones between C and c have real independence or, to use a current but certainly musically unsuitable expression, 'equal rights'" (Schenker 1935/1979, 13; as cited in Ewell 2020, 9). For that reason, some colleagues prefer to ignore Schenker's music-theoretical contribution. But his influence is such that a theory of tonality (including the theory presented here) is barely imaginable without him. Similarly, the Western canon is hardly imaginable without the music of Richard Wagner, whose antisemitism—as appalling as it is to us today—does not prevent us from performing, enjoying, and admiring his music.

Carol Krumhansl

The late-twentieth-century academic discipline of psychology brought forth a new set of ideas and methodological tools with which to revisit earlier music-theoretical ideas. In that context, an American psychologist, Carol Krumhansl, developed a cognitive approach to MmT based on the idea of key profiles (Krumhansl and Kessler 1982; Krumhansl and Shepard 1979; see figure 1.1). The key profile quantified a well-known music-theoretical concept: the relative stability of scale steps in a musical scale in the context of a major or minor key. But it went well beyond existing, introspective, music-theoretical assumptions about stability.

Krumhansl demonstrated empirically that the shapes of major and minor key profiles were remarkably independent of the actual tones or chords included in a passage, provided the passage clearly instantiated a given musical key (Krumhansl and Kessler 1982; Krumhansl and Shepard 1979). Instantiating a key was somehow the same, psychologically, as instantiating a key profile. The profile shape was also remarkably independent of musical expertise.

Both those observations supported the idea that key profiles reflected cognitive structures that exist in the musical minds of Westerners. The profiles demonstrated the psychological reality of major and minor tonalities of mainstream music theory—at least for Westerners with extended experience of Western music. It is one thing to claim that diatonicism has mental reality (e.g., Huovinen 2006); Krumhansl went a step further, assigning different degrees of stability to diatonic scale steps and affirming the psychological reality of those distinctions.

Krumhansl's key profiles raised issues of cross-cultural generalization. In various collaborations, she applied the empirical methods used to establish Western key profiles to the music of North India, Bali (Indonesia), and Finland (North Sami yoiks). Her approach enabled a new examination and separation of quasi-universal and culture-specific principles while remaining sensitive to the inherent limits of comparative approaches. The results revealed contrasting cognitive representations of musical pitch in different cultural contexts that depended on musical experience and expanded the music-psychological project in the direction of ethnomusicology. The results also suggested a possible musical quasi universal, namely the tendency for the stability of a scale step to depend in a simple way on how often it is heard in a musical passage or its total duration (Krumhansl 2004).

In the process of investigating the perception and cognition of MmT, Krumhansl (1990) developed methods and concepts that were widely imitated and came to be regarded as disciplinary benchmarks. It became clear that key profiles had considerable potential as a foundation for modeling quantitative data on perceived relationships between tones, chords, and tonalities. Her multidimensional scaling solutions of data on the perception of musical tones and chords in tonal contexts were similarly informative. She was the first to quantify plausibly the tonal stability of scale tones and diatonic triads in diatonic contexts and the distance between two keys in a modulation. She established quantitative relationships that enabled the development and testing of an early key-finding algorithm, inspiring many later variants (Leman 2012; Toiviainen and Krumhansl 2003; P. G. Vos 2000). Her experiments on the relatedness of pairs of tones and pairs of chords in tonal contexts demonstrated that progressions from unstable to stable are perceived as more related than progressions in the opposite direction.

Some of Krumhansl's data on the perception and cognition of MmT are amenable to psychoacoustic modeling. The key profiles can be modeled surprisingly accurately, simply by counting notes in the scores of the cadential progressions that she used to instantiate tonality in experiments (Butler 1989). Andrew Milne (2015) developed a model of the key profiles that considered only spectral pitches. Parncutt (1989) showed that predictions could be improved by accounting for variations in virtual-pitch salience. Such a model can also include memory decay and primacy/recency effects (Parncutt 1994; Parncutt and Bregman 2000), allowing for empirical data on modulation from one tonality to another to be explained (Huron and Parncutt 1993). These points are consistent with the idea that a passage of music in MmT is a prolongation of the tonic triad (Parncutt 2011).

Not all of Krumhansl's data can be modeled in this way. Her data on the stability of a chord within a tonality (ratings of chords in harmonic-hierarchy experiments; Krumhansl 1990, 171–172) quantified the relationship between a chord and a prevailing major or minor tonality. In each trial of those experiments, the listener heard a predominant-dominant-tonic cadence (a "strong key-defining context") followed by a single chord and was asked "how well the final chord fit, in a musical sense, with the preceding key-defining context" (Krumhansl 1990, 169).

One might expect data of that kind to be explicable in terms of pitch commonality between the chord and the tonic triad. But things are not that simple. The theory of pitch commonality predicts that a iii chord goes well with a I chord because the two chords have two pitch classes in common. But as Krumhansl's data and everyday music theory show, a V chord goes better with a I chord than a iii chord does, although V-I has only one pitch class in common (e.g., in the progression GBD-GCE, the common tone is G).

In this way, Krumhansl's data confirmed that listeners internalize the central importance of tonic, dominant, and subdominant chords in MmT, consistent with the theories of Rameau and Riemann. Her findings are evidence that cognitive structures of musical pitch exist psychologically, and that they affect or determine the perception of musical pitch relationships. To understand that in a psychohistoric approach, we need to think about how sensitivity to the hierarchy of tonic, dominant, and subdominant may have arisen historically (see chapter 20).

The epistemological and ontological status of Krumhansl's multidimensional scaling solutions may be questioned. They conveniently summarize a complex data set but do not necessarily confirm the existence of underlying cognitive structures. In an ecological approach, it is unclear whether a multidimensional scaling solution represents physical structures that exist in the directly perceived environment or cognitive structures that in some sense are independent of that environment. Said another way, it may be more promising to regard cognition as happening not in the head but rather in the body and the environment, assuming that "(1) cognition is situated; (2) cognition is time-pressured; (3) we off-load cognitive work onto the environment; (4) the environment is part of the cognitive system; (5) cognition is for action; (6) offline cognition is body based" (Wilson 2002, 625).

The academic and historical context of Krumhansl's theory included long-term developments in the discipline of psychology during the twentieth century that tended to reject the introspection of nineteenth-century psychology, including that of Wilhelm Wundt and Gustav Fechner. The relativism of psychological introspection was challenged by both behaviorism (Skinner 1938) and cognitivism (Chomsky 1959). The discipline of cognitive psychology became quantitative, experimental, non-introspective, and non-behaviorist (Hatfield 2002). Cognitive psychologists saw the human mind and brain as analogous to a computer with input stimuli and output behaviors and experiences.

Krumhansl's doctoral supervisor, Roger Shepard, was a leading figure in the cognitive psychology of the 1960s, 1970s, and 1980s. He made major contributions to the understanding of spatial relations and mental rotation and the development of multidimensional scaling—a mathematical method commonly used to investigate cognitive representations. That research turned out to be compatible with representations of tonal relations in music such as Riemann's *Tonnetz* (cf. Shepard 1982). It also paved the way for a fruitful interdisciplinary interaction between music theory and cognitive psychology.

Shepard (1964) attracted the attention of the music-cognition community with the idea of pitch circularity, which he demonstrated using tones whose partials are spaced at octave intervals across the audible spectrum with a bell-shaped amplitude envelope (later called Shepard tones or octave-complex tones). As the pitch class of such a tone is continuously or repeatedly increased, the tone seems to rise indefinitely, without actually getting higher—an illusion used artistically by composers such as Claude Risset (Braus 1995). But the illusion disappears when we learn how it works. A Shepard tone comprises several partials at octave intervals, only some of which can be separately heard. Usually, attention is focused on a virtual pitch in the middle of the musical range. As the pitch class is raised, it is possible to hear a new tone enter in the bass, or a high tone to disappear. Shepard tones were indispensable for Krumhansl's research on octave-generalized aspects of musical structure, and later for my own research.

Ernst Terhardt

An electrical engineer and psychoacoustician from Munich, Germany, Ernst Terhardt developed a theory of pitch perception that shed new light on the perception of musical tones, chords, and tonal relationships. The theory could be applied to any everyday sound, including speech, music, and other environmental sounds. The theory's generality made its explanations of Western harmony convincing from a scientific viewpoint, while at the same time raising controversial cultural issues in a humanities context (see chapter 23).

Terhardt's experience and skill as a jazz pianist gave him detailed knowledge of musical chords and their roots. That may have inspired his theory of pitch perception, according to which most pitches that we hear in everyday life are virtual, including all voiced speech sounds, and virtual pitches are subharmonics of salient spectral pitches (audible partials). On the telephone, the pitch of every phoneme produced by the remote speaker is virtual. It is also a pitch at the missing fundamental: a typical telephone does not transmit the fundamental frequency of regular speech. Similarly, most musical pitches are virtual, whether the corresponding fundamental is physically present and audible or not. By clarifying these issues, Terhardt gave Rameau's theory of chord roots a new scientific foundation.

Like his supervisor and colleague Eberhard Zwicker, Terhardt was influenced by the academic discipline of cybernetics (Zwicker and Feldtkeller 1967/1999). In a cybernetic approach to animal or human behavior (Umpleby 1997), the organism interacts with the environment as it perceives it, in a circular relationship. When listening to music, for example, we constantly predict what will happen next; we then respond differently depending on whether what we predicted actually happens. To understand an interactive system of that kind, a cybernetically oriented engineer understands that it is not enough to observe and measure the behavior. You also need to be able to model it, and your model's success in predicting the data is a measure of how well you understand the system.

The theory of virtual pitch (Terhardt 1972) was conceived of as a black box—a central concept in cybernetics, related to other concepts such as mechanism, machine, stability, feedback, information, communication, regulation, co-ordination, and control (Ashby 1956). A black box is a complex system whose contents are unknown. All we know is the mathematical relationship between input and output. The pitch algorithm of Terhardt et al. (1982b) modeled quantitative relationships between physical properties of diverse sounds and corresponding perceived pitches. The details of the algorithm did not necessarily correspond to the physical workings of the ear and brain due to fundamental uncertainties about the mind–brain relationship and the complexity of temporal and spectral effects in the neurophysiology of pitch. Assumptions about brain processes were excluded on the assumption that to include them would have been unrealistic. Terhardt regarded that limitation as a strength. In general, it is easier to solve a complex problem when its domain and complexity are limited. One should first clarify the input–output relationship of the black box; only after that might one reasonably wonder about the status of its detailed contents.

The theory was nevertheless influenced by physiological considerations in two cases. First, the auditory periphery (specifically, the basilar membrane) was understood to separate high from low frequencies, creating a kind of running spectrum. The membrane is lighter and tighter at the end that responds to higher frequencies. How auditory spectral analysis actually happens physiologically, involving a complex combination of temporal and spectral effects, was not explicitly modeled. Second, patterns in the auditory scene are learned and recognized by neural networks (Laden 1993). Again, the process involves a complex combination of temporal and spectral effects that were not modeled.

As the example of pitch perception illustrates, a cybernetic model of the brain, or more generally of the perceiving organism or body, is not primarily about anatomy and physiology in World 1. It is more about the brain as an information-processing machine in World 3. As such, Terhardt's approach could be regarded as cognitive—but he avoided that term, being skeptical of the research tradition of cognitive psychology, with its reliance on fuzzily defined cognitive structures that cannot be directly observed.

The academic discipline of cybernetics differs from cognitive psychology in various ways. According to W. Ross Ashby (1956), "cybernetics . . . is a 'theory of machines,' but it treats, not things but *ways of behaving*. It does not ask 'what *is* this thing?' but '*what does it do?*' Thus it is very interested in such a statement as 'this variable is undergoing a simple harmonic oscillation,' and is much less concerned with whether the variable is the position of a point on a wheel, or a potential in an electric circuit. It is thus essentially functional and behaviouristic. *The truths of cybernetics are not conditional on their being derived from some other branch of science*" (1, italics in original).

If that is true, Terhardt was justified in applying cybernetic principles to questions of pitch perception while at the same time ignoring parallel developments in cognitive psychology, as well as physiologically based approaches to the psychoacoustics or pitch perception. The point, from Terhardt's perspective, was to reduce the problem to its essentials and solve it mathematically, creating testable predictions. The resultant algorithm was not perfect (some fine-tuning would possible, if the many relevant experiments were carefully repeated and corresponding adjustments made), but given the long list of empirically investigated and documented effects with which the algorithm was consistent (listed by Terhardt et al. 1982a), it was a remarkable achievement.

This book focuses on Terhardt's theory of pitch perception for several reasons. Apart from making music-theoretically interesting predictions, the theory clearly separates the physical from the subjective and regards pitch as purely or inherently subjective, consistent with methodological paradigms in psychoacoustical practice. It focuses on quantitative relationships between physical sound parameters and subjective measures—an appropriate way to understand musical structure as perceived or experienced.

As we listen to music, we consciously hear some pitches and not others, and some pitches sound more salient or stable than others. If we want to understand musical tonality, we need to understand why that is so. Pitch salience predictions are integral to Terhardt's approach, and under certain circumstances, they can predict the stability of scale steps in musical contexts. For any steady-state input, the model predicts a series of possible pitches and their relative saliences. That output can readily be compared with empirical data, whether from listening experiments or music database analyses.

Fred Lerdahl

Another important influence on the present work was the *Generative Theory of Tonal Music* by Lerdahl and Jackendoff (1983). In that book, a composer and a linguist got together to present a rule-based system for the analysis of a passage of tonal music, reducing it to hierarchical tree structures called grouping structure, metrical structure, time-span reduction, and prolongational reduction. These processes were assumed to reflect cognitive processes in Western music listeners—the spontaneous construction of musical understanding.

Lerdahl approached these issues from his perspective as a composer. He wanted to understand how we spontaneously perceive a tone or chord according to its function in its tonal context—something that composers and performers of tonal music are constantly doing. He suspected that a cognitive model of that process could make a big contribution to research on music cognition, and that Noam Chomsky's (1957) theory of generative grammar could provide a suitable foundation. His approach uniquely combined music-theoretical, compositional, and linguistic insights. Lerdahl and Jackendoff (1983) inspired many empirical studies in the fledgling international discipline of music psychology and became one of its most cited sources.

Looking for empirical confirmation for his theory, Lerdahl got involved in empirical research. In his (1988, 2001) simple model of tonal pitch space (see table 21.1), hierarchical depth correlated with Krumhansl's empirical key profiles. To better account for empirical data, Lerdahl and Krumhansl (2007) introduced an additional psychoacoustic predictor called *surface tension* and understood it to depend on scale degree, chordal inversion, and nonharmonic tones—but not directly on sound spectra. The authors concluded that listeners often compare unconnected time windows, hearing hierarchically rather than sequentially "through the prolongational analysis from global to local levels in the tree structure" (334). Whether the average concertgoer actually does that is an interesting question. Comparable studies came to the opposite conclusion (Bigand and Parncutt 1999; Bigand et al. 1996): listeners tend to focus on what is happening in a relatively short time window of a few seconds and not on the longer-term structure of the music (cf. Tillmann and Bigand 2004).

Taking a broad perspective, on the one hand we can see that nonsequential hierarchical processing of the kind investigated by Lerdahl and Krumhansl must exist psychologically and must therefore be considered an essential aspect of MmT perception. Without such processing, the musical forms (e.g., rondo or sonata) and tonal drama of classical works such as Beethoven's sonatas (Rosen and Ricci 2002), which played such an important role in music history and also influenced Lerdahl's own compositional practice, would make no sense. It's hard to imagine that such structures, which are so important for understanding musical form, might exist only in musical scores and the minds of composers and not in the minds of listeners. On the other hand, it has been remarkably difficult to confirm the psychological reality of such structures in empirical studies (see chapter 21).

Allen Forte

Of Forte's many contributions to music theory, I will focus on his pitch-class theory. His book on that topic, published in 1973, became a standard reference. Inspired by composer Milton Babbitt, Forte's approach had emerged within the small but growing North American discipline of music theory, influenced by a broader academic culture of mathematical positivism (Davis 1995). David Forrest and colleagues (2021) described how it happened: "In the mid-1950s, American composer-theorists began reformulating the conceptual basis of the nineteenth-century European tradition by proposing new explanatory models of musical structure influenced by the developments in contemporary science, philosophy, and linguistics. Their investigations resulted in their advocacy of a music-theoretical discipline centered at universities and composed of professional theorists or composer-theorists who had mastered an essential tripartite core of knowledge: Schenkerian theory and analysis, history of theory, and the pioneering subfields addressing post-tonal music, twelve-tone theory, and set theory" (from the introduction).

Forte's theory of pitch-class sets systematically enumerated all possibilities within given constraints. Whether he studied mathematics and whether his theory was directly inspired by mathematical group theory (or whether the relevance of group theory only later became apparent), his book certainly revealed his mathematical talent. Before introducing a new mathematical idea, he usually offered a musical example, ensuring that the entire theory was grounded in music. He also pioneered computer-based methods in music research—as Bregman, Krumhansl, and Terhardt also did in different ways.

It was an impressive academic achievement to present a new theory of such flexibility and potential that generations of North American music theorists would learn to understand and apply it. Joseph Straus (2015) commented: "Forte's most central and lasting contribution, and the one most closely identified with him, is his mapping of the terrain of the pitch-class set. That theorists generally acknowledge the centrality of the pitch-class set in not only the modernist music of twentieth-century Europe and the Americas, but also other musical repertoires before, during, and after, and that they generally think about pitch-class sets in particular ways, are marks of Forte's durable influence" (3).

Pitch-class set theory was originally developed to understand the structure of atonal music. It later became clear that the theory had a broader domain of application. The present book incorporates pitch-class set theory within a psychologically oriented theory of MmT—but only for the purpose of labeling pitch patterns. Pitches are labeled only relative to each other, in accordance with the mathematical principle of transpositional equivalence. Other mathematical relationships are harder to hear and are therefore not considered. They include intervallic inversion (inverting each interval within the octave), retrograde (playing a melody backwards), and symmetry (symmetrical sets include [02468T], [0369], [048], and [06]). Complementary relations (the complement of [047] being [1235689TL]) and set multiplications, while inspiring for many composers, are even farther removed from musical perceptibility. If the main features of MmT can be explained without recourse to complex mathematics or imperceptible transformations (such as Callender et al. 2008), the principle of parsimony implies that those complexities should be avoided—at least for the purpose of establishing the foundations of MmT.

Pitch-class set theory assumes initially that the twelve pitch classes are equally important, with no tonal centers. But even in the post-tonal repertoire, pitch classes often differ considerably in perceptual salience or stability for a mixture of sensory and cultural reasons. This book puts those sensory and cultural factors in the foreground on the assumption that they affect how music works and how we experience it more directly and simply than the mathematical properties of pitch-class sets. I claim that central structural features of MmT can be accounted for by combining relatively simple analyses of pitch-class sets with relevant psychological, musical, and historical insights.

10 Scales and Intervals

Having reviewed diverse approaches to MmT and its interdisciplinary foundations, we can now return to some basic questions. Is a musical interval a frequency ratio or a pitch distance? Are musical intervals absolute and universal, or are they culturally embedded and learned? How are musical scales structured, and what is the origin of the chromatic and diatonic scales? For millennia, Western music theory has been dominated by mystical Pythagorean concepts of intervals and scale steps as ratios of small integers. Ancient and medieval theorists favored Pythagorean ratios, which were limited to factors of two and three; major 3rds were considered dissonant (64:81). Renaissance theorists such as Zarlino assumed Just tuning and intonation, in which intervals are simple ratios and major 3rds have the ratio 4:5. Musical ratio theory is an important part of cultural heritage and the history of ideas but misleading in an empirical psychological approach. Empirical studies suggest that musical intervals are inherently approximate pitch distances that deviate systematically from theoretical ratios and, like most things about music, are learned from oral-aural tradition. Their optimal tuning is a compromise among partly conflicting criteria. Whereas diatonic scales maximize the number of octave and 5th intervals between scale steps, the tuning accuracy of each interval and each scale step is limited by JNDs. This psychocultural *approach has interesting implications for a theory of MmT.*

Psychohistoric Origins

A musical scale is a set of tones or pitches that are used to make melody, polyphony, and harmony. It is a resource for making music. It is also an abstraction, relative to the more concrete phenomenon of melody. If a given scale can be transposed to any starting pitch, its important parameters are the intervals between the tones.

To understand what scales are, we need to think about where they come from. For an individual listener or performer, a musical scale is learned from music. Musical interval sizes are similarly learned from music. The ideal size or intonation of a musical interval is that which is most familiar from music performances. That is part of the oral-aural tradition of music, which includes all audible details of musical performances.

Lewis Rowell (2015) offered examples from Indian music. For Western music, Anne McLucas (2010) explained:

> Oral tradition in music, then, refers to those aspects of music that are passed down by humans teaching one another the art form, whether in person or by means of recordings, radio, television, or other non-written means. It also refers to the kinds of composition that occur completely without referent to written notation, that is, the creative as well as the re-creative forms of oral tradition. (2)
>
> Different parameters of the musical experience may stand out when looked at from the perspective of oral tradition. The melodic and harmonic aspects so prized by Western musicologists and theorists may give way in analysis of orally transmitted music to aspects of rhythm and timbre that are less easily expressed in notation, either because of their complexity, as with timbre and the details of rhythm, or because of their all-pervasive repetitive nature, as with many danceable pieces. (157)

As the word "interval" implies, the most psychologically salient feature of a musical interval is its perceived width. In a psychoacoustical approach, that is not necessarily exactly the same as its physical width (frequency ratio or width in cents) due to pitch shifts (see chapter 6). When music students with good musical listening and performing skills hear isolated harmonic (simultaneous) intervals and are asked to label them (Plomp et al. 1973), confusions depend mainly on width: major 3rds are confused with minor 3rds more than with minor 6ths, although a minor 6th is an intervallic inversion of a major 3rd. Most of the incorrect responses in Plomp's study were a semitone away from target responses, but there were also occasional confusions between perfect 4ths and 5ths and rarer confusions between major 2nds and minor 7ths. Results were similar for pure and complex tones. Although the intervals were tuned to the equally tempered scale, the results would have been similar using Just or Pythagorean tuning.

Cazden (1954) argued that "the source of the musical scale is neither Nature nor abstract numbers, it is the history of the art of music. This is the only assumption that can account for its known variation in historical time and in different cultures. Scales are doubtless influenced, but they are not determined, by certain elementary psycho-acoustic phenomena, notably the special perceptual configurations of the octave and the fifth" (301). Along similar lines, Alexander Ellis (1885) commented that "the Musical Scale is not one, not 'natural,' nor even founded necessarily on the laws of the constitution of musical sound, so beautifully worked out by Helmholtz, but very diverse, very artificial, and very capricious" (as cited in Hui 2012, 125).

In principle, a scale step can lie at any point on a frequency or pitch continuum. In practice, adjacent frequencies are far enough apart to allow them to be easily and categorically distinguished in music performance and perception. When Curt Sachs (1962) wrote somewhat condescendingly of "tumbling strains," citing Australian Aboriginal song, he seemed to be referring to melodies in which scale steps are defined relatively approximately, or in which there are relatively few scale steps per octave: "The most fascinating of the oldest melody patterns may be described as a 'tumbling strain.' Its character is wild and violent: after a leap up to the highest available note in screaming fortissimo, the voice rattles down by jumps or steps or glides to a pianissimo respite on a couple of the lowest, almost inaudible notes; then, in a mighty leap, it resumes the highest note to repeat this cascade as often as necessary" (51). But approximate tuning of scale steps is the rule—not the exception. The question is not whether the scale steps are approximate but *how* approximate they are. The inherent uncertainty in the tuning of scale steps is a musical universal that is based on the perceptual universal of JNDs and their context dependencies.

Ellis (1885) documented variations in scale tuning in different musical cultures. His conclusions undermined Western thinking about scales: "Ellis had found that pitch systems did not emerge from acoustic principles occurring naturally in the world. Instead, they were artifices fashioned diversely from place to place through direct human intervention and choice" (Stock 2007, 308). Whereas in the nineteenth century many agreed with Ellis's rejection of ratio theory, his prescient message was unfortunately not understood by all contemporary theorists of Western music, whether in humanities (e.g., Riemann) or sciences (e.g., Helmholtz). Many continued to theorize about scales and intervals using number ratios.

The inherently approximate nature of musical intervals and scale steps can be confirmed in a thought experiment. Imagine a sound recording of any music in the world that is comprised of tones that musicians or musicologists agree belong to scales. Measure the frequency of each tone. That is not as easy as it sounds—the tones have to be separated from the sonic mix, and the measurement needs to be made in a temporal window that starts a certain time after the onset and has a certain duration. Since the perceived pitch depends on several harmonics (see chapter 13), inharmonicity can be a problem. In any case, the measurement should be done in a consistent way, taking advantage of the empirical methods and theories of psychoacoustics. If the musicians whose tone frequencies are being measured are free to change their intonation during performance, making small adjustments to fundamental frequencies by ear, then the measured frequencies of each tone, as it occurs in different musical contexts, will typically vary more than the physical uncertainty of the measurement procedure. Whereas that physical uncertainty might be a few cents, the musical variation (changes in the frequency of each scale step as it recurs) will typically be a few tens of cents. That uncertainty depends on two main factors that are difficult to separate: quasi-random variation (including limited ability to intone accurately) and deliberate variation. The existence of expressive intonation is evidence that scale-step frequencies or pitches are inherently imprecise.

The tuning of scales in oral-aural tradition follows different principles that can provide clues about how scales emerge and develop. One principle applies to more stable anchor tones if they span consonant intervals such as 4ths and 5ths (diatonic principle). Another principle applies to the tones between the anchors if they tend toward equal spacing (equitonic principle). If both principles apply, we may speak of a hybrid scale. The same principles show how musical scales might be hierarchically organized—similar to the hierarchical organization of MmT.

Rytis Ambrazevičius and colleagues (2015) found that "the equidistant principle of scale composition interacts with the constraint of consonances" (23): "[In] 'framework' scales, a rough, loosely intoned equitonics is anchored in a frame of consonances (mostly fifths or fourths). Equitonics can also be specific to 'dissonant' cultures, with modifications according to the psychoacoustic requirements of sonority, roughness, etc." (23–24). They offered two contrasting examples:

> Susanne Fürniss (2000) noticed that the scales of polyphonic (four-part) vocal performances of Aka pygmies are based on consonances of a polyphonic framework. They form steady trichordal reference points while the intermediate scale degrees are intoned more freely (see also Arom and Voisin 1998, 268). (23)

> The śruti system indeed exists, but it is not totally based on "simple ratios." Krishnaswamy (2003, 13) concludes that a hybrid scheme most likely appears in

> Carnatic instrumental music. The tuning of the strongest consonances in characteristic accords shows some resemblance to simple ratios such as 3/2 or 4/3, but this is not valid for the other tuning procedures. (32)

These arguments are examples of psychohistoric theorizing. The sizes of intervals and their musical functions are held to depend on both psychology and history. To understand what a musical interval is and how it works in a musical context, we need to combine historical and psychological approaches. Psychological theory and experimentation can shed light on how intervals are perceived and performed today, but the dependence of interval sizes on musical context may be better explained in a historical approach that considers the historical (or prehistoric) development of musical structure and tuning/intonation. In such an approach, problematic theoretical or mathematical issues of tuning disappear, and theories based on ratios become obsolete. Tuning and intonation instead become practical and aesthetic matters that depend on numerous partly contradictory constraints.

The Pythagoreans

No account of the nature and origins of intervals and scales would be complete without reference to the Pythagoreans—the ancient Greek followers of Pythagoras, a mysterious figure about whom little is directly known. The Pythagoreans believed in the music of the spheres (or harmony of the spheres), according to which both musical intervals and the movements of the planets were determined by ratios of small integers. In some interpretations, the planets were held to produce an inaudible, spiritual form of music called *musica universalis* that represents a causal connection between the universe and the human soul. The mathematical mysticism of the Pythagoreans, like that of Empedocles (fifth century BC; Kingsley 1995), prevailed for millennia.

The Pythagoreans believed that musical intervals *are* ratios in the sense of ontology (being, existence, reality)—not merely that they *correspond* to ratios, as many believe today. They confirmed their belief empirically by subdividing a vibrating string (the monochord) into equal parts. Halving the length raised the pitch by an octave. Dividing it by three raised the pitch by an octave plus a 5th, and so on. In a Pythagorean approach, musical interval ratios are examples of Platonic forms: they represent what intervals are, or should be, ideally.

We should be careful not to take the Pythagoreans too seriously. Thomas Britz (2022) described them as "a mystic cult of mathematicians who saw many numbers as having mystical, philosophical and even ethical significance." Along similar lines, Carl Huffman (2018) explained that:

> Pythagoras presented a cosmos that was structured according to moral principles and significant numerical relationships and may have been akin to conceptions of the cosmos found in Platonic myths . . . the planets were seen as instruments of divine vengeance ("the hounds of Persephone"), the sun and moon are the isles of the blessed where we may go, if we live a good life, while thunder functioned to frighten the souls being punished in Tartarus. The heavenly bodies also appear to have moved in accordance with the mathematical ratios that govern the concordant musical intervals in order to produce a music of the heavens, which in the later tradition developed into the harmony of the spheres.

Cazden (1954) was similarly disparaging: "Musicians are not generally aware that the 'law of Nature' doctrines, from Pythagoras to our day, while cast in a seeming scientific guise, do not result from or lead to a scientific approach to music or to Nature at all. These doctrines represent in reality a species of mysticism dealing with magic numbers. In their current forms, the Nature theories represent a retreat to mediaeval, pre-scientific modes of speculation and not a proper philosophy and method of science" (289).

The first to write extensively about the Pythagorean movement was the more down-to-earth Aristoxenus (Zhmud 2012). He tended toward what we might today call ecological materialism, with a focus on how real people create and perceive real music (although he also contributed to musical ratio theory). His proposal that the ear (not mathematics) is the ultimate judge of good or bad tuning has repeatedly been supported in modern research on intonation in music performance. Looking back, the music theory of Aristoxenus seems closer to the modern humanities and sciences than that of the Pythagoreans.

But Pythagorean thought was persistent. A millennium later, the sixth-century philosopher Boethius, guided by an immaterial but omnipresent Christian God, still regarded music of the spheres as the real music and human music as a mere approximation. Another millennium later, the scientific revolution led to the discovery of spectral analysis, after which many understood musical intervals not only as ratios of string lengths, but also as frequency ratios: ratios between the fundamental frequencies of harmonic complex tones.

Just versus Pythagorean

Most ratio theory in the Western history of music theory can be reduced to two approaches: Pythagorean and Just (or pure). The terminology is ambiguous: both Just and Pythagorean approaches are Pythagorean in the sense that both regard intervals as ratios of small integers. The Pythagorean approach is more characteristic of ancient Greece, whereas the Just approach was more common in the Renaissance.

Pythagorean intervals are combinations of octaves (ratio 2:1) and perfect 5ths (3:2). A Pythagorean major 3rd is four perfect 5ths minus two octaves or a ratio of $(3/2)^4/2^2=81/64$. That's 4.08 semitones or 408 cents—relative to 12-EDO, the tuning of modern keyboards (neglecting octave stretch for the moment). In Just tuning, major 3rds have a ratio of 5:4 or 3.86 semitones—22 cents narrower than Pythagorean. In short, Pythagorean major 3rds are slightly wider than four semitones in 12-EDO, and Just major 3rds are slightly narrower.

In music performance, it is not generally possible to distinguish between Just and Pythagorean versions of the same interval. Intonation typically varies over a range that includes both—even at high levels of musicianship (cf. Devaney et al. 2011; Duke 1985; Karrick 1998; Loosen 1993; Mason 1960; Morrison 2000; O'Keefe 1975; Sundberg 1982). Within a section of a high-level student choir (i.e., among singers singing the same note), the standard deviation in fundamental frequency under ideal conditions can be as small as 10–15 cents (Ternström and Sundberg 1988).

There is no evidence for the musical existence of two different ratios for the major-3rd interval (Just versus Pythagorean) or for a bimodal distribution in which major 3rds tend toward one or the other of the two ratios. There may, however, be a weak and inconsistent tendency to reduce the size of major 3rds in slow music with long non-vibrato tones (to reduce perceived roughness) and an opposite tendency in fast music (perhaps to clarify tonal functions or movement implications).

The empirical findings suggest that the real major 3rd is neither Just nor Pythagorean. Neither ratio is consistently preferred relative to four semitones in 12-EDO, which lies between them. Confronted by these observations, a ratio theorist might argue that idealized ratios are distorted when transformed in music performance—for reasons to do with physical constraints, musical contexts, or expression. Such distortions exist, of course—but relative to which of the two ratios? The question seems to be unanswerable. So, a theory of that kind is unfalsifiable (Parncutt and Hair 2018).

Pythagorean tuning allows for every tone in a major or standard diatonic scale to be exactly defined and, on that basis, any other tone in the chromatic scale—at least in an abstract world of pure mathematics. Enharmonic equivalents are tuned differently, sharps being sharper than enharmonically equivalent flats (C♯ is sharper than D♭). Some theorists have proposed that Pythagorean tuning is ideal for Gregorian chant on the assumption that diatonic scales are created most simply by combining perfect octaves and 5ths or that Pythagorean tuning derives from ancient Greek theory (Barbour 2004; Bower 2002; Krahenbuehl and Schmidt 1962). But to my knowledge, there is no empirical evidence that chant was or is performed closer to Pythagorean than to Just.

Renaissance theorists such as Zarlino preferred Just intonation, which is limited to factors of two, three, and five: all intervals are combinations of octaves (2:1), 5ths (3:2), and major 3rds (5:4). Contrary to the Pythagorean system, accidental sharps in Just are flatter than enharmonically equivalent flats. Again, there is no empirical evidence, to my knowledge, that Renaissance polyphony was or is intoned closer to Just than to Pythagorean.

It is not mathematically possible to tune all intervals between all tones of a diatonic scale within the Just system, for example when tuning the harpsichord (Sukljan 2018). At least one interval must be tuned non-Just. The major-2nd interval has a ratio of 8:9, and the perfect 5th has a ratio of 2:3. If we create a major 6th by stacking a major 2nd on a perfect 5th, the resultant ratio is 16:27, which is not Just (a Just major 6th is 3:5). On the modern piano, if the major 2nd between C and D is tuned 8:9, and the major 6th between C and A is tuned 3:5 (the intervallic inversion of the minor 3rd, which is 5:6), then the perfect 5th between D and A is not 3:2 (Zweifel 1994). Any attempt to tune Just, or at least partially Just, results in shifting scale steps, which then sound out of tune. In that sense, there is no such thing as a Just diatonic scale in musical practice, whereas Pythagorean tuning is at least theoretically possible.

Consider a piece of music that is confined to a seven-tone diatonic scale. The music can be performed with Pythagorean tuning on a regular keyboard by tuning the white keys Pythagorean and ignoring the black keys. More complex music that modulates to any of the twenty-four major and minor tonalities can be played in Pythagorean tuning on a keyboard that has split black keys with two different tunings for enharmonic equivalents (A♯ sharper than B♭). It may also be necessary to split some white keys to distinguish B♯ from C, C♭ from B, E♯ from F, and F♭ from E. Double sharps and flats cause additional problems, but their frequencies are at least clearly defined.

Beyond Ratios

Interval ratio theory has limited usefulness in a modern, empirically based psychological approach to music theory. In practice, musical intervals may correspond approximately to ratios of small integers, but they also often deviate systematically from them. Larger intervals (not only octaves) are consistently stretched relative to both theoretical ratios and 12-EDO (Rakowski 1994; Ward 1954), whereas for small intervals (mainly minor 2nds, but also to a smaller extent major 2nds and minor 3rds), the opposite tendency has been observed (Rakowski and Miśkiewicz 1995).

Simultaneous and successive intervals between harmonic complex tones may tend under some circumstances to be tuned such that some upper partials coincide, creating simple ratios (Villegas and Cohen 2010). But that is only one of several tuning criteria. The optimal size of an interval in context also depends on the stretch of large intervals, sharp intonation of soloists relative to accompaniment, anticipation of melodic movement, and individual differences in psychoacoustic pitch shifts. Most of all, musicians learn interval sizes from music itself (Kopiez 2003). Tuning standards are passed from one generation to the next "by ear" (in oral-aural tradition). In Western culture, the tuning of the piano may be acting as a psychological standard for tuning in any tonal style.

Standard pentatonic and diatonic scales are based on perfect 5th and 4th intervals. Those intervals are learned pitch distances—not ratios, although they are certainly close to 3:2 and 4:3, respectively. If listeners are sensitive to (and prefer) the consonance of perfect intervals (a culturally arbitrary criterion), the standard scales have few alternatives. These simple observations answer the question "Why are standard pentatonic and diatonic scales the most common in Western music?" without the need to consider ratios (e.g., Zweifel 1996).

The circle of 5ths is a music-theoretical structure with a long history. Ancient Greek music theorists, regarding musical intervals as simple ratios and music theory as a branch of mathematics, realized that one of the twelve

5ths must be mistuned to complete the circle, because $(3/2)^{12}$ is not exactly equal to an integer power of two. A whole number of ideal 5ths can never equal a whole number of ideal octaves; $(3/2)^{12} = 129.7$, but $2^7 = 128$.

That may be an interesting mathematical point, but its practical musical and psychological significance is limited. In musical practice, scales are collections of pitch categories—not exact pitches. The tuning of all intervals, including octaves and 5ths, is inherently flexible. The adjustment that piano tuners make to the perfect 5th interval to solve the problem (reducing its size from an exact 2:3 ratio of 702 cents to a tempered 5th of 700 cents) is imperceptible in musical contexts and, in that sense, musically irrelevant (Ward and Martin 1961). Joos Vos (1986) found that under musically atypical ideal conditions, purity ratings for two simultaneous harmonic complex tones spanning a perfect 5th were only affected if mistuning exceeded 2–10 cents. In real music, the variation in interval size is usually much larger. In a psychological approach that puts the experience of listeners (including performers, improvisers, and composers) in the foreground, the pitch-class circle and the chromatic scale are effectively closed and complete.

Music is not the only phenomenon that has been explained mathematically since ancient times. A well-known theory of visual aesthetics invokes the golden ratio or divine proportion ($\phi = 1.618$). The golden ratio plays an important role in certain geometric figures and is related to the Fibonacci sequence. The idea fascinated the Pythagoreans, Euclid, da Vinci, Kepler, and Euler. But architecture, paintings, and human bodies are not more beautiful if their dimensions correspond to the golden ratio (Stieger and Swami 2015). Moreover, and contrary to rumor, ancient Greek builders and architects were seldom guided by the golden ratio (Foutakis 2014). Incidentally, the ratio of the sides of DIN-A4 paper is $1{:}\sqrt{2} = 1{:}1.41$, which is much smaller than the golden ratio—but that does not make A4 paper any less beautiful, and US letter is not even less beautiful because it lies even further from the golden ratio (1:1.29).

Stretched Octaves

The width of an octave is not only inherently approximate, but also stretched or enlarged. If it is not stretched, it may sound out of tune. If all octaves are played exactly 2:1 in a piece of music, very low tones sound sharp, and very high tones sound flat. In the violin concerto repertoire, high natural harmonics sound flat, but when physically measured, they tend to be sharp relative to 2:1 (Guettler 2002). In listening experiments, octaves are preferred if they are 15–20 cents too wide (Dobbins and Cuddy 1982; Fransson et al. 1974; Houtsma et al. 1987, Demonstration 15).

If an octave is not a simple number ratio, no interval is a simple number ratio. But ratio theory continues to thrive, despite the clear empirical counterevidence. In one modern approach, Timothy Hubbard (2022) proposed that the Pythagorean comma can explain the stretched musical octave in the following way. The Pythagorean comma is obtained by moving around the circle of 5ths, adding together twelve 5ths (by multiplying their ratios) and subtracting seven octaves: $(3/2)^{12}/2^7 = 3^{12}/2^{19} = 531441/524288 \approx 24$ cent $\approx 1/4$ semitone. That corresponds approximately to empirical measures of octave stretch in music (Dobbins and Cuddy 1982), suggesting a link between the two. Hubbard explained the link in terms of "a desire to return to the tonic or other central pitch" or be "consistent with the idea of movement through pitch space" (677). He implied that when listeners perceive an octave interval in music, they imagine (or somehow cognitively process) a twelve-step (!) voyage around the circle of 5ths, in which each 5th is tuned to within 2 cents of 3:2, and each octave is tuned to within 2 cents of 2:1.

If Hubbard was wrong, how else can octave stretch be explained? We need to think about the origin of the musical octave from an ecological-evolutionary or psychophysiological viewpoint. An important function of hearing is speech processing. The auditory system is constantly responding to voiced speech sounds—harmonic complex tones. The octave is the perceived distance between the 1st and 2nd harmonics (or the 2nd and 4th, or the 3rd and 6th). Perception of octaves within harmonic complex tones is enabled by the basilar membrane in the inner ear, but the basilar membrane is not a perfect spectral analyzer. The information it sends down the auditory nerve is slightly distorted due to the enormous frequency and amplitude range of hearing. As a result, the perceived octave between the 1st and 2nd harmonics of a typical harmonic complex tone tends to be slightly stretched (Terhardt et al. 1982b). The stretch can be explained in either a spectral or a temporal approach (Hartmann 1993); in both cases, it involves the neurophysiology of the auditory periphery.

An octave is a musical phenomenon—not a mathematical one. In a circular operational definition, an octave is the pitch distance or frequency ratio between musical tones that are perceived to lie an octave apart. That being the case, we can establish the size of an octave by measuring musical sound. Empirical findings suggest that the perceived octave between the 1st and 2nd harmonics of a single complex tone is stretched relative to a 2:1 ratio due to pitch shifts by roughly the same amount as a typical performed octave in music—suggesting the two have a common origin, or the perceived octave represents the origin of the performed octave.

There is also a physical (acoustical) reason for musical octave stretch. Octaves on the piano (or any instrument with freely vibrating strings such as the guitar) are physically stretched due to the stretching of intervals between partials. The effective length of a piano string is different for each partial, because the string is not perfectly elastic: higher partials cause the string to bend more at the bridge. When piano tuners line up the 2nd partial of a lower tone with the 1st partial of a higher tone, the result is a slightly stretched octave (Giordano 2015). In such cases, a stretch of 20 cents is typical, but the stretch is smaller in the central range and larger at low and high frequencies. A piano whose octaves are not stretched in this way sounds clearly out of tune (D. W. Martin and Ward 1961).

12-EDO

The standard against which tuning in Western music is most appropriately measured in practice is neither Just nor Pythagorean. It is a division of the octave into twelve equal semitones (12-EDO), such that each semitone corresponds approximately to a ratio of the twelfth root of two ($1:\sqrt[12]{2}$) or about 1:1.06 (a 6-percent change in frequency). In addition, the octave should be slightly stretched, causing all the semitones to be stretched by the same amount, but that stretch is not normally included in the definition of 12-EDO. Based on what we know about intonation in real music, 12-EDO is the most realistic and convenient reference for measuring the size of intervals.

In music performance, the size of an interval can vary considerably. Consider the most common interval between successive tones: the major 2nd. Ancient theorists argued over whether its ratio was 8:9 or 9:10, but in performance—even by the best modern ensembles with the most sensitive tuning—tuning varies over a range that includes both these ratios. Moreover, nothing special happens, psychologically or musically, in the immediate vicinity of the ratios themselves. Like all intervals, major 2nds can be relatively large or relatively small—nothing more. In everyday musical contexts, the interval of a minor 2nd can vary across a relatively wide range—say, 0.5–1.5 semitones—and still be recognized as a minor 2nd (cf. Devaney et al. 2011). Given the inherent uncertainty in the tuning and intonation of all scales, Western music being no exception, it is appropriate to define the chromatic scale as approximate 12-EDO.

Why is the piano keyboard based on 12-EDO and not 13-EDO, 11-EDO, or another number? That is one of the few questions that ratio theory can clearly and plausibly answer. In separate studies, Ramon Fuller (1991), Dirk de Klerk (1979), and Sebastian Von Hoerner (1976) showed mathematically that certain equal divisions of the octave come closer to Just intervals than others. The most promising divisions have five, seven, twelve, nineteen, thirty-one, forty-three, or fifty-three equal parts. To understand why MmT is based on 12-EDO and not one of the other numbers in this list, we need to think about how real music works:

- Two musical tones that are a semitone (100 cents) apart are usually perceived as clearly and categorically different in pitch—even in relatively complex musical contexts and considering typical limitations on the intonation of the voice and melodic instruments. In the psychology laboratory, under ideal conditions and using trained listeners, the JND for the frequency of successive pure tones in the central range can be as small as 3 cents, but the smallest microtonal intervals in musical use are in the range 33–50 cents (for a literature review, see Parncutt and Cohen 1995).
- The number twelve evidently approaches the largest the number of pitches per octave that an expert musician can hold in imagination in typical musical contexts—a kind of upper limit of human cognitive processing capacity. There is no clear evidence for this assertion, but it is interesting to compare George Miller (1956).
- The difference between Just tuning and 12-EDO for musically important intervals such as major 3rds is small enough to lie within normal ranges of intonation in vocal or instrumental performance.
- The twelve-tone scale enables a large number of scalar subsets to be created and enumerated in pitch-class set theory (Forte 1973). Those subsets are usually unequally spaced, which helps listeners to get their bearings relative to tonal references or tonics (Trehub et al. 1999).

If 12-EDO is an appropriate standard against which tuning and temperament can be measured, it follows that Just and Pythagorean tuning are not.

That casts a shadow of doubt over the word "temperament" itself. The word suggests mistuning relative to simple frequency ratios, on the assumption that those simple ratios are standard or ideal. But neither Just nor Pythagorean tuning can be considered standard or ideal if there is no evidence that musicians consistently prefer one over the other.

The idea of measuring interval sizes relative to 12-EDO was introduced to the discipline of comparative musicology (an approach to musicology based on intercultural comparison—in particular, comparison to Western music) by Alexander Ellis in the colonialist nineteenth century. Ellis proposed measuring musical intervals in cents, or hundredths of a semitone. His project to measure the scales of different cultures systematically was promising but only partly successful. Today, despite the efforts of Ellis and his followers, we cannot draw upon a systematic collection of common non-Western tunings because of the wide range of tunings of any given interval in any given cultural context. Western music is typical in that regard: operatic singers depart from 12-EDO by a wide margin in both vibrato (often plus/minus a semitone) and deviations between the center of the vibrato (perceived as the intended pitch) and the intonation of the accompaniment (Prame 1997).

In an innovative attempt to address this issue, Simha Arom and Frédéric Voisin (1998) had musicians in Central Africa and Java tune an electronic keyboard with natural-sounding timbres to find out what intonations they intended or preferred—what interval sizes were stored in their minds. To my knowledge, this promising technique has not yet been applied systematically to musical styles in different geographical regions—possibility due to ethical considerations or simply intonational variability. Arom nevertheless claimed to have measured interval sizes to within 10–20 cents.

Today, most Western music is played in or near 12-EDO. That includes music written long before the idea of 12-EDO crystallized in the early eighteenth century with the first book of J. S. Bach's *Wohltemperiertes Klavier*. Even then, Bach was not promoting exactly equal semitones; *wohltemperiert* is not the same as *gleichstufig temperiert* (Rasch 2011). Anyway, the idea was not new in the eighteenth century: much sixteenth-century lute music was performed in something approaching 12-EDO (Ozmo 2016). The extent to which Bach's clavichord might have been (almost) equally tempered in the modern sense is unclear, but we do know from the music itself that all twenty-four major and minor keys were eminently usable, even if there were slight tuning variations from one key to another (Rasch 2011). Small tuning variations of that kind are not relevant for the theory presented here. Quarter-tone tuning in Middle Eastern musical traditions (Persian *dastgah*; Arabic *maqam*) can be similarly approximate.

One can argue that the chromatic scale, defined as approximate 12-EDO, is much older than Bach. The division of the perfect 4th into two whole steps and one half step in ancient Greek theory is consistent with 12-EDO. During the next few centuries, Chinese tuning theory independently developed something close to 12-EDO based on the circle of 5ths (McClain 1979). The ancient Greek tetrachord was four tones, spanning a perfect 4th interval or 3:4 ratio. In the diatonic genus, and starting from the higher tone of the 4th and moving downward, the tetrachord comprised two whole-tone intervals (understood as 8:9 or 204 cents each) and one semitone (243:256 or 90 cents). These three intervals multiply to 3:4 and add to 498 cents (cf. Barbera 1977; Gollin 2004). Although the 243:256 ratio was smaller than that of the semitone in today's 12-EDO and corresponds to Pythagorean tuning (producing slightly sharp leading tones), it still lies well within the usual range of tuning in today's music performance. From that, we may guess that a lot of ancient Greek music practice conformed approximately to 12-EDO—similar to MmT today.

Octave Equivalence

The music-theoretical principle of octave equivalence says that tones an octave apart have the same harmonic function—at least from the point of view of Western tonality. But octave equivalence is not a musical universal, as is often claimed:

- In Indonesian gamelan tunings, even if octave equivalence seems to apply, the octaves may be stretched more than Western octaves (Carterette and Kendall 1994)—possibly due to the familiarity of gamelan players with the nonharmonic spectra of gamelan gongs (McLachlan, Marco, and Wilson 2013).
- Some Australian indigenous melodies seem to lack octaves (Will 1997), but that does not mean the performers or listeners are insensitive to them. Instead, octaves may simply be unimportant or irrelevant in this musical style. Another possibility is that the category boundaries for the perception of musical pitches and intervals are different in Australian indigenous

and Western music, causing musically trained Western listeners to get the impression that the music lacks octaves.

- Nori Jacoby and colleagues (2019) remarked that "sung correlates of octave equivalence are undetectable in Amazonians, suggesting effects of culture-specific experience . . . octave equivalence may be culturally contingent, plausibly dependent on pitch representations that develop from experience with particular musical systems" (3229). Specifically, "it is possible that perceptual octave equivalence only emerges in the presence of an octave-based musical system with a large melodic range" (3241).

Even if octave equivalence is not a musical universal, sensitivity to octaves can be considered a perceptual universal, given the importance of octaves in the harmonic series that every human perceives within voiced speech sounds when tracking the fundamental frequency of speech. Laurent Demany and Françoise Armand (1984) found on the basis of a careful empirical study that "two pure tones forming an octave interval have some degree of perceptual equivalence for three-month-old human infants" (63). However, their finding could be accounted for by familiarity either with the harmonic series within voiced speech sounds or with Western music, both of which can be heard both before and after birth. The perceptual octave effect can be quite weak; Howard Kallman (1982) presented successive pure tones spanning chromatic intervals to musicians and asked them to rate similarity. He observed almost no effect at the octave. Responses depended more on interval size.

Despite the ambiguity of the empirical findings, it is reasonable to assume that everyone in the world is capable of perceiving the similarity of successive harmonic complex tones spanning octave intervals, in which the audible harmonics of the first tone line up with those of the next, by comparison to mistuned octaves—even if octave equivalence plays no role in their music. Regardless of cultural background, every human has a basilar membrane that enables a running spectral analysis. Similarly, everyone in the world is capable of perceiving the consonance (harmonicity and/or smoothness) of simultaneous octaves by comparison to slightly larger or smaller intervals (cf. Plomp and Levelt 1965). The ability to recognize harmonicity in running spectra logically includes the ability to recognize octaves (frequency ratios of approximately 2:1) as the most common and hence most important indicator of spectral harmonicity. All of that is presumably true across cultures, even if octave relationships play no role in the music of a given culture or if they play a different role.

Octave equivalence in Western and other musics may be considered psychohistoric in the sense that neither psychology nor history alone can account for it, but only a combination of the two. Octave equivalence depends not only on perceptual properties of tones and intervals as we perceive them (the psychological aspect), but also on the musical contexts in which octave intervals occur and the history of perceiving and performing octaves in diverse musical contexts. Octave equivalence may be learned from exposure to octave intervals within the spectra of individual harmonic complex tones in speech, music, and everyday sounds, in parallel with harmonic fusion (Demany et al. 2021).

In Western music theory, tones an octave apart are equivalent, but tones a 5th apart are categorically different. The distinction may be clearer for musicians who can easily recognize pitch patterns due to their musical training, and their understanding of music theory and music notation in which octave equivalence plays an important role. Experiments on similarity judgments of musical tones at different intervals (e.g., Parncutt 1989) do not imply a categorical difference between octaves and 5ths for nonmusicians.

Perfect Intervals

In the history of Western music theory, octaves, 5ths, and 4ths between simultaneous and successive tones are *perfect consonances*. But for simultaneous tones, the 4th is paradoxically often regarded as a dissonance. When the 4th occurs within a chord and the lower tone of the 4th is in the bass, the chord is dissonant and requires special treatment or resolution (e.g., a major or minor triad in 2nd inversion).

Despite the ambiguous C/D of the simultaneous perfect 4th interval, melodic perfect 4ths and 5ths evidently played an important role in the prehistory of standard pentatonic and diatonic scales. In oral tradition, intonation is learned from performance and passed from one musician or generation to the next. Quasi-diatonic scales may have emerged when consonant melodic intervals (octaves, 5ths, 4ths) came to be preferred between nonadjacent scale tones (Gauldin 1983).

A possible explanation involves pitch commonality. The pitch of a musical tone is ambiguous (Terhardt 1974a), and tones are heard to go well with each other if they have pitches in common (Parncutt 1989). Pitch commonality involves both spectral pitches (audible partials) and virtual pitches (possible fundamentals; see chapter 13). Successive tones at perfect octave, 5th, and 4th intervals have both spectral and virtual pitches in common (see figures 13.1 and 13.2). The perceptual coherence of a scale may depend on the degree to which its tones are related by perfect intervals.

The idea that pitch commonality drove the prehistoric emergence of scales in oral tradition can hardly be tested. We have no recordings of ancient music; we only have recordings of traditional music that we guess was not influenced by MmT. But if the idea is valid—and it is certainly straightforward—we can apply it in an attempt to explain why some scales, understood as sets of pitch categories, are more common than other scales—at least within Western music. Familiar scales from music theory (church modes, the major scale, variants of the minor scale, the blues scale) tend to maximize the number of perfect 4th or 5th relationships. After the octave, that is the strongest harmonic relationship—the one with the highest pitch commonality between harmonic complex tones.

The principle applies in particular to standard pentatonic and diatonic scales, whose steps correspond to adjacent points on the circle of 5ths. Starting from C, the first five tones on the circle (C, G, D, A, E) make up a standard pentatonic scale, as do the five tones starting with the second element (G, D, A, E, B), and so on. The first seven tones (C, G, D, A, E, B) make a standard diatonic scale, as do the seven tones starting with the 2nd element (G, D, A, E, B, F♯), and so on. In a psychohistoric account of the origin of such scales, pentatonic and diatonic scales developed as 5ths were favored between non-adjacent tones due to their pitch commonality.

In an ecological approach, the process did not rely on musicians and listeners recognizing scales as such. Instead, they recognized melodies, which were preferred if they conformed to scales. Scales may have been recognized later at a more abstract level (perhaps by trained musicians). Recognizing a scale meant somehow understanding the pitch pattern as a whole; it also meant recognizing the more common (usually smaller) intervals within that context. The process did not necessarily involve the development of a cognitive representation of the circle of 5ths.

Asymmetry and Evenness

The intervals of the seven-tone diatonic scale, considered as a subset of 12-EDO, are unequal (separated by whole tones and semitones). The scale cannot be transposed onto itself without changing one of the tones unless the transposition is through an octave. The scale's asymmetry makes it easier for listeners to get their bearings and hear the music relative to tonal references or tonics. Whereas Sandra Trehub and colleagues (1999) suggested that humans might have an "inherent processing bias favoring unequal-step scales," one could also say in an ecological approach that it is inherently easier for any perceiving machine or organism to get oriented relative to an unequally spaced scale because the unequal interval sizes allow one to locate the tonic or other pitch reference unambiguously. For similar reasons, scalar asymmetry facilitates the learning of new melodies (Pelofi and Farbood 2021).

A related property of diatonic scales is evenness. The whole tones and semitones are distributed almost equally around the pitch-class circle (Clough and Douthett 1991). As a result, the size of most intervals corresponds roughly to the number of scale steps, give or take a semitone: intervals of a diatonic 2nd are one or two semitones, 3rds are three or four, 4ths are five or six, 5ths are six or seven, 6ths are eight or nine, and 7ths are ten or eleven (see table 3.1). Each diatonic interval has two common variants a semitone apart, which limits variation and simplifies the mapping of diatonic onto chromatic. If one interval between neighboring tones is smaller than average, the intervals on each side of it are usually larger than average, restoring evenness. Within such a diatonic scale, when pitch patterns are transposed to different starting pitches, the exact sizes of individual intervals may change, but the relationship between larger and smaller intervals stays the same.

The diatonic scale can be constructed by stacking six perfect 5th intervals and transposing back down by octaves. The result is a scale with seven tones per octave: CDEFGAB. It is interesting to ask why the scale has seven tones and not six or eight. A scale of six tones, CDEFGA, would be inconsistent: two of its minor 3rds (DF and EG) would be split (by E and F), whereas AC would not be split. It is more logical to split either all minor 3rds (producing a standard diatonic scale) or none of them (producing a standard pentatonic). Similarly, a scale of eight tones, such as CDEFGAB♭B, would be inconsistent because the minor 3rd AC would be split into three, whereas DF and EG would be split into two.

Performed Intonation

Tuning is frequency adjustment before performance, such as the tuning of a piano or a guitar. Intonation is frequency adjustment during performance—in real time, from one tone to the next, as in a choir or string quartet. Musical intonation strays in both quasi-random and systematic-intended ways from theoretically correct frequencies or pitches: various studies of Western performance suggested that intonation tends to be closer, on average, to Pythagorean or 12-EDO than to Just (e.g., Mason 1960).

Good intonation is about honing in on the familiar average sizes of musical intervals. It also involves keeping the pitch of scale degrees constant when they occur at different times and in different musical contexts, but it may mean changing those pitches if the musical style includes rule-based microtonal variations (Ahlbäck 2018). In ensemble performance, good intonation also involves responding appropriately and sensitively to intentional and unintentional changes in intonation by other performers. Good intonation is inseparable from other musical skills and other aspects of familiarity with relevant musical styles.

During the course of a performance on instruments with variable intonation (including voices), there is no exact standard of pitch or frequency. Instead, the tuning of more stable tones such as tonics tends to be more exact, and the tuning of less stable tones such as leading tones tends to be more variable (Ambrazevičius 2005). The intonation of a given scale step varies over a range that seems smaller to the listener than to the acoustician making physical measurements (cf. Devaney et al. 2011). The range of variation is greater in everyday singing by nonmusicians, in which aesthetic judgments can be remarkably independent of intonation; in that case, "interval tuning is imprecise in both production and perception" (Pfordresher and Brown 2016, 11).

Musicians and listeners intuitively apply several conflicting criteria to decide whether a pitch or an interval is "in tune" in the context of a musical performance. The criteria include:

- Context-dependent learned interval sizes. Interval sizes are not only learned from music, they can also depend on musical context (Ahlbäck 2018).
- Interval stretch (Rakowski 1994). Different people prefer different amounts of stretch, and upright pianos have more stretch than grands (Railsback 1938).
- The relationship between foreground and background (solo versus accompaniment). Solos are often intoned sharp relative to accompaniment (Kantorski 1986).

- Anticipation of pitch movement. Rising leading tones are intoned sharp, and falling leading tones are intoned flat. In both cases, intonation is more variable than for other scale steps (Fyk 1997). Rytis Ambrazevičius (pers. commun.) found a similar principle in Lithuanian traditional singing: Scale degrees adjacent to tonal anchors tended to gravitate toward them, an interval of about two semitones shifting toward about one-and-a-half semitones.
- Emotional expression. This can affect intonation in many ways, perhaps by imitating emotional speech, but there are also typically musical gestures, such as sliding toward target pitches (P. Johnson 2004).
- Style. Slow, sustained music without vibrato may tend in the direction of Just intonation if that makes beats less audible. Singers who think Renaissance polyphony should be sung with Just intonation may deliberately reduce the size of their major 3rds.
- Pitch shifts. In principle, all pitches in a piece of music are slightly shifted due to perceptual distortions, for which performers may compensate. Pitch shifts depend on sound level or masking and are smaller for complex than for pure tones (Terhardt 1975).
- Idiosynchronies of musical instruments. High natural harmonics on string instruments may sound flat because octaves are not stretched (or not stretched enough; Guettler 2002).
- Individual psychological and physiological differences. These apply for instance to the size of pitch shifts.

Good intonation is generally a compromise among such conflicting criteria. Consistent with that idea, Anders Friberg and colleagues (2006) developed, adjusted, and tested a set of performance rules for computer-controlled expressive performance. The rules included:

- *High sharp*: Stretch all intervals in proportion to size.
- *Melodic intonation*: Intonate according to melodic context.
- *Harmonic intonation*: Intonate according to harmonic context.
- *Mixed intonation*: Intonate using a combination of melodic and harmonic intonation.

In melodic intonation, semitones are performed smaller than in 12-EDO. In harmonic intonation, beating in chords is minimized, which sometimes means tuning in the opposite direction (toward Just rather than Pythagorean), leading to questions of mixed intonation.

In addition to the listed criteria, intonation in musical performance is inherently uncertain. Musicians (including singers) have limited control over the pitch of their instruments, leading to quasi-random variation around approximately defined target frequencies. Eric Prame (1997) analyzed twenty-five tones sung by sopranos in each of ten commercially available recordings of Franz Schubert's *Ave Maria*. Whereas the pitch of the tones was heard near the middle of the vibrato (as confirmed by Shonle and Horan 1976), the mean extent of vibrato according to Prame was 0.7 of a semitone in both directions. So, it spanned 1.4 semitones. Larger vibratos were measured for Verdi opera arias than for Schubert. The difference between relatively sharp and relatively flat singers was remarkable: "The mean difference across artists between sharpest and flattest intonation was 54 cent" (Prame 1997, 618). Moreover, "the intonation of tones with greater vibrato extent tended to be sharper" (619).

Although the JND for the frequency of two successive pure tones, when measured in a psychoacoustic experiment and under ideal conditions, is much smaller than a semitone (typically 5 cents, with large individual differences), it is larger in complex musical contexts, especially when those contexts feature audible mistunings (often of the order of a semitone) due to the technical demands of musical sensorimotor coordination (e.g., choral singing). Because our ears are used to variation around target pitches in music, we are surprisingly tolerant of it. That can explain why the tuning variations that are revealed in physical measurements of music performance tend to be greater than the mistunings that we think we are hearing. Musicians who are courageous enough to have their own intonation measured may be surprised at how poor it is from a physical viewpoint.

But it is not the physical viewpoint that counts. What counts is perception. Besides, musicians may deliberately and audibly change the pitch of a tone during the course of the tone for expressive reasons, starting a little flat and gradually rising to a target pitch. The tonal references relative to which intonation is being adjusted may also change, from melodic intonation (where tonal references are successive) to harmonic intonation (where tonal references are simultaneous). If such adjustments are performed effectively, no one will describe the deviations as "mistuning" or complain about poor intonation.

The importance of beating (amplitude modulation from almost coinciding partials) for tuning in music performance is smaller than many musicians believe. The sound of beating is familiar from the "wa-wa" sound of a single note on a badly out-of-tune or "honky-tonk" piano. If the beats are too fast to be heard individually (more than twenty per second) and slower than about 250 per second (Terhardt 1968), the sound is rough. One criterion for good intonation in music performance is to minimize or avoid audible beats and roughness, but that roughness or beating may be only perceptible if the tones are long and sustained and lack vibrato. In music, it is common either for tones to die away quickly (like piano tones do) or for their fundamental frequencies to change slightly during the course of the tone (as often happens in vocal, woodwind, brass, or bowed-string music).

Categorical Perception

Perception of MmT involves psychological pitch and time categories that correspond roughly to notated pitches and durations. Given the variability and context dependence of intonation in music performance, the psychological categories that make up a Western musical scale are typically about a semitone wide (i.e., a central pitch plus or minus a quarter tone).

The German word *zurechthören* means to hear a tone correctly, even though it was not played correctly, or to guess which tone a musician intended to play and hear the sound as if that tone had been played. We do that all the time when we hear music that is sung or played on instruments with variable intonation. Within the range of *zurechthören*, both the performer and the audience perceive a tone as in tune; professional performers are usually more sensitive to mistuning than their audiences.

Melodies tend to focus on a small set of approximately defined pitches called scale steps, such that musical pitch is perceived in categories. They do not usually slide up and down like speech does (portamento). The ultimate origin of psychological pitch categories in music is unclear. Various possible psychohistoric scenarios might have led to their emergence. All of them are speculative, since we have no way of going back in time and performing psychological experiments:

- Musical pitch categories may enable and facilitate melody transmission in oral tradition. They may allow melodies to be memorized and passed from one generation to the next with a minimum of information content (even if the music is not written down). A modern analogy is the MIDI file, which takes up much less space on a computer hard disc than a corresponding MP3 or .WAV file. But the idea is problematic. People in oral traditions also remember and reproduce subtle variations in intonation (McLucas 2010): they remember much more than just the "notes."

- Musical pitch categories may be a consequence of the physics of musical instruments, including the finger holes of flutes or techniques to tune string instruments that involve avoiding beating. Musical instruments therefore tend to produce specific frequencies rather than pitch slides. According to this idea, pitch categories emerged in prehistory when early singers imitated instruments (E. Schubert and Wolfe 2013).
- Pitch categories may be related to category boundaries for the mistuning of harmonics within complex tones. Modern psychoacoustic experiments have shown that the spectral pitches of partials within complex tones are perceived categorically: a partial is perceived either as a harmonic of a given fundamental or as a separate sound, depending on the degree of mistuning. The ear is surprisingly tolerant of mistunings of partials within individual tones—perhaps because many tones in the physical world have almost-harmonic partials. Bells are an example: usually, some partials in the spectrum correspond to an incomplete, mistuned harmonic series, and others do not. Another example is the piano: the partials of piano tones are stretched relative to the harmonic series, the octave between the 1st and 2nd partials being perceptibly bigger than 2:1, the perfect 5th between the 2nd and 3rd bigger than 2:3, and so on. Moreover, the ear cannot hear pitch very exactly in very short sounds (due to the uncertainty principle—a hard physical limitation; see chapter 5). A harmonic partial within a harmonic complex tone can therefore be heard as part of the tone, even if it is noticeably mistuned, and the maximum mistuning typically lies between a quarter tone and a semitone (B. C. J. Moore et al. 1986). The categorical pitch perception of partials in complex tones could explain why the pitches of tones in melodies are also perceived categorically.
- We tend to categorize everything that we perceive in different ways. In everyday life, we categorize the distal stimulus (people talking, dogs, bells, trucks, the wind in the trees, rivers flowing). In music, we categorize the proximal stimulus or musical surface (the pitch, timbre, loudness, and perceived duration of voices and instruments). The two perceptual processes are evidently related. A focus on specific pitches—held and repeated during a melody—attracts our attention to the proximal stimulus and is typical of music listening.

11 Chant and the Leading Tone

The tonic of a scale is often the higher tone of a semitone interval. The origin of the rising leading-tone effect can be traced to structural properties of religious (Gregorian) chant, although chant itself does not include leading tones. Christian chant was an oral-aural tradition. Repertoires were first notated on a musical staff (like a graph of pitch against time) in the eleventh century, and polyphony emerged in the twelfth and thirteenth centuries. I propose that scale steps (individual sung tones) in chant vary in consonance, depending on how well their audible partials match the prevailing diatonic scale—a form of pitch commonality. Consonant tones were sung more often. The scale step fa *was more consonant than* mi *because the audible harmonics of* fa *better matched the prevailing diatonic scale. That made* fa *more stable than* mi*, which can explain the tonicizing function of rising leading tones in later polyphony.*

Background: Speech versus Song

Both speech and song convey a mixture of lexical and prosodic meaning. The meaning of speech is mainly lexical (like that of words in a dictionary), whereas that of song is more prosodic (depending on changes in pitch, loudness, timbre, and articulation). Prosodic aspects of speech communicate important meanings such as approval and intentionality (Esteve-Gibert and Prieto 2018). In music, specific pitch-time contours occasionally have lexical meanings, a well-known case being the leitmotif in Wagner's operas.

Although language and music vary in their details from one culture to the next, they also have quasi-universal features. One of them is the arbitrary nature of relationships between sound and meaning in language. The sound of a word such as "typewriter" or "laptop" has nothing to do with its meaning, and so the connection is arbitrary. The auditory system of a child quickly learns an enormous number of such arbitrary relationships. Only occasionally does the meaning of a word depend on its sound, the "meow" of a cat being an example (onomatopoeia).

The similarity of speech and song, and their clear separation in only one species (human), suggest that they have a common origin in earlier acoustic-gestural communication (Stumpf 1911/2012) or "musilanguage" (S. Brown 2001). The musilanguage of our distant primate ancestors was possibly comparable with, or based on, motherese or baby talk (Dissanayake 2001). Speech and song may have separated in human evolution during the emergence of what archeologists call *human behavioral modernity*—a complex collection of human behaviors that also include art and religion. Very approximately, behavioral modernity must have started 100,000 years ago (Parncutt 2019b).

Speech is about everyday communication, whereas music is more special, confined to rituals or ritual-like situations. In a broad definition of ritual that includes both religious ceremonies and playful caregiver–infant interactions or lullaby singing, we can say that music communicates abstract (non-lexical) ritual meanings. It also emotionally marks important social changes (birth, marriage, death, healing, new identity, new leader), with implications for social cohesion and psychological health (Lloyd et al. 2008).

As mentioned, both speech and song involve pitch-time patterns called prosody or melodic contour. In song, attention is focused on pitch contours and their emotional connotations, whereas in speech, attention is focused on phonemes (timbres) and lexical meanings. In that regard, both are hierarchically organized. Both have rhythm, but only musical rhythm is perceived to be isochronous. People can usually tap at almost equal time intervals to musical rhythm, which they can also do together, tapping in synchrony (phase locked). It is harder for a group to chant regular speech together, because speech is rhythmically less regular and more variable.

A human speaker normally produces four or five phonemes per second (Tendera et al. 2019). The rate of tones in melodies spans a wider range, from very slow to very fast (say, two to ten per second; Palmer 1997). The average pitch interval between successive melodic tones in the music of the world is approximately a whole tone (or a frequency ratio near 8:9 or 9:10; P. G. Vos and Troost 1989); the range of a melody is usually limited to about an octave. Melodic phrases tend to rise at the start and fall at the end (Huron 1996), just as the fundamental frequency of speech goes up and down with breathing (inclination), but prosodic changes from one word or sound to the next (due to expressive intention—important for understanding the message) may be bigger in language than in music (e.g., S. E. Miller et al. 2010).

Melodic and linguistic quasi universals may depend on universals of human biology and physiology: what fundamental frequencies, noises, and acoustic resonances can be produced comfortably by the human vocal tract, the dependence of voice frequency on subglottal pressure, how fast and accurately the sounds can be produced and perceived, or how sound memory works. Auditory processes such as temporal and spectral discrimination, Gestalt perception, and auditory scene analysis are also relevant.

Gregorian Chant

These generalizations about song apply equally to European plainchant and specifically to Gregorian chant, which emerged in Europe in the ninth century, based on oral traditions of Jewish and Christian song and ritual dating back to the fourth century (Möller and Stephan 1991). Chants were learned by ear and classified into modes, which helped singers memorize the repertoire: "classification of material is generally a sign that it was intended to be memorized" (Berger 2005, 5). From the twelfth century, as student singers started to learn polyphonic composition, "musicians were less concerned with learning basic rules and more with memorizing various alternatives for setting melodic formulas" (Berger 2005, 6).

Many chants are confined to six tones, called the hexachord (*ut, re, mi, fa, sol, la*). The intervals between adjacent tones in a hexachord are whole tone, whole tone, semitone, whole tone, whole tone. From the eleventh century, students learned how to match the hexachord to the chants in their repertory. Today, a hexachord can be played on the white keys of the piano in two places, starting on C (CDEFGA) or G (GABCDE). If transposed to start on F, a black key is introduced (FGAB♭CD).

The only semitone in the hexachord lies between *mi* and *fa*. In later music, and before today's conventions for marking accidentals emerged, tones that acted as *mi* were sometimes marked by predecessors of the sharp symbol, such as the square-looking *b quadratum* corresponding to B♮ in the hard hexachord GAB♮CDE. Tones that acted as *fa* were sometimes marked by predecessors of the flat symbol, such as the rounder-looking *b rotundum* corresponding to B♭ in the soft hexachord FGAB♭CD (cf. Henderson 1969).

Some modes happened more often than others. Willi Apel (1958, 142–165; as cited in W. E. Thomson 1999) commented that in the tonal chant repertory, "the *protus* [D modes] and *tetrardus* [G modes] occupy a considerably more prominent place than the other two, there being approximately 750 chants in the *protus* [D modes], 450 in the *deuterus* [E modes], 250 in the *tritus* [F modes], and 900 in the *tetradus* [G modes]" (38).

But there is more to mode classification than the final (which was D, E, F, or G in the medieval eight-mode system) and the authentic/plagal distinction (i.e., higher/lower ambitus relative to the final). Modes also have different reciting tones (also referred to as psalm tone, tenor, confinalis, dominant, or repercussion), which can make different modes sound more or less similar. Another problem is the occasional use of B♭. If a B♭ appears in a D mode, as a minor 6th above the final D, the result sounds like a transposition of an A mode, in which F is a minor 6th above A.

Ambiguities of that kind mean that Apel's numbers are only approximate. Nevertheless, some simple explanations come to mind for his observations about the frequency of occurrence of modes. The reason why the E and F modes were less common than the D and G modes may be due to dissonances within the scale relative to the final: the E-F semitone in the E modes, and the F-B tritone in the F modes. These intervals may have been perceived as dissonant or at least unrelated—even when the tones were not sounded simultaneously. In the white-note diatonic scale, G and A seem more consonant as finals because their upper and lower neighbors are a whole-tone away rather than a semitone—a more consonant interval. Final G is flanked by F and A, and final A by G and B. Because B is relatively dissonant (due to the tritone between B and F), final G has an advantage over A.

Melodic C/D

Both simultaneous and successive sounds can be heard as consonant or dissonant (C/D). Under ideal conditions, such as when chant is sung in a resonant space with no background noise, about ten harmonics are audible in each tone, the 9th and 10th harmonics being about a whole tone apart. That has interesting implications for the C/D of successive whole-tone and semitone intervals:

- When the interval between the fundamentals is a whole tone, harmonics eight, nine, ten, and eleven of the lower tone almost line up with harmonics seven, eight, nine, and ten of the upper tone, producing a feeling of pitch commonality. For example, in the interval G2-A2, both tones have harmonic partials near G5, A5, B5, and C#6. The interval is perceived as consonant.

- If the interval is a semitone, the audible spectra do not overlap. There is no pitch commonality, and the interval is perceived as dissonant.
- If the interval is a perfect 5th, the 3rd harmonic of the lower tone lines up with the 2nd harmonic of the upper tone, producing pitch commonality. If the interval is a perfect 4th, the 4th harmonic of the lower lines up with the 3rd of the upper. In both cases, there is perceptible pitch commonality. In both cases, the interval is perceived as consonant.
- If the interval is a tritone, there is again a lack of pitch commonality, and the interval is perceived as dissonant.

These arguments can explain why whole-tone intervals are so common in melodies, including chant. They are relatively easy to intone (sing in tune) because part of the audible spectrum of one tone lines up with part of the spectrum of the other. By comparison, the semitone interval seems arbitrary; it is not anchored to anything in the stimulus. The same comparison can be made between perfect 4ths and 5ths, on the one hand, and the tritone, on the other.

The effect can be observed in similarity judgments of successive complex tones, as perceived by nonmusicians (who presumably do not recognize specific intervals). Similarity is greater for an interval of two semitones than for an interval of one semitone—but only when the tones are harmonic complex (not when they are pure). The same applies to the tritone when compared with perfect 4ths and 5ths (Parncutt 1989, 120, figure 5.5a).

The preference for whole-tone intervals in Western chant is unsurprising, given that the whole-tone interval is the world's most common melodic interval (P. G. Vos and Troost 1989). That raises the possibility of an alternative explanation. In many musical styles (e.g., those based on pentatonic scales), the whole-tone interval may be the smallest for which the two pitches are perceived to be categorically different, especially considering the importance of song for the prehistorical development of music and the imprecision involved in vocally generating intervals (Pfordresher and Brown 2016). From the available evidence, it is unclear which of the two explanations—pitch commonality or categorical pitch perception—is preferable.

A third explanation involves the amount of information we can store in short-term memory, which according to George Miller (1956) is seven items.

That seems to explain seven-tone diatonic scales, but only on the assumption that tones an octave apart are equivalent. In fact, octave equivalence may have been less important for medieval chant than for later polyphony and MmT. In any case, if one tone in a melody is transposed up or down by an octave, and the others stay where they are, the melody's contour is totally changed, which radically changes the melody's character (Dowling 1978).

Counting Tones

Why did some modes happen more often than others? In a reductionist approach, we might start by counting individual tones, regardless of mode, and ask how often each scale step happens.

The biggest chant corpus available for computer-based statistical analysis is the *Liber Usualis*, a book of Catholic chants that was transcribed from oral tradition and edited by monks from Solesmes Abbey in late nineteenth-century France. The corpus can be analyzed online, courtesy of Distributed Digital Music Archives & Libraries Lab (DDMAL; Schulich School of Music, McGill University, Montreal, Canada; ddmal.music.mcgill.ca). Over the centuries, the repertory could have been influenced by Renaissance and Baroque polyphony and hence by MmT (Dodds 2003). In that case, it does not necessarily reflect chant performance in the Middle Ages.

The distribution of scale-step prevalence in the *Liber Usualis* is shown in figure 11.1(a). The most commonly notated tones are G and A, and the least common (apart from B♭) is B. That is no surprise for chant theorists, who know (referring to the twelve-mode system of Glarean) that modes beginning on G and A are relatively prevalent, whereas the mode on B does not exist in practice due to the tritone above the final. The figure shows that tones C, G, and A happen more than twice as often as B; for a psychologist wondering about statistical significance, that is a big effect.

To check these findings, I counted scale steps by hand in another large collection of Gregorian chants. The corpus of John Bryden and David Hughes (1969) was obtained from "five selected manuscripts . . . the Antiphonals of Lucca and Worcester, and the Graduals of Benevento, St. Yrieix, and Sarum" (vol. 1, vii–viii). The results are shown in figure 11.1(b).

The most striking difference between the two analyses is the reversed relationship between C and D. In Bryden and Hughes (1969), D happens more often than C, whereas in the *Liber Usualis*, C happens more often than D. The latter is surprising, given the medieval four-mode system in which only

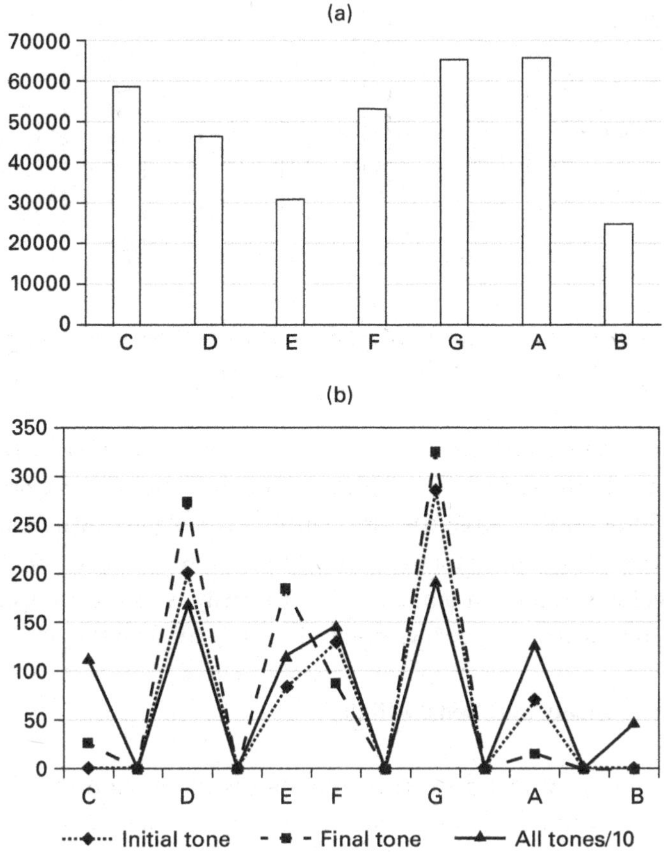

Figure 11.1
Frequency of occurrence of diatonic scale steps in Gregorian chant, calculated in two contrasting ways: (a) by analyzing the *Liber Usualis* using the search machine of DDMAL and (b) by analyzing randomly selected chants from J. R. Bryden and D. G. Hughes (1969) by hand. In (b), results for "all tones" (full line) have been divided by ten for ease of comparison.

D, E, F, and G were recognized as finals. The prevalence of the tone A in *Liber Usualis* is also surprising.

The reason for the difference evidently involves tone repetition: some tones are repeated more often than others, as shown in figure 11.2. Tone repetition was included in *Liber Usualis* as made available by DDMAL, but not in the Bryden and Hughes (1969) collection. Each final in the medieval modal system (D, E, F, and G)—of which G was the most prevalent, followed by D—had its own tenor or reciting tone/note, which was often repeated. For

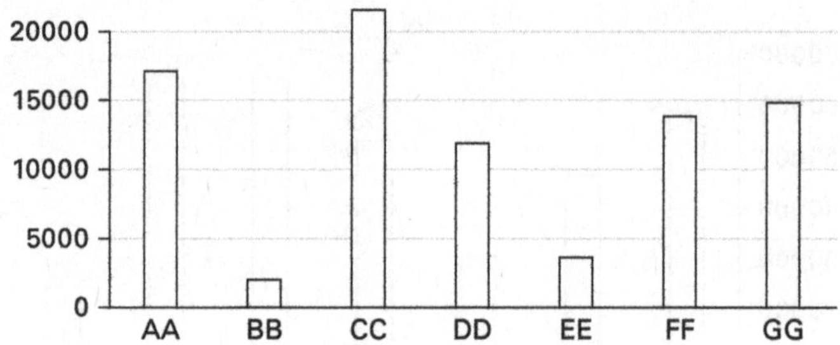

Figure 11.2
Tone repetitions in *Liber Usualis* according to DDMAL: how often a given tone is immediately followed by a repetition of the same tone.

the higher-pitched authentic modes, the tenor was A, C, C, and D, respectively. As a rule, the tenor was a 5th above the final, but it was a 6th higher for the mode on E to avoid the dissonance of B. For the plagal modes, with their lower ambitus, the tenor was a usually a 3rd lower: F, A, A, and C, respectively (avoiding B for the mode on G). The analysis of the *Liber Usualis* in figure 11.1(a), with peaks at A, G, and C, is consistent with these points.

Pitch Commonality and Tonal Affinity

Both parts of figure 11.1 are consistent with the assumption that individual scale steps vary in consonance (Parncutt 2019a). The tone G sounds consonant because all its audible partials (assuming they are harmonics one to ten) correspond to diatonic scale steps: G, D, B, F, and A in different octave registers. Psychoacoustic experiments suggest that the mistuning of the 7th harmonic relative to 12-EDO (about 1/3 semitone) is seldom noticeable in typical musical contexts. The tone B sounds dissonant because about four of its audible partials do not correspond to diatonic scale steps (F♯ in two octave registers, and D♯ and C♯ in one each). For similar reasons, C sounds more consonant than B, and F sounds more consonant than E, the difference being smaller for F and E. In medieval terms, *fa* is more consonant than *mi*—and is therefore sung more often.

A simple model based on the number of audible partials corresponding to scale steps—a form of pitch commonality—predicts that G should be the most prevalent tone. But such a model cannot explain why the tones A and G are about equally common in the DDMAL analysis. As mentioned, a possible reason involves reciting tones. Or perhaps the tone A is more common because the subtonic–final relation G-A (at the end of a chant on A) avoids semitone intervals by the widest possible margin: the scale tones adjacent to both G and A lie a whole tone away on both sides. Similarly, the mode on F does not necessarily happen more often than the mode on E, even if scale degree F happens more often than the scale degree E. The final on E may sometimes be preferable because its subtonic lies a more consonant major 2nd below it, whereas the tone below final F is a semitone below.

A theory of scale-step consonance in chant, in which tones happen more often if their audible partials belong to the prevailing diatonic scale, recalls Helmholtz's (1877) theory of tonal affinity (*Tonverwandschaft*). In the interpretation of Benjamin Steege (2012):

> Helmholtz's theory of tonal affinity required just two seemingly intuitive assumptions, which were nevertheless of a distinctly speculative psychological character: first, what is normally thought of as a singular tone sensation is in fact composed of a manifold bundle of less noticeable discrete sensations, which "the ear" tends to associate by unconscious habit and to identify as constituting a single perception, and second, that despite this unconscious synthesis, listeners tend, even under ordinary circumstances, to retain some intuitive "feeling" for the presence or absence of the constituent sensations, the partial tones. On the basis of the potential greater or lesser degrees of coincidence between the partials of any two tones, Helmholtz supported that a tone would sound inherently more related (*verwandt*) to some tones than to others. (130)

In this way, Helmholtz formulated two ideas that were central to Terhardt's approach to pitch and music. First, virtual pitch (Helmholtz's "single perception") depends on harmonic patterns within more complex spectral-pitch patterns (his "manifold bundle of less noticeable discrete sensations"); the relationship involves a learned process of pattern recognition ("unconscious habit," "unconscious synthesis"). Second, two sounds sound related if they have pitches in common. Helmholtz's idea of *Tonverwandschaft* was confined to common spectral pitches ("greater or lesser degrees of coincidence between the partials"). The idea was developed by Terhardt et al. (1986), who considered not only common spectral pitches, but also (and primarily) common virtual pitches.

Early chant singers were presumably unaware of variations in the C/D of individual tones, although some realized they were avoiding B due to the tritone to F. They could not have imagined a theory of pitch commonality

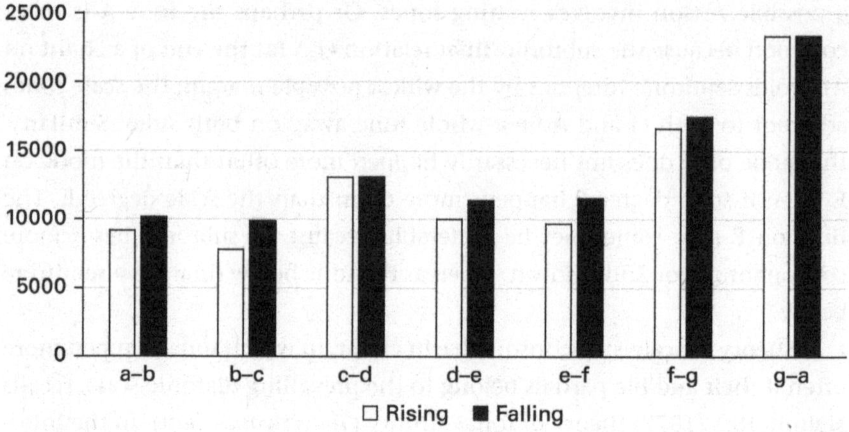

Figure 11.3
Stepwise movement in *Liber Usualis* according to DDMAL. The first white bar on the left shows how often the tone A (in any register) is followed by the tone B in a rising step in the entire corpus. The first black bar on the left shows how often the tone B (in any register) is followed by the tone A in a falling step.

between individual tones and their diatonic context if they were unaware of the audible harmonics within each sung tone. Instead, they preferred some chants over others, or considered some chants more suitable for certain purposes (and hence certain texts) than others. Those preferences may have involved intuitions about the C/D of individual tones. Figure 11.1 suggests that as the chant tradition evolved over the centuries, more consonant tones tended to be sung more often, as chant melodies that emphasized more consonant tones were sung more often.

Counting Intervals

Given that most melodic motion in chant (and most melody) is stepwise, it is interesting to consider melodic progressions systematically, from any tone to its upper or lower diatonic neighbor. Figure 11.3 counts all stepwise movements in *Liber Usualis*. The figure shows that melodic seconds between some scale steps happen about twice as often as between others—another big difference that invites explanation.

Consider first the difference between falling and rising versions of the same interval. In general, falling steps happen slightly more often. Melodies of all kinds tend to start with a rising leap followed by a series of falling steps—a kind of implication–realization relationship (Narmour 1990). The effect is small in chant due to the relative absence of larger leaps.

The peaks at G-A and A-G are striking. Tones G and A may be common either because their partials match the prevailing scale or because they are surrounded by relatively consonant whole-tone intervals (F-G and A-B). The only stepwise motion in which neither tone has a semitone neighbor is G-A. Scale steps B, C, E, and F may be avoided because they are next to dissonant semitones (B-C and E-F). Interval C-D is more common than D-E; of the three tones (C, D, and E), E is the most dissonant, having two audible non-diatonic harmonics (G♯ and F♯), whereas C and D have only one each (B♭ and F♯, respectively).

F-G is more prevalent than A-B, presumably because B is the most dissonant diatonic tone, with the largest number of audible partials that do not correspond to diatonic scale steps. Assuming octave equivalence, the audible partials of B include three non-diatonic pitch classes (F♯, D♯, and C♯), whereas E has only two (G♯ and F♯). The dissonance of B can explain why A-B and B-C are the least common melodic seconds. Of the two, B-C happens less often (at least for rising intervals) because the B-C semitone is more dissonant than the A-B whole tone.

These explanations are provisional and have not been explored systematically, nor have the predictions been quantified in a model. Their explanatory power is evidence for the validity of a theory of scale-tone prevalence in chant that is based on pitch commonality—either between each tone (including its audible partials) and the scale or between successive tones.

The Leading Tone

In 1855, the Belgian Jesuit composer Louis Lambillotte argued that Gregorian tonality, and in particular the *mi-fa* semitone, was full of what Fétis had called "appellative" elements (Christensen 2019). But he was listening to chant with nineteenth-century ears and incidentally had no qualms about harmonizing chant with modern chord progressions.

Even if there are no leading tones in chant, a statistical analysis of chant can give us a clue to the origin of leading tones in early polyphony. Returning to figure 11.1, if we focus on the two semitone intervals (E-F and B-C), it is clear that the higher tone of the semitone happens more often than the lower tone. The reason for the difference probably involves the non-diatonic harmonic partials of B and E. The difference is bigger for B-C because the audible harmonics of B clash more audibly with the prevailing scale. Modern empirical research in music psychology has shown that melodic tones that happen more often are more likely to be perceived as stable points of reference (Lantz et al. 2020; Oram and Cuddy 1995). We therefore expect F and C to be perceived as stable references in chant, and B and E to be relatively unstable. Consistent with this prediction, Guido used colored lines for *do* (yellow) and *fa* (red) in his graphic notation (van Waesberghe 1951). The colors not only attracted attention to the semitone intervals, but also marked the more stable of the two tones in each case.

These considerations can explain why leading tones in MmT tend to rise rather than fall. In the notated polyphony of the fourteenth century (composers such as Vitry, Machaut, Landini, and Ciconia; see Jürgensen 2011), some tones did not conform to the prevailing diatonic scale; their pitches were determined by the rules of *musica ficta* (not by notated accidentals, as in later music). These new tones created new semitone intervals against their diatonic neighbors, which contributed to tonal expectations: "Renaissance music is highly directional, and shows clear tonal goals that are often articulated by means of cadences" (Stern 1981, 5).

The only semitone in the hexachord *ut-re-mi-fa-sol-la* is *mi-fa*. Medieval listeners may have heard non-diatonic tones as either *mi* or *fa*, depending on whether the more stable neighbor was a semitone higher or lower. Therefore, non-diatonic tones in medieval and Renaissance polyphony tended to rise by a semitone to a diatonic tone, making the target tone more stable and creating a temporary reference pitch (cf. Jürgensen 2011). In that way, they tonicized the more stable tone (in Schenkerian terminology).

Jennifer Bain (2003) explained that "because descending semitones are completely absent at final cadences, their tonal effect is one of relative weakness" (333). In addition, Bain (2005) pointed out that "implied semitone motion can contribute fundamentally to the projection of tonal structure in a song" (60) and "this acute concern in medieval notational and pedagogical systems with the placements of semitones suggests very strongly that they are important to tonal organization" (73). In particular, "semitone placement can contribute to the strength or weakness of a cadence" (75). Such remarks are consistent with a chant-based theory of the origin of rising leading tones, and can explain why we take for granted that "leading tones" will rise unless otherwise stated.

Ernst Kurth (1917) argued that although the leading tone is usually the 3rd of V in MmT, the origin of the rising leading-tone effect must be sought elsewhere: "The semitone is a symbol of increased intensity of force. In this final 'chromatic' step to the octave of the tonic [at the conclusion of an ascending major scale], the 'working-off' of the leading note is completed, i.e. of its vital force, manifested to an enhanced degree . . . (The explanation of the effect of the leading note, to be found in very many harmony textbooks, according to its place in the dominant chord, rests on a confusion between cause and effect)" (G. Chew 1983, 37, his translation; the text in parentheses is from Kurth).

Imagine a piece of early music that does not conform exactly to a diatonic scale, so it cannot be transposed onto the white keys of the piano. The underlying scale might for example be what later theorists called harmonic minor, or natural minor with a sharpened leading tone. There are three semitone intervals in the harmonic minor scale: $\hat{2}-\hat{3}$, $\hat{5}-\hat{6}$, and $\hat{7}-\hat{1}$. If early polyphony was based on chant, and if familiarity with chant influenced how early polyphony was perceived, any non-diatonic tone that did not belong to the prevailing diatonic might have been perceived as either *mi* or *fa* if it was heard to be a semitone above or below an existing scale step. The tone would be perceived as *mi* if its main diatonic neighbor was a semitone higher in pitch and as *fa* if it was lower. In the melodic minor, we might expect scale steps $\hat{3}$, $\hat{6}$, and $\hat{1}$ to be tonicized in this way. For $\hat{1}$ and $\hat{3}$, that is no problem—they are part of the tonic triad. The tonicization of $\hat{6}$ creates ambiguity that is resolved by the emotionally laden falling-semitone resolution to $\hat{5}$ (cf. Cooke 1959, as criticized by Clarke 1989b).

A psychoacoustic theory of melodic tonicization by leading tones might be divided into two parts. First, there is a general tendency to resolve dissonances through small steps, independent of direction. That is consistent with the principle of grouping by pitch proximity in auditory scene analysis, as well as the trill threshold (G. A. Miller and Heise 1950; Shonle and Horan 1976). Steve Larson and Leigh Vanhandel (2005) referred to musical magnetism, while for Jamshed Bharucha (1996), the effect was one of melodic anchoring; a music theorist might refer to the "rule of the semitone." Second, rising leading tones tonicize more strongly than falling semitones because in chant and medieval polyphony, *fa* is more stable than *mi*.

Implications

The ideas in this chapter have interesting implications for musical intonation. Pythagorean theorists assumed that chant was ideally tuned, with "pure" octaves (1:2), 5ths (2:3), and 4ths (3:4). Whether that was true in practice, we do not know. At the time, equipment for measuring intonation in live performance did not exist; reliable measurements have only become possible in recent decades (and are still problematic; Devaney et al. 2011).

We can now revisit this question theoretically. Listeners may perceive the audible harmonics of chant relative to the prevailing diatonic scale. Individual tones may be perceived as consonant if the audible harmonic series above them belongs to that scale. The audible harmonics include the 5th, which lies a Just major 3rd above the 4th. If the theory is correct, the major 3rds between C and E, F and A, or G and B in Gregorian chant may ideally be Just. That would be consistent with evidence that the relative prevalence of scale steps in chant depends on how well the audible harmonics match the diatonic scale.

But one can argue, equally convincingly, that chant was tuned to Pythagorean ratios. The traditional argument that tuning depended mainly on 4th, 5th, and octave intervals remains valid. That being the case, the ideal tuning of chant may have been a compromise between Just or Pythagorean. Therefore, any such compromise may be acceptable for historically informed modern chant performances, with 12-EDO being one such compromise. (A major 3rd is 386 cents in Just tuning, 408 cents in Pythagorean tuning, and 400 cents in 12-EDO.) But discussions of this kind may be irrelevant if the intonation of real singers in real performance is not accurate enough to distinguish between Just and Pythagorean.

This chapter has presented a new approach to the structure of medieval chant. The theory is consistent with the idea that musical pitch structure generally depends on the audible partials of each sound—even if listeners are not aware that they are hearing those partials. A partial may be considered audible if a listener can hear the difference between a sound including the partial and the same sound in which only that partial is missing. More generally, a psychological theory of musical pitch structure should ask which partials in musical sounds are audible (as spectral pitches) and which are not. It should then ask how the conscious perception of pitch in music might depend on those audible partials. Terhardt's theory of virtual pitch offers a way forward. It allows us to ask: Which spectral and virtual pitches are audible in a musical passage? How do the virtual pitches depend on spectral pitches? Which of these pitches are consciously perceived? How can the answers to these questions shed light on music-theoretical questions?

12 Accidentals and Enharmonics

Music theories tend to take conventions of music notation for granted. To understand the process of score-based music analysis in both psychological and historical contexts, we need to review standard music notation and the assumptions upon which it is based. Why is music notation like it is and not different? Where did its conventions come from, and what others might have been possible? Accidentals are sharps and flats relative to a prevailing key signature. In what sense is C♯ the same as, or different from, D♭? What is the scientific status of enharmonic equivalence, and what is its role in a psychoacoustical theory of musical structure? Can we imagine a theory of MmT without accidentals? In a quasi-general, psychologically founded theory of MmT, prior assumptions about musical pitches should be minimized. A convincing theory should initially treat all twelve pitch classes as equal and then show how they become unequal in tonal contexts. A chromatically based approach frees the researcher to focus on music as performed and perceived rather than music as written and read.

History of Notation

Standard Western music notation uses a staff whose lines and spaces represent successive tones of the diatonic scale—an idea attributed to the eleventh-century monk Guido of Arezzo. Whereas his staff had four lines, today we usually use five. Modern notation also features accidentals (sharps ♯ and flats ♭) that change the pitch by one semitone (up or down, respectively) relative to the prevailing scale, which is defined by the key signature. Accidentals were not part of Guido's original concept but were added later.

The notational conventions that we use for MmT today have been stable since about the eighteenth century. They are the result of countless past experiments and decisions about both notation and the music itself. "In the course of its history, systems of musical notation were invented and came to mesh with the age-old systems for making, remembering, and disseminating music through performance alone, and at the same time to contribute to the transformation of those systems" (Treitler 1992, 132).

New notational ideas tend to be introduced gradually and in different variants. When an idea becomes internationally accepted, the reason is primarily practical: successful notational conventions allow composers to communicate clearly with musicians, and musicians with each other. That helps musicians realize their musical ideas across cultural and linguistic boundaries. Misunderstandings and ambiguities are always possible, but they become less common as a notational system becomes more standardized.

Music notation cannot do everything. Today's standard conventional notation is a compromise that evolved within multiple constraints, one of which is appearance (visual aesthetics are also important). But a piece of music is successful because of how it sounds—not how it looks on paper. While standard notation may be an ideal way of passing a composer's intentions to a performer, at least within the boundaries of MmT, it does not necessarily reflect how music is perceived or understood by listeners. For that reason, it does not necessarily represent an appropriate foundation for a psychological theory of musical structure. Musical meaning depends primarily on sound, for which notation is always an imperfect representation. That is especially true when one considers unnotated details of timing, dynamics, and timbre that are passed on in oral-aural traditions (Cook 2013).

Enharmonic Spelling

When composers choose between two possible spellings for a give note, such as D♯ or E♭, they may unintentionally create implications that music theorists enthusiastically discuss over coffee at conferences, centuries later. The composer in question may merely have chosen what seemed to be the most readable solution from two available options. That being the case, we may ask how important enharmonics really are for music theory and music psychology.

To answer that question, it helps first to formulate the main principles of enharmonic spelling. Surprisingly, a formulation of that kind is hard to find in the literature. The following is a brief attempt (cf. Parncutt 1999). All three principles tend to minimize the number of accidentals, making the manuscript more concise:

- *Scalar principle*: If the immediate context of the note in question conforms to a standard diatonic scale (a transposition of the white keys on the piano), notate it as part of that scale, in the usual way, using the usual key signatures.
- *Harmonic principle*: If a note is perceived to lie a perfect 5th or major 3rd above or below a stable pitch, notate the interval accordingly. Do not notate the 5th as a doubly augmented 4th or diminished 6th, and do not notate the 3rd as a doubly augmented 2nd or diminished 4th.
- *Melodic principle*: If a note is perceived to function as a leading tone, notate the leading semitone as a minor 2nd and not as an augmented unison.

Here is a simple example. Mozart sometimes wrote a D♯ just before the 3rd (E) of a C-major harmony (a rising chromatic appoggiatura). Why didn't he write E♭? In this case, it is not possible to apply the scalar principle because the tones in question are not confined to a single standard diatonic scale. The melodic principle can explain why the appoggiatura is notated D♯, and the harmonic principle yields the same result if B is the reference pitch, D♯ being a major 3rd above B. In that case, harmonic and melodic criteria agree. If, however, the stable reference pitch is understood to be G, the appoggiatura is notated E♭—a major 3rd below G—and the following E bears a natural symbol (♮). The tone is also E♭ if it falls to D, according to the melodic principle.

Seen in this way, enharmonic ambiguities arise from competition between melodic and harmonic principles or between different reference pitches. The choice between two enharmonic options may depend on the relative stability of possible pitch references, creating an interesting link between rules of enharmonic spelling and Krumhansl's key profiles.

If we apply the above principles strictly, the result can be a relatively pedantic style of enharmonic spelling. Pedantry has its advantages: users of music notation programs such as Sibelius, Finale, and Musescore know that carefully following such rules facilitates automatic transposition into different keys.

But pedantry can also be problematic. When in 1850 Franz Liszt notated measure 81 of *Liebestraum* No. 3 (figure 12.1), he chose to write an A-major triad, although the above rules clearly imply B♭♭, to clarify the Neapolitan (♭II) relationship to the tonic A♭ major. Perhaps he wanted to make the

Figure 12.1
The last seven measures of *Liebestraum* No. 3 by Franz Liszt. From musescore.com.

music easier to read and memorize. Pianists would think of the chord as A major regardless of how it was notated.

Today, we can enter the score of this *Liebestraum* into a music notation program, just as Liszt wrote it. We can then transpose the score up four semitones to the key of C major. The A-major chord then becomes a C♯-major chord, and musicians will agree that it should be flipped to D♭. But that does not make Liszt's original enharmonic choice incorrect if he simply wanted to make his score easier to read.

In my experience, Frédéric Chopin followed the above rules more closely. If he had been writing this piece, he might have chosen B♭♭ at this point, considering it to be theoretically more correct or because it clarified voice-leading relationships (Parks 1976). We presumably do not know what either Chopin or Liszt thought about such things, but we can guess that the main issue from their viewpoint was to ensure the pianist depresses the intended piano keys and not some other ones. Seen that way, conventional Western music notation forces composers to choose between enharmonic equivalents, whether they like it or not. In psychological jargon, they make a two-alternative forced choice.

On the whole, correct (or conventionally expected) enharmonic spelling is important for efficient musical communication—if not for the credibility of the composer or arranger. Misunderstandings of enharmonic conventions can give a bad impression, like unintended parallel 5ths in classical harmony. But as the above examples and principles show, the rules of enharmonic spelling are often inherently ambiguous. Moreover, from a psychological perspective, relationships among enharmonic spelling, intonation, and meaning are inherently unclear.

It is often unclear which specific interval relationship determines a spelling, or why the relationships highlighted by one spelling are more important than the relationships that would have been highlighted by another. If the tone between D and E is called D♯, and it is heard relative to the interval C-G in a C-major context, it is not clear whether the spelling was determined by the melodic interval D♯-E or the harmonic interval B-D♯. Nor is it necessarily clear why the interval G-E♭ (implied if the tone were called E♭) is considered less important. In fact, all tones in a polyphonic passage stand in multiple interval relations to all other tones, and all those relationships can contribute to the perception and meaning of the passage. All tones have multiple tendencies to move in different directions based on the listener's learned expectations. Complexity of that kind cannot be encapsulated by a single accidental symbol.

Do we need rules of enharmonic spelling? The answer is surely yes. However formulated, they are an important part of a long tradition of written musical communication and must be understood and respected. But they are not necessarily relevant for understanding the perception of musical structure—and without perception, one may argue, musical structure (considered as part of World 2, the world of subjectivity) does not exist.

Shirlaw (1957) complained that the piano "provides but a single key for two entirely different sounds, such as g♯ and a♭" (266). That was a misleading statement. Any key on the piano can sound entirely different in different contexts: compare the sound of the tone D within a B-minor triad with that of the tone D within a B♭-major triad. Imagine singing those two different Ds as an alto in a four-part texture and having two entirely different "D-experiences." Clearly, the issue is not one of enharmonic spelling per se. Besides, as Shirlaw surely knew, G♯ is higher than A♭ in Pythagorean tuning and lower in Just. If the single key were split into two, how should they be tuned?

On that basis, one might argue on the contrary that a music theorist or performer interpreting a musical score should take care *not* to be influenced by its enharmonic spellings (beyond perhaps checking whether the usual rules of spelling have been followed), just as a participant in a psychological Stroop experiment should understand the word "green" to mean "green," even if it is written in red ink. That experimental participant cannot completely separate the word from its color (hence the Stroop effect), just as theorists can hardly prevent their judgments from being influenced by enharmonic spellings. But a "science of harmony" (as Shirlaw called it) should not depend on notational conventions or enharmonic spelling. Instead, it should depend on psychoacoustics and the cultural history of musical listening.

Empirical measurements of performance intonation (see chapter 7) suggest that in ensemble performance (voices, strings, winds), the tone G♯ is neither consistently sharper nor consistently flatter than A♭ (assuming that the difference between G♯ and A♭ is clearly defined, which it is not). One could test this claim directly by having a passage performed that includes the chord progression C-E7-F-Fm-C (or similar, transposed to different keys), in which the G♯ in the E chord is compared with the A♭ in the Fm chord. The order of chords could also be reversed: C-Fm-E7-Am. Existing research on intonation suggests there would be no difference, but the question is interesting all the same. According to Cazden (1954):

> In Just or Pure Intonation, a sharp tone is invariably lower than its enharmonically equivalent flat, but in Pythagorean tuning this is reversed, and the sharp tone is invariably higher. Now musicians consistently play the sharps distinctly higher than the corresponding flats, and they approach very closely the Pythagorean values. In practice they are about as far away as they can get from the much-vaunted Pure intervals, even though many of them still do not know this. And everyone listening to music prefers the Pythagorean tunings, and does not like Just or Pure values when these are offered, save in occasional isolated chords. These things have been convincingly demonstrated in careful experiment for the past two centuries, though the facts have not stopped the daily eulogies of the glorious Just Purities by those who copy their prayers from the old books. (303)

Music theorists often tacitly assume that enharmonic choices say something salient about musical structure and meaning. In an analysis of the *Andante* of the overture to Mozart's *Don Giovanni*, Roger Kamien (1986) noticed "a subtle enharmonic relationship between the augmented sixth chord in m. 10 and the V^4_2 of ♭II in m. 20: the $g\sharp^1$ in the top voice of the augmented sixth is transformed into the A♭ in the bass of the applied V^4_2 chord" (173).

To understand this comment, note that a "V^4_2" chord is a third inversion dominant 7th, and "applied" means secondary dominant—a dominant relative to the following chord. The comment refers to an augmented-6th chord notated B♭-D-F-G♯ in measure 10, and a secondary (applied) dominant 7th notated A♭-B♭-D-F in measure 20. The two chords comprise the same four pitch classes. In that sense, they are the same chord, but they are voiced differently and appear in different contexts. The chords are notated differently because in the first case, the G♯ rises to A, whereas in the second case, the A♭ falls to G. In each case, Mozart followed notational convention and chose a notation that minimized the number of accidentals, eliminating the need for natural symbols (♮). The interesting point is not the enharmonic choices but rather the two different resolutions of the G♯/A♭, which are what make the chord sound different—not the chord itself.

I am writing in this way to simplify and clarify the discourse. There are some basic questions lurking in the background: Why does the music comprise these tones and not others? How does the listener understand these patterns of pitch? The history of music theory and the more recent history of music cognition suggest that these questions are very complex. To answer them, we need to clear away the debris and try to get to the heart of the matter.

In that spirit, allow me to pose a different question. What if the music notation of Mozart's time had been based on 12-EDO? In that case, the G♯/A♭ in question would not have been notated as an accidental. It would have simply been the tone between G and A, and it would have had its own name. Parncutt and Stuckey (1992) called the black notes on the piano VW XYZ (grouping them into twos and threes), so G♯/A♭ was Y. But the idea was not new (e.g., Vincent [1894] had proposed something similar) and involved yet another arbitrary choice.

The discussion is clarified if we label the chord relative to the chromatic scale. It is a [047T] chord built on ♭$\hat{6}$ ([8]) relative to the tonic D. The [047T] chord, incidentally, has the clearest root of all chords used in MmT—even clearer than the root of the major triad (see chapter 15). It also has the highest harmonicity, as well as relatively high roughness due to the tritone interval between [4] and [T].

In one case, a dissonant tone in a dissonant chord resolved in one direction, and in another case, in the other direction. That is not an unusual phenomenon. Any chord can be put into different contexts and its tones can progress or resolve in different directions. A C-major triad can appear in any of the twenty-four major and minor keys, and its three tones can resolve in different directions. A famous case is the Tristan chord: a half-diminished 7th [036T] that is placed in an unusual context and therefore resolves in an unusual way. In such cases, enharmonic spellings or relationships are not necessarily relevant for questions about how the music works as sound, or what the music sounds like. Instead, they can be seen as unavoidable idiosyncrasies of conventional notation.

In a music-psychological example, John Sloboda (1991) found that an "enharmonic change" in a piece of music evoked emotional responses, including tears and shivers (see his table IV). But it was not the enharmonic change itself that evoked the response. Enharmonics are a feature of notation—not of music. Besides, an enharmonic change can often be removed from a score, or shifted to another location, by flipping some accidentals and taking advantage of double accidentals. What really evoked the emotional response was the surprising nature of the modulation—the tonal distance between the old and the new key.

Doing Without Accidentals

A psychologically based theory of musical structure, for music composed within the twelve-tone chromatic scale, should be independent of enharmonic spelling. That raises a more general question: What would the theory, composition, and performance of music be like in a chromatic world without accidentals and enharmonic equivalents?

For the past few centuries, musicians have been trying to reform music notation by basing it on the chromatic rather than the diatonic scale (Read 1987). So far, no proposal has achieved musicological acceptance and respectability. That is surprising, given evidence that MmT is ultimately based on the chromatic scale, of which the diatonic scale is a subset. Alternative music notations failed for several reasons:

- The ability to read conventional notation, with all its complexity, is a hard-won skill that would be threatened by a standard alternative. A competing music system would have to be learned from scratch.
- Musicians and nonmusicians alike love the appearance of conventional notation—almost as much as they love the sound of the music. Its beauty is part of musical culture, and it motivates conservative resistance to change.
- It would be difficult to get musicians to agree about specific elements of an alternative notation. "Notation inventors" lack the necessary political power.

But bilingualism is common in both language and music (bimusicality; Hood 1960). Musicians can and often do learn two contrasting notation systems and translate from one to the other. Many guitar players read both conventional notation and tablature (D. E. Thompson 2011), and many Japanese musicians read both traditional notations (for gagaku, jōruri, or koto) and Western notation (Tokita 2014).

Modern computer technology suggests a way forward. It is now possible to develop software to convert music in any computer format (e.g., MIDI or MusicXML) into different possible notations, so that an amateur learning a new system would have an unlimited supply of music in that system. That might happen in the near future, radically changing the way we think about music theory. This book quietly anticipates that change of perspective. Although the music examples in this book are in conventional notation, the accompanying theory and analysis assumes that the twelve tones in the chromatic scale are equal, and enharmonic choices are irrelevant. The music is analyzed as if it was written in a notation system based on the chromatic scale (see the author's internet page on Keyboard Trigram notation).

Salient Flats

In cases where enharmonic choices are relatively clear, such that different composers would have notated the same passage the same way, the theory of musical pitch perception presented here permits a new insight. Accidental flats (flats relative to a key signature) tend to be more perceptually salient than enharmonically equivalent sharps (Parncutt 2014). Insofar as a flat is perceived to lie a major-3rd interval below a diatonic scale step (e.g., A♭ is a major 3rd below C) and a sharp is perceived to lie a major 3rd above a diatonic scale step (e.g., G♯ is a major 3rd above E), accidental flats are more likely to be perceived as chord roots than accidental sharps. The theory of Terhardt (1982) and Parncutt (1988) suggests that the psychological salience of accidental flats is reinforced by their upper major 3rds (octave equivalent to the 5th harmonic), whereas accidental sharps are not reinforced in this way. Instead, they boost the salience of the pitch a major 3rd below.

Consider a modulation (key change) in which the starting and ending keys are notated with different key signatures. Modulations can be big or small, close or remote. The size of a modulation in music theory is sometimes understood as the difference in the number of sharps or flats in the key signature. For example, a modulation from F major to E♭ major means adding two flats. A modulation from G major to F major means removing one sharp and adding one flat—equivalent to adding two flats. Both modulations are understood to have the same size or traverse the same distance.

W. F. (Bill) Thompson and Lola Cuddy (1989) showed that flat-side modulations (those in which the number of flats in the signature increases or the number of sharps decreases) seem psychologically bigger or more remote than sharp-side modulations. For example, a modulation from G major to F major (two steps toward the flat side) sounds more noticeable or significant than a modulation in the opposite direction, from F major to G major (two steps toward the sharp side).

Classical composers tended to start a piece by consolidating the home key and then modulating to the dominant (from a major key) or relative major (from a minor key). Modulations of that kind are held in music theory to have "passing" character, suggesting that the two keys in question did not sound very distant from each other. Modulating to the subdominant sounds more surprising. If the circle of 5ths (or line of 5ths; D. Temperley 2011) has a cognitive representation, both standard music theory and the empirical results of Thompson and Cuddy suggest that the representation is asymmetrical: key changes sound more striking when they move toward flatter keys.

Psychoacoustic theory offers an explanation for the asymmetry that is based on direct perception of the musical stimulus. The new tones that are added to the scale in a flat-side modulation are more perceptually salient than the tones added in a sharp-side modulation. If we modulate from G major to F major, the new tones are F♮ and B♭. Both can be chord roots in the new key, the chords being F major and B♭ major (I and IV in the key of F). If we modulate in the reverse direction, from F major to G major, the new tones are B♮ and F♯—tones that rarely function as roots in a G-major context. More often, these tones function as 3rds of major triads (G major and D major, respectively, or I and V in G major).

IV Perception of Sonority

13 Pitch

The word sonority *encapsulates aesthetic and structural aspects of tone simultaneities in MmT. Related to* texture, *sonority is a subjective property of musical tones, chords, passages, and pieces. The perception of tone simultaneities depends on the perception of individual tones. The pitch of a pure tone or partial is* spectral, *whereas that of a complex tone (voiced speech sound, musical instrument, environmental sound) is* virtual. *A virtual pitch does not correspond to a specific partial but is derived from multiple spectral pitch contours by spectral/temporal pattern recognition. A virtual pitch is heard when audible partials describe an incomplete harmonic series as if they came from the same periodic source. The main virtual pitch of a harmonic complex tone in the central musical range is determined by higher harmonics; it is not shifted by shifting the fundamental. Since virtual pitch may correspond to both temporal and spectral aspects of a signal, it is difficult to predict from physical measurements of neural activity. A pattern-recognition model of pitch perception can explain the nature and origin of musical chord roots and tonics.*

Measuring Pitch

Frequency and pitch are very different things. The frequency of a periodic sound is the number of vibrations per second. The pitch of a sound is its perceived height on a one-dimensional scale from low to high. In everyday life, the number of times per second something happens has nothing at all to do with the height of something above the ground. A periodic sound does not in any physical sense have height; nor can one perceive individual vibrations within a regular steady tone in the range of human hearing (20 Hz–20 kHz).

It is clear how frequency is measured in physics. Count how often a periodic pattern repeats per unit of time. If the frequency is constantly changing, limit the counting to a temporal window and count within several overlapping windows.

Pitch is not so easy to measure. Like any other quantitative psychological variable, it needs to be operationalized in a standard experimental procedure. Different experimental measures of pitch can only be compared if the method is standardized. Misunderstandings in the pitch-perception literature may arise from differences in experimental methods; sometimes, those differences involve unstated philosophical assumptions.

Pure tones are appropriate standard references for pitch experiments because they are physically simple. There are few parameters to adjust and few psychoacoustic dependencies. The pitch of a pure tone is relatively clear and unambiguous (pitch salience is relatively high). Whereas the timbre of a pure tone changes with frequency (from a low rumble to a high squeak), the pitch–timbre relationship is always the same for a pure tone of given amplitude and a given listener.

In a typical standard procedure, a listener alternately hears a short test sound and a short pure reference tone, with short time gaps between them, and adjusts the frequency of the pure tone until the pitch of the test sound matches the pitch of the pure tone (Terhardt 1988, 312). Here, "short" means a few hundred milliseconds (typical of speech and music), and the SPL of the pure tone is understood to remain constant. In this way, the pitch or pitches measured in a sound can be measured in equivalent hertz.

In such experiments, psychoacousticians traditionally ignore individual differences. Instead, they average results over several listeners on the assumption that individual differences are relatively small. Perhaps they should not, given differences between spectral and fundamental listeners (Seither-Preisler et al. 2007).

Using this method, one can show that the pitch of a pure tone depends not only on its frequency, but also on its intensity and on the presence of other simultaneous sounds (pitch shift). One can also show that the pitch of a harmonic complex tone is ambiguous: depending on how tones are presented in a random sequence, listeners may match the pure reference tone to different pitches within the same complex. Those pitches do not necessarily correspond to partials.

Different Kinds of Pitch

Some music psychologists regard pitch as comprising two psychological dimensions—pitch height and pitch class—but the idea is problematic (see chapter 14). Instead, we can distinguish two kinds of pitch: spectral and virtual (Terhardt 1972). In a complex tone (i.e., a tone comprising different partials of different frequencies), spectral pitch is the pitch of a single partial, and virtual pitch is the pitch of the whole tone.

When we hear a pitch, it is not immediately apparent whether it is spectral or virtual. Both spectral and virtual pitches are associated with diverse timbres (using the word "timbre" in the sense of a subjective experience—what a tone sounds like). We may for example hear a church bell with two pitches: a strike tone with a virtual pitch, and a hum tone with a spectral pitch (Terhardt and Seewann 1984). Without theoretical knowledge about strike and hum tones, it is unclear which is which. The terms "spectral" and "virtual" refer to the origin of the pitches—not what they sound like. They refer to what each pitch depends on and how it might be predicted. Spectral pitches depend directly on spectral frequency, whereas virtual pitches involve pattern recognition.

Rather than distinguishing two kinds of pitch, it is sometimes more intuitive to refer to two kinds of tone sensation. A pure tone sensation is the experience of a listener who is consciously perceiving a pure tone; a complex tone sensation is the experience of a listener who is consciously perceiving a complex tone. A spectral pitch is the pitch of a pure tone sensation, whereas a virtual pitch is the pitch of a complex tone sensation (Parncutt 1989).

In a typical harmonic complex tone, there is both a virtual and a spectral pitch at the fundamental. We are typically aware only of the virtual pitch because it is more salient. The tone's (virtual) pitch and timbre usually depend on the upper partials more than they depend on the fundamental. If the harmonic complex tone has no fundamental, the main virtual pitch typically corresponds to the missing fundamental.

An everyday example is a voice heard on the telephone. Telephones usually do not transmit acoustic frequencies below about 300 Hz. When the phone rings and we answer, not expecting anyone in particular, we quickly assess whether the speaker has a higher voice (probably a woman or child) or a lower voice (probably a man). To make that assessment, we perceive the virtual pitch at the missing fundamental. In everyday nonmusical perception, most pitches are virtual and perceived in the same way (Terhardt 1974a). The telephone example illustrates that virtual pitches are a normal, effortlessly perceived part of everyday life.

Spectral Pitch

From the viewpoint of acoustics and the auditory scene, the simplest sound is a pure tone, in which the relationship between air pressure and time is sinusoidal. Spectral pitch is the pitch of a pure tone, or the pitch of an audible partial within a complex tone when the tone is perceived analytically.

Spectral pitches are subjective experiences of the listener, whereas spectral frequencies are physical measures. Spectral pitches and spectral frequencies differ in two ways: small deviations called pitch shifts, and variations in audibility. Masking can cause a partial to become aurally irrelevant: the spectral frequency exists but the spectral pitch does not.

Terhardt et al. (1982b) assumed that a spectral pitch is perceptually salient, and hence more likely to be noticed separately ("heard out") or to contribute to virtual pitch, for three reasons:

- It has relatively high SPL relative to masked threshold. That's how overtone singers make one overtone in their voice audible: by changing their vocal tract resonances to increase the sound level of that specific overtone. If the SPL is already well above masked threshold, increasing it further will increase loudness but not pitch salience.
- It lies near the center of the frequency region within which the ear is most sensitive to spectral pitch. The spectral dominance region lies between about 300 and 2,000 Hz, with a peak near 700 Hz; it corresponds roughly to the range of the first two speech formants (cf. B. C. J. Moore et al. 1985; Ritsma 1967; Terhardt et al. 1982b). S. S. Stevens et al. (1947) estimated that the frequency region for speech transmission is confined to 250–7,000 Hz.
- It is not masked, having a relatively high SPL or being relatively distant in frequency from other partial tones or noises.

Virtual Pitch

A virtual pitch is the pitch of a complex tone when perceived holistically. In Gestalt psychology, most perception in everyday life is holistic. In an approach called holism, perceptual experiences are regarded as intrinsically holistic (Wagemans et al. 2012).

Virtual-pitch perception involves pattern recognition in the auditory scene. The physiological foundations are complex, involving a mixture of innate and learned processes, and temporal and spectral patterns (Assmann and Summerfield 1990; Vliegen et al. 1999). But the same applies to the pitch of a pure tone: it also depends on both spectral and temporal information (B. C. J. Moore 1989; Oxenham 2013). The neural foundations of virtual pitch are located in the auditory cortex (Bendor and Wang 2005; Lütkenhöner et al. 2010; Ross et al. 1996)—specifically, Heschl's gyri (Krumbholz et al. 2003; Zatorre 1988); Catherine Liégois-Chauvel and colleagues (2001) and Andrew Oxenham and colleagues (2004) investigated cortical tonotopic representations. For music-theoretical purposes, we can ignore the physiology of virtual pitch, which is complex and poorly understood, and focus instead on the listener's subjective auditory experience.

Helmholtz (1856) clarified the physical difference between pure tones (*einfache Töne*) and complex tones (*Klänge*) and between combination tones and beats, reconciling differences between Seebeck and Ohm (Steege 2012). A pure or complex tone is something that exists in sound before it hits the ear. Combination tones and beats are created by the ear, and they are two quite different phenomena. Combination tones are distortions that produce new frequencies that can be heard as new pitches. The vibrations exist as additional pure tones on the basilar membrane, even if they do not exist in the sound outside the ear. Beats are amplitude modulations; beat frequencies do not evoke pitch. A pure tone whose amplitude is varying has a pitch corresponding to the carrier frequency; the beating is perceived as variation in loudness. If the modulation is fast, the individual beats can no longer be heard, and the sound becomes rough.

The relationship between spectral and virtual pitch normally has nothing to do with combination tones (which are inaudible at usual sound levels for both speech and music; Plomp 1965) or beats (which do not evoke new pitches). The relationship is difficult to investigate neuroscientifically because both spectral and virtual pitch belong to World 2, and the relationship between Worlds 1 and 2 is the famously intractable mind-body problem.

Helmholtz explained why we are usually unaware of spectral pitches in everyday sounds: "Whether the voice of a dog contains the higher octave or the twelfth of the fundamental tone is without practical interest and not an object of our attention. Thus the overtones are assimilated into those peculiarities of a tone, not to be further specified, which we call its timbre" (Steege 2012, 75; translated from Helmholtz 1858). Put another way, our sensory organs are constantly sending information to our brain, but we only become aware of a small fraction of that information. Those percepts of which we are consciously aware tend to correspond to environmental affordances, and their pitches are virtual.

To understand speech, the brain must track the fundamental frequency, allowing the listener to focus on the speech signal and ignore the background noise. The fundamental frequency contour communicates prosodic meanings that augment the lexical meaning of the words being spoken. To follow it, the brain needs to recognize harmonic patterns in the running spectrum of the auditory scene and separate them from the background noise.

We might therefore predict that the most salient virtual pitches lie in the fundamental frequency range of speech and singing (about 65–1,000 Hz). Calculations using the algorithm of Terhardt et al. (1982b) are consistent with that prediction (Huron 2001). The range is close to that of Bach's clavichord (C2–C6), and virtual-pitch salience peaks near about 300 Hz (D4). The most salient spectral pitches lie in a higher region that is centered on about 700 Hz and roughly symmetrical in log frequency.

The (virtual) pitch of a harmonic complex tone is generally ambiguous. In holistic perception, it may be an octave higher than the fundamental, the even-numbered harmonics being perceived as a tone within a tone. To demonstrate the effect using a digital sound editor, slice a high note out of an operatic tenor solo. Let's say the note is A4, which is theoretically 440 Hz; in fact, it may be a little higher because solo intonation tends to be slightly sharp relative to the accompaniment (Kantorski 1986). Now, compare the pitch of the tenor's voice in the recording with two successively presented pure tones: one at the same fundamental frequency (say, 445 Hz) and one an octave higher (say, 990 Hz). If both seem to match the original quite well, which is *the* pitch? That is one kind of ambiguity, and there is another rather simple one: if a complex tone is heard analytically, individual partials can be heard out (the spectral pitches). Again, it is unclear which is *the* pitch.

It is possible to hear a pure tone sensation and a complex tone sensation at the same time with the same pitch. In other words, a spectral pitch can coincide with a virtual pitch, and we can be aware of the coincidence. It can happen when our attention is attracted to the fundamental of a regular harmonic complex tone—the pure tone at the fundamental frequency. Imagine a sound demonstration in which a complex tone alternates with a pure tone and the frequency of the pure tone is gradually changed (Houtsma et al. 1987; Terhardt 1998). As the pure tone approaches each partial, we start to hear that partial as a separate tone within the complex tone—we "hear out" the harmonic. The same thing happens when the pure tone approaches the fundamental: we start to hear the pure tone at the fundamental as a different tone with the same (spectral) pitch as the (virtual) pitch of the complex tone as a whole but a different timbre (that of a pure tone sensation rather than a complex tone sensation). The demonstration confirms that there are two pitches at the fundamental: a spectral pitch corresponding to the partial at the fundamental, and a virtual pitch corresponding to the tone as a whole associated with a different timbre.

Another way to demonstrate this effect is to present two harmonic complex tones in alternation that are identical except that the fundamental is missing from one of them. The comparison attracts attention to the fundamental when it appears in the complete tone. We hear the fundamental as a separate pure tone—with the same pitch as the complex tone but different timbre.

Virtual pitches can be evoked by patterns of unresolved higher harmonics—partials that are inaudible in the sense that listeners cannot attend to them. The effect can be demonstrated by presenting a well-known melody from tones comprising only high harmonics that cannot be resolved and asking a listener to name it (Moore and Rosen 1979). A monist—someone who believes that only the physical world exists—might argue that the spectral pitches in question do not exist, but the virtual pitch does. A dualist, who also believes in the existence of subjectivity, might come to a different conclusion. If a spectral pitch affects subjective experience in any way, it is aurally relevant in that sense. Aurally relevant partials form a spectral-pitch pattern that, like the spectral pitches it contains, belong to World 2—even if the listener is incapable of analyzing the pattern into its parts (i.e., even if the spectral pitches are unresolved).

The results of Roelof Ritsma (1967) and others are consistent with the following principle: A tone's most salient virtual pitch is a subharmonic of its most salient spectral pitch(es). Consistent with that idea, Houtgast (1976) found that even the pitch of a pure tone sometimes corresponds to a subharmonic, especially if the pure tone is presented within noise that could have masked a tone at the subharmonic. Dik Hermes (1988) developed a pitch model based on subharmonic summation, and the success of Terhardt's pitch model is further evidence for this principle. Similarly, the root of a chord tends to correspond to a subharmonic of one or more chord tones.

The predicted spectral and virtual pitches of an isolated pure tone on C4 and an isolated harmonic complex tone on C4 according to these principles

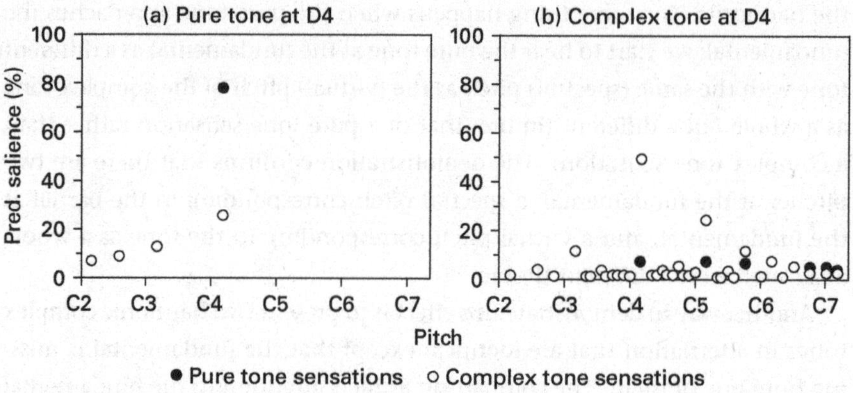

Figure 13.1
Pitches evoked by (a) a pure tone and (b) a harmonic complex tone according to Parncutt's (1989) adaptation of the pitch algorithm of Terhardt et al. (1982b). Filled circles are spectral pitches (pure tone sensations), and open circles are virtual pitches (complex tone sensations). The spectrum of the harmonic complex tone on the right is musically typical (the amplitude spectrum is specified in Parncutt 1989). Pitch-salience values have been scaled to predict the probability of noticing.

are shown in figure 13.1. Part (a) of the figure features weak subharmonic virtual pitches of a pure tone that have been observed empirically in the presence of background noise (Houtgast 1976) and are not predicted by Terhardt et al. (1982b). The pitch ambiguity of the complex tone (in particular, at octave and 5th intervals) is consistent with empirical studies on the similarity of successive pure and complex tones, including Parncutt (1989), Gerhard Stoll and Parncutt (1987), and Terhardt et al. (1986). Rising octave intervals between successive harmonic complex tones are more similar than falling, possibly due to an implication–realization relationship (similar to that described by Narmour 1990): the second-most-salient pitch in the first sound is realized as the most salient pitch in the second sound. For the same reason, *falling* octave intervals between successive *pure* tones are more similar than rising, as shown in figure 13.2.

Spectral-Pitch Shift

The perceived pitch of a pure tone (how high it sounds) can be shifted slightly (so that it becomes slightly sharp or flat) without changing its frequency, in

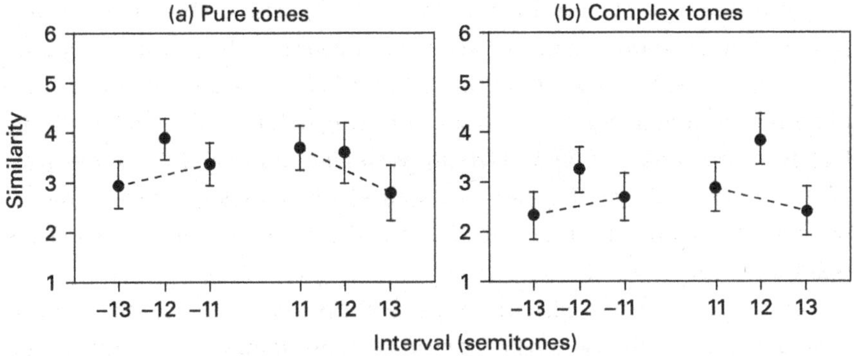

Figure 13.2
Mean similarity judgments of (a) pure and (b) complex (sawtooth) tones spanning major-7th, octave, and minor-9th intervals (eleven, twelve, and thirteen semitones). Rising intervals are assigned positive values on the horizontal axis; falling intervals are assigned negative values. Dots are mean similarity ratings over thirty listeners. Error bars are 95 percent confidence intervals. Adapted from Parncutt (1989, figure 5.5b).

two ways: by changing its amplitude (the tone seems to go slightly sharp or flat if suddenly made much louder or quieter), or by partially masking the pure tone with other, simultaneous sounds (the tone's pitch moves away slightly from the pitch of the masker). Pitch shifts of this kind have also been called pitch bias (e.g., Chuang and Wang 1978).

Many empirical studies have independently confirmed and measured spectral-pitch shifts (Cheveigné 1999; A. Cohen 1961; Hartmann and Doty 1996; Morgan and Galambos 1943; W. B. Snow 1936; S. S. Stevens 1935; Stoll 1985; Verschuure and Van Meeteren 1975; Walliser 1969). Whereas individual differences have been observed in the size of the shifts, the direction of the shift is usually the same for different listeners.

Despite repeated empirical confirmation of this phenomenon, some researchers remained skeptical. Robert Peters and colleagues (1983) had three listeners match the pitches of partials in harmonic complex tones comprising harmonic numbers 1–7 or 4–7 (the latter having a missing fundamental). Not expecting to find shifts, they found that "the pitches of components in context were again found to be very close to those of components in isolation" (abstract), implying the existence of small deviations.

Spectral-pitch shifts originate in the physiology of the cochlea—the complex and imperfect way in which it generates a running spectral analysis over a very wide range of sound levels (Evans 1981). The cochlea has no easy task: a difference of 120 dB between the quietest and loudest perceptible sound corresponds to a factor of one million in air pressure and one trillion in power. From that viewpoint, it is unsurprising that there are pitch shifts. In fact, it is surprising that pitch shifts are as small as they are. In an ecological-evolutionary approach, the reason they are so small involves the importance of frequency patterns for the recognition of environmental sound sources (see chapter 6).

We can confirm the existence of pitch shifts informally. Listen to a pop song that starts with a solo bass line. In the terminology of cognitive music psychologists such as Krumhansl (1990), the solo bass line instantiates the song's tonality in the mind of the listener. During the introduction, suddenly and drastically change the loudness. Compare the pitch at very low sound level with the pitch at very high sound level. Alternatively, listen to a solo bass line at a low sound level, although it was originally recorded at a high level. The pitch shift becomes clear when a higher instrument or voice enters. The following original recordings may serve as examples: *My Girl* and *Papa Was a Rolling Stone* by The Temptations, *Pumped Up Kicks* by Foster the People, and *Could You Be Loved* by Bob Marley.

Many people have experienced pitch shifts without realizing it. Imagine trying to sleep with a party going on in another apartment or down the road. Mostly, you hear only the bass line: boom, boom, boom. The pitch of those booms allows you to imagine the tonality. Suddenly, someone opens a door, and you start to hear the melody. It sounds like the singer is singing in a different key. That is because you are hearing the bass line more quietly than the people in the house are hearing it. For the people in the house, the music is loud and the bass sounds in tune. For you, the music is quiet, and the bass sounds out of tune.

Virtual-Pitch Shift

Virtual pitches may also be shifted, but not as much as spectral pitches (Stoll 1985)—perhaps because a virtual pitch typically depends on several aurally relevant partials (spectral pitches), only some of which are shifted by changing SPL or introducing maskers. Virtual pitches may be shifted for two reasons. First, they depend on shifted spectral pitches; if a virtual pitch is a subharmonic of a spectral pitch, and the spectral pitch is shifted, the virtual pitch should be shifted by the same amount. Second, the subharmonic intervals that are applied in Terhardt's pitch model are slightly stretched, because the spectral pitches in the harmonic spectra from which they are learned are shifted.

The second idea can explain why octaves in music are stretched relative to a 2:1 ratio. The octave intervals within the auditory system's harmonic template (between harmonics 1 and 2, 2 and 4, 3 and 6, and so on) are stretched. As a result, the interval between a salient spectral pitch and a corresponding virtual pitch is generally not exactly harmonic. The pitch of a harmonic complex tone with a missing fundamental is often slightly flat relative to the pitch of a pure tone at the same frequency (Terhardt et al. 1982b).

There are also pitch shifts in complex tones with physically stretched harmonics, such as piano tones. In that case, the shifts are both physical and perceptual. If such a tone is perceived holistically, in the usual way, the perceived pitch may change when attention is attracted to the lowest partial.

Experiments on pitch shifts in complex tones are sensitive to variations in method. Stoll (1984) and Allan Vurma and colleagues (2010) claimed that a sung speech vowel with a brighter timbre (e.g., /i/) has a slightly higher pitch than a harmonic complex tone with a duller timbre (e.g., /u/). In each experimental trial, two vowels were presented in succession, and the listener was asked which was higher in pitch. But there was a methodological problem: the tones being compared varied in both pitch and timbre. If the aim was to measure pitch, the results were confounded by timbre, and asking the participants to "ignore any differences of timbre between the stimuli of a pair" (Stoll 1984, 139) could hardly solve the problem. The solution is to hold the timbre of the comparison tone constant, or only to consider trials in which the comparison tone was constant (Chuang and Wang 1978).

No current model of pitch perception reliably and generally predicts pitch shifts of pure tones as a function of intensity and masking, and corresponding shifts in virtual pitches. Terhardt et al. (1982b) claimed to predict the direction of such pitch shifts but not their magnitude. That is reasonable, given large individual differences in the perception of pitch shifts. Temporal models of pitch perception have difficulties dealing with mistuned partials, such as octaves that are not exactly 2:1. Hartmann and Doty (1996) accounted for perceptual pitch shifts in physically shifted (mistuned) harmonics by a combination of spectral (excitation pattern interaction) and temporal (neural timing) effects, but their solution was not generalized to accept any spectrum as input.

Mistuned Harmonics

The literature on pitch shift is confused by different usages. In the present book, and following Terhardt, I use the term "pitch shift" only for small changes in the (spectral) pitches of pure tones as a function of SPL or masking and corresponding changes in the (virtual) pitches of complex tones. As we will see in this section, other researchers used the term differently.

Inspired by Schouten's (1940) theory of the residue, De Boer (1956) investigated the pitch of complex tones whose partials are equidistant on a linear frequency scale but not harmonic. He investigated the shift in virtual pitch when the frequencies of the partials of a harmonic complex tone are increased linearly and found that "midway between two harmonic situations, the pitch is indeterminate" (536). For example, if a complex tone with three partials at 800, 1,000, and 1,200 Hz is shifted upward so the spectral frequencies become 850, 1,050, and 1,250 Hz, the virtual pitch at the missing fundamental rises from 200 to about 210 Hz. In that case, the first tone is harmonic, but the second is not; the auditory system tolerates the small deviation from harmonicity. If the frequencies are further increased to 900, 1,100, and 1,300 Hz, the pitch becomes ambiguous: the three partials could be perceived either as harmonic numbers 4, 5, and 6 of $f_0 \approx 1{,}100/5 = 220$ Hz, or as harmonic numbers 5, 6, and 7 of $f_0 \approx 1100/6 = 183$ Hz. Schouten et al. (1962) showed that the virtual pitch of such approximately harmonic complex tones could not be due to difference tones. He termed the difference between the perceived pitch and the constant spacing of the partials (here, 200 Hz) the first effect of pitch shift. The effect can be heard on the Auditory Demonstrations CD of Houtsma et al. (1987, Demonstration 21 "Shift of virtual pitch").

These results suggest that a virtual pitch is generally an approximate subharmonic of a salient partial. Here, 850/4=213, 1050/5=210, and 1250/6=208. A virtual pitch corresponding to 210 Hz is approximately subharmonic to all three spectral pitches. In a spectral pattern-recognition approach, a spectral-pitch pattern does not have to be exactly harmonic to be recognized by the auditory system as harmonic. A temporal approach can also allow for deviations, but less directly.

These ideas contributed to the formulation of the pitch algorithm of Terhardt et al. (1982b). Terhardt and Manfred Seewann (1984) used the same idea to predict the pitches of church bells, whose spectra are nonharmonic but comprise approximately harmonic subsets. Consistent with pitch theory, J. H. (Berry) Eggen (1986) found that the strike note of a bell depended on the octave (2nd harmonic), the 12th (3rd harmonic), and the upper octave (4th harmonic).

There is also a second effect of pitch shift, in which the pitch of a harmonic complex tone becomes slightly lower if the tone is made slightly inharmonic by stretching the spectrum. Mike Greenhough and colleagues (1973) found that "the magnitude of the second effect depends on the loudness of the stimulus" and "the magnitude of the effect is a function of the position of stimulation of the basilar membrane" (277).

Effects of this kind may be explicable by both time- and frequency-domain models. In Terhardt's approach, which can be described as a spectral-pitch-domain model (spectral pitches being subjective, whereas frequencies are objective), virtual-pitch shifts depend on spectral-pitch shifts, which in turn depend on SPL and mutual masking. They also depend on learned pitch distances between partials in harmonic complex tones such as speech vowels.

Yet another kind of pitch shift (actually, a frequency shift) involves physically mistuning a partial within a harmonic complex tone. Empirical studies have shown that a mistuned harmonic contributes to the virtual pitch at the fundamental for mistunings of up to 3 percent or half a semitone (Darwin et al. 1994; Lee and Green 1994; B. C. J. Moore et al. 1985). That is unsurprising in an approach based on the learning and recognition of pitch patterns within speech and environmental sounds; in everyday life, the pitches of partials are only approximately defined and can be shifted by changes in amplitude and masking. Moreover, some quasi-harmonic sounds have physically stretched partials relative to a harmonic series (piano tones being an example).

These uncertainties suggest that the brain uses a kind of harmonic template to recognize harmonic complex tones. Given that the elements of the template are only approximately defined, the expression "harmonic sieve" is appropriate (Roberts and Brunstrom 1998). These ideas are merely images that help us to understand the process; what actually happens in the physical world is largely concealed within biological neural networks.

Experiments on the perception of mistuned harmonics are relevant for music theory. The 7th partial of a harmonic complex tone (7th harmonic) is about 1/3 of a semitone flat relative to a minor-7th interval, when all semitones are equal. If the 7th harmonic is raised 1/3 of a semitone to align with 12-EDO, it still contributes to the virtual pitch at the fundamental. In that regard, the difference between a Mm7 chord in 12-EDO and the so-called natural 7th chord (4:5:6:7) is smaller than music theorists of the past have assumed. The natural 7th may sound out of tune to our 12-EDO-influenced ears, but the 7th still contributes to the clarity of the root. Tuning and intonation are in any case inherently and generally approximate (see chapter 7).

There are additional implications for music theory. A musical chord may fuse (or almost fuse) if it is almost the same as a major triad or Mm7 chord, one voice being shifted up or down by up to a semitone. Applying this principle systematically yields a list of chords in which fusion is relatively strong, but not quite as strong as in major triads and Mm7 chords. Shifting the root of a major triad up a semitone produces a diminished triad; shifting it down a semitone produces a minor triad. Shifting the 3rd up or down produces a suspended-4th chord or minor triad, respectively. Shifting the 5th produces a major triad with diminished 5th (tritone) or an augmented triad. All such chords happen relatively often in tonal music (the last two being less common for other reasons: the dissonance of both the tritone and the major-2nd interval between the major 3rd and diminished 5th above the root, and the root ambiguity of the augmented triad). Shifting the tones of an Mm7 chord in a similar way produces chords such as minor 7th, major 7th, diminished 7th, 7th with suspended 4th, 7th with diminished or augmented 5th, and others. This procedure (which is similar to a procedure proposed by Rameau) produces the main sonorities of MmT, with some exceptions (e.g., the half-diminished 7th, which can be created by shifting *two* tones of a Mm7 chord through a semitone). Of course, the degree of fusion of a musical chord also depends on other factors, such as voicing (chordal inversion, spacing, doubling), musical context (immediately preceding and following sounds), and listener (some are more likely to listen holistically, others analytically).

Temporal or Spectral?

Both spectral and virtual pitches, and the relationship between them, depend on both temporal and spectral aspects of sound, or of neural responses to sound. Temporal aspects are aspects that can be seen in the waveform, such as period, waveform shape, changes in the waveform from one repetition to the next (hence, departures from periodicity), and amplitude modulation (beating). Spectral aspects can be seen in the running spectrum or spectral representation—primarily, the changing frequencies and amplitudes of the partials and noise components. Although several researchers have compared the performance of spectral versus temporal models of pitch or have assumed that one of the two is superior to the other (M. A. Cohen et al. [1995] preferring a spectral model, Patterson et al. [1995] a temporal model), results of such comparisons have been indecisive—in part due to the mathematical equivalence of periodicity and harmonicity.

From an evolutionary or ecological viewpoint, pitch perception is constrained both physiologically and environmentally: "pitch perception is critically shaped by the constraints of natural environments in addition to those of the cochlea" (Saddler et al. 2021, 1). The brain therefore takes advantage of both spectral and temporal cues (B. C. J. Moore 1989; Klapuri 2004): the "non-linear place-time model" of Peter Assmann and Quentin Summerfield (1990) produced "the most accurate estimates of the f_0's of pairs of concurrent synthetic vowels and comes closest to predicting the identification responses of listeners to such stimuli" (680). A combined temporal-spectral approach may enable the brain to separate voices in everyday situations of extraordinary complexity (the cocktail party effect; Bregman 1990) as well as multivoiced music (Bello et al. 2006). When Aniruddh Patel and Evan Balaban (2001) aimed to investigate the relationship between pitch and the "timing of stimulus-related cortical activity" (839), they commented that "because of frequency-dependent time delays in the cochlear, either [spectral or temporal] method of extraction (or a mixed method) could contribute to cortical activity in which the timing varies with the perceived pitch of the stimulus. These findings also do not rule out subtle spatial changes in the magnitude of cortical activity during pitch perception, because these may be reflected in aspects of the brain signal that our study was not designed to detect" (843). In general, then, pitch can correspond to (or somehow emerge from) many different aspects of brain activity, of which any given empirical study has access to only a part.

In this regard, the term "spectral pitch" can be misleading. Spectral pitches do not depend only on the spectrum or frequency-domain representation. Like virtual pitches, they also depend on the temporal pattern of electrical responses in the auditory nerve (B. C. J. Moore 1989). It is also misleading to call Terhardt's model "spectral." A sound spectrum is part of World 1, but his model focuses on the World-2 relationship between spectral and virtual pitch. The model is only spectral insofar as the calculations are carried out in the pitch domain for convenience, assuming a one-to-one correspondence between the pitch and the frequency of a pure tone of constant SPL.

Brian Moore (1989, 1990) presented a model of pitch perception that incorporated both spectral (tonotopic) and temporal (timing) information. Like Terhardt's, it began with acoustic input to a filter bank: each hair cell on the basilar membrane may be regarded as an auditory filter that responds to a limited range of frequencies. Spectral pitch was assumed to depend on a mixture of spectral information (position on the basilar membrane) and temporal information (in auditory nerve fibres). The next stage involved the recognition of periodic or harmonic patterns in an approach that differed from Terhardt's: "In [Terhardt's] model each partial gives rise to potential pitches corresponding to subharmonics of that partial, and the perceived pitch is that for which the number of coincidences is greatest. In our model these subharmonic pitches correspond to the interspike intervals of 1, 2, 3, 4, 5 . . . ms, and these correspond to pitches of 1000, 500, 333.3, 250, 200 . . . Hz" (B. C. J. Moore 1989, 186). Note the mathematical equivalence of periodicity and harmonicity in this case.

Terhardt's Algorithm

In this book, I am following the approach of Terhardt et al. (1982b), for the following reasons:

- It assumes that the relationship between spectral and virtual pitches is partly learned and can be affected by musical experience (Preisler 1993). For example, gamelan and Western musicians hear different pitches in nonharmonic gamelan sounds (McLachlan, Marco, and Wilson 2013; Parncutt 1989).
- It acknowledges the existence of pitch shifts and tries to model them. It assumes plausibly that virtual-pitch shifts can be predicted from spectral-pitch shifts (due to changes in SPL and masking) and learned harmonic patterns (slightly shifted or distorted spectra of harmonic complex tones).
- Its predictions are limited to aurally relevant aspects of the sound signal. Unresolved partials are included in the calculation insofar as they are aurally relevant.

For an accessible description of how Terhardt's algorithm deals with different combinations of frequencies in an input spectrum and how it applies Bregman's principle of harmonicity in different cases, see Huron (2016, 27–32). The algorithm is based on simple, widely accepted assumptions:

- Only aurally relevant partials influence pitch. Spectrally unresolved upper harmonics may or may not be aurally relevant.
- Spectral pitch salience is a measure of a partial's aural relevance. It depends on SPL relative to masked threshold and frequency relative to the spectral dominance region, the ear being more sensitive to spectral pitch in a log-frequency region centered on about 700 Hz.
- The greater the number and salience of spectral pitches matching a harmonic template, the more salient is the virtual pitch at or near the fundamental frequency.
- Aurally relevant partials that correspond to lower harmonics of a virtual pitch have a greater effect on its salience than those corresponding to higher harmonics.
- Partials can be mistuned relative to the harmonic template by up to about a semitone (6 percent) and still contribute to virtual pitch at the template's fundamental; mistuning gradually reduces the contribution.

The small number of assumptions makes the model easy to falsify, and the lack of references to auditory physiology makes the model less arbitrary.

Terhardt's algorithm was in part inspired by Helmholtz. Here is an extract from Helmholtz's (1856) paper on combination tones: "Now, the ear is accustomed to hearing the musical tones of musical instruments, the human voice, etc., *always in the same composition repeatedly*, so that they become to the ear specific and familiar sense perceptions, about whose composition it has no occasion to reflect, indeed little more than we habitually make clear to ourselves that the sensory intuition of a physically extended object is composed of two different retinal images in both eyes. Indeed, in the everyday practice of our senses, we only attend to our sense perceptions at all to the extent that they enable us to recognize objects and events of the outside world" (Steege 2012, 73; italics in original).

The composition of musical instrument tones and voice sounds to which Helmholtz was referring is the harmonic series formed by the audible partials, which according to Terhardt becomes familiar to the auditory system in early life. That familiarity enables the ear to track the fundamental frequency and distinguish different timbres. But we are not aware of these recognition processes. Instead, we are aware of the sound source as constructed by our perceptual system. Philip Glotzbach and Harry Heft (1982) described how our perception constructs the current state of the outside world:

> Sensory input remains ambiguous, or at least incomplete. As a result, the organism must construct hypotheses to deal with it. (109)

> [In] traditional perceptual theories . . . animals are forced to gamble their well-being by constructing perceptual hypotheses of the "most likely" distal environmental conditions based on incomplete proximal cues. But one can ask: if the environmental regularities are available, why could not species have evolved the capacity to detect them directly, without having to resort to mediating probabilistic constructs? (110)

Our sensory systems are constantly constructing a model of the outside world based on available real-time sensory inputs and previous experience of environmental objects and the stimuli they usually produce. That principle applies to any aspect of perception in any sensory modality. The construction happens spontaneously and effortlessly, and indeed without awareness that construction is happening. Instead, we are aware of the output of the process: the world as constructed by our senses. In vision, for example, we are not aware of differences between the images picked up by the left and the right eye. Instead, we are aware of the distance of an object from us and the object's 3D features, which the distance between the eyes enables us to perceive.

One might argue that Helmholtz was implying (or even predicting) an ecological approach to perception, in which it is primarily the interaction between the perceiver and the environment that determines what we attend to. But his approach also anticipated today's cognitive approaches by dividing perception into two separate unconscious stages: sensation and inference (Barlow 1990). He also correctly implied that there is no contradiction between ecological and cognitive approaches.

Subjectivity of Pitch

Skeptical monists may doubt the existence or relevance of virtual pitches. Here are three possible reasons:

- If only the physical World 1 exists, virtual pitches do not exist by definition, although they may somehow exist in the brain as emergent properties of neural networks that cannot yet be reliably predicted.

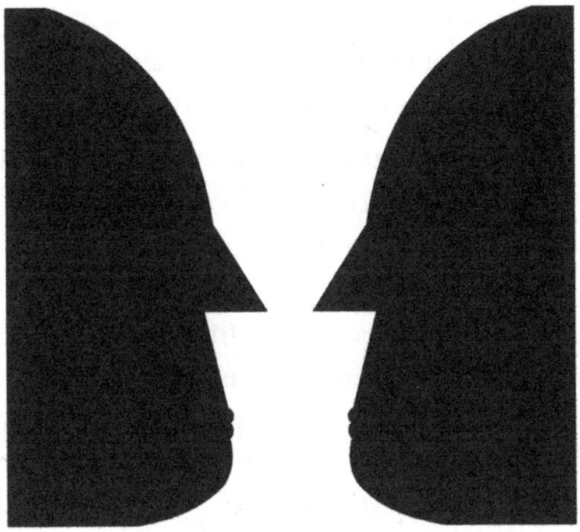

Figure 13.3
The Rubin vase. The figure can be seen as either a white vase or two black faces. (From Wikipedia: Rubin vase; Gestalt psychology.)

- Virtual pitches tend to "disappear" (become inaudible) when the listener focuses attention on them, just as holistic aspects of a pattern tend to disappear when we focus our attention on an element. In vision, the boundaries of the ghost triangle in the Kanizsa illusion and the dots at intersections in the Hermann grid illusion may disappear when we attend to them; different physiological bases have been proposed (Schiller and Carvey 2005). In the sound of a church bell, the first pitch we hear is the strike tone, which is a virtual pitch that depends on recognition of a harmonic pattern among the bell's audible partials. If we then focus our attention on the hum tone (the most salient spectral pitch), the virtual pitch becomes less salient—comparable with switching attention between the white Rubin vase and the black pair of faces in figure 13.3. Readers are invited to listen to an Internet recording of Vienna's biggest church bell, the *Pummerin*, whose strike tone is about F3 and whose hum tone is about D♭3. As the sound dies away, the (virtual) F3 becomes less prominent, and the (spectral) D♭3 becomes more so.
- There are big individual differences in pitch perception. People tend to be either fundamental listeners or spectral listeners (Seither-Preisler et al. 2007). Presented with harmonic complex tones with missing fundamentals, overtone listeners are more likely to hear spectral pitches (the pitches

of upper partials) consciously, whereas fundamental listeners are more likely to hear the pitch at the missing fundamental. Spectral listeners may be more likely to doubt the existence of virtual pitches.

Despite these uncertainties, there are good reasons to believe that virtual pitches exist. Beyond the empirical and theoretical evidence reported in this book, there is everyday experience of speech: we effortlessly perceive fundamental frequency trajectories, even when speech is masked by background noise (conversation, ocean waves, road traffic, music; J. H. Song et al. 2011). That is a practical consideration, and there is also a philosophical one: if most pitches heard in speech and music were virtual, it would be surprising if they did not exist. By the same token, if all of humanity was engaging with certain nonessential behaviors, calling them "music," and explaining their value in terms of subjective experience, it would be surprising if that experience did not exist.

The standard experiment for measuring pitch (described above) is objective in the sense that the results cannot be biased by the experimenter. A computer plays sound over headphones and automatically shuffles the order of trials for each new participant. It makes no difference if the experimenter accidentally smiles when listeners should reply "yes" and frowns when they should reply "no." If the results show that listeners consistently perceive a pitch, then it exists.

Pitch is inherently subjective, in two different ways. First, there are individual differences in the salience and shift of individual pitches. Second, pitch is part of Popper's World 2. To measure pitch, a human listener has to adjust the frequency of a pure tone until its pitch matches that of a test sound. That process relies on consciousness and attention, and hence on subjectivity.

To understand the inherent subjectivity of pitch better, compare it with another inherently subjective entity: timbre or sound quality. Skipping over attempts to define or operationalize timbre in the literature for the moment, timbre might loosely be defined as "what a sound sounds like" by comparison to known models in different sensory modalities (e.g., a bright light, a rough stone, chocolate, or a clarinet). Like pitch, timbre depends on various physical properties of a tone—not only spectral envelope, as is often assumed. Unlike pitch, timbre cannot be measured on a one-dimensional scale, but it does have perceptual dimensions that can be measured; they may correspond physically to spectral energy distribution or spectral fluctuation (Grey 1977). Those are good reasons to regard timbre as quite a different thing from the temporal and spectral envelopes upon which it depends—just as there are good reasons to regard pitch as different from frequency.

We can also regard a perceived tone as a subjective entity, of which pitch as an attribute. In both cases, a pitch is a *quale*—a unit or instance of subjective, conscious experience. The quale in question can be seen either as a property of a physical phenomenon (the tone—an oscillation in air pressure) or as a property of another quale (the tone sensation—a subjective entity).

Whereas some qualia are ineffable, pitch (as defined) can be reliably measured—contradicting claims in the literature that qualia cannot be measured. A standard psychoacoustic experiment allows us to measure pitch reliably. The result is usually presented as an average across several listeners on the assumption that they are experiencing almost the same thing. I will assume that these global claims are true regardless of how the term "qualia" is understood in detail, or what detailed questions about (musical) qualia are of concern to philosophers (e.g., Schiavio and van der Schyff 2016).

The discipline of psychoacoustics (more generally, psychophysics) can be seen as a form of quantitative phenomenology. Subjective experiences are measured, and results from different participants are averaged to improve reliability. The approach is comparable with what Daniel Dennett (2003) called heterophenomenology or a "third-person approach to human consciousness" (19)—phenomenology of another, not of oneself. Dennett was interested in qualitative description, whereas we are here dealing with quantitative measures. The advantage of Dennett's idea, like the critical phenomenology of Max Velmans (2007), is that it enables a more objective description of subjectivity than mere introspection. The disadvantage of heterophenomenology is that ineffable aspects of experience are excluded.

This discussion implies that reliable information about subjective entities is possible, which in turn justifies believing in the existence of those entities. The reliable information in question includes replicable findings of psychoacoustic experiments if we ignore the uncertainty in the measurements; after all, physical measurements are also generally uncertain.

To clarify the inherent subjectivity of pitch further, imagine an experiment on pitch perception in monkeys. In each trial, the monkeys hear two successive tones. The pitch of the second tone is either higher or lower than the first. The monkeys are trained by operant conditioning to respond differently to rising and falling tone pairs. In experiments with New World monkeys, Xindong Song and colleagues (2016) reported that "the subject had to detect the F0 change and respond by licking at a feeding tube placed in front of its mouth during the response window ("hit") to receive a food reward" (785). The research team found that the relationship between spectral and virtual pitch in monkeys is much the same as it is in humans. So far, so good.

Now imagine a new experiment to test the perception of pitch shifts or pitch ambiguity in monkeys. Pitch shifts are small changes in pitch due to physical intensity or masking (typically, less than a semitone), whereas pitch ambiguity means pitch can flip across larger intervals (e.g., an octave). Monkeys and humans hear complex tones in a similarly holistic manner. But in the case of monkeys, we have no access to their subjective impression of pitch because we have no access to their private world. As in humans, their impression could be affected by diverse aspects of neural processing that correspond in complex ways to both time-domain and frequency-domain aspects of sound. Proponents of 4E cognition would agree that a monkey's impression of pitch also depends on interactions between the brain and the rest of the body or between the body and the environment.

Like humans, monkeys have enormous numbers of neurons in their brains, and the number of synapses joining them together is greater again. Trying to understand how that complexity relates to pitch judgments of monkeys is like trying to solve the mind-body problem in humans. This comparison between humans and monkeys is valid regardless of whether monkeys have consciousness and, if so, of what kind.

The point applies equally for both spectral and virtual pitch. Just because spectral pitch corresponds to an approximate position on the basilar membrane and/or to information contained in neural timing patterns does not mean that spectral pitch is a physical parameter or somehow more physical than virtual pitch. Like virtual pitch, spectral pitch can only be determined by an experimental procedure that involves subjective appraisal and comparison by a conscious observer. Like virtual pitch, spectral pitch is purely subjective in nature.

Neuroscience of Pitch

The physical states and processes of the human brain that correspond to experiences of pitch are complex and difficult to investigate (Plack et al. 2014). Deborah Hall and Christopher Plack (2009) found evidence for neural processing corresponding to pitch in the lateral Heschl's gyrus, planum temporale, temporo-parieto-occipital junction, and prefrontal cortex. For perceiving and producing pitch in speech, Martin Meyer and colleagues (2004) discussed the roles of the superior temporal gyrus, dorsal striatum (putamen and caudate nucleus), right Rolandic operculum (premotor cortex), left inferior frontal gyrus (Broca's area), left inferior precentral sulcus, and left Rolandic fissure.

Different brain areas can contribute to pitch perception in different ways, depending on the linguistic, musical, or environmental function of pitch (cf. Gandour et al. 1998). There are also individual differences: different brains can process pitch in different ways. That being the case, we might pose an interesting armchair question: Imagine that we knew everything that there was to be known about the physical state of a given brain, including the state of every neuron and every synapse, from one moment to the next. Would we be able to predict perceived pitch from physical measurements? Given the inherent complexity of the mind–brain relationship, the answer is unclear.

I prefer not to jump to conclusions about such a difficult problem, and I agree with Terhardt that a music-theoretically interesting model of pitch perception should follow cybernetic principles and not depend on neurophysiology. It is possible that a future neuroscientist will achieve a breakthrough that will make the preceding arguments obsolete. That would be great, but in the meantime, we should remain scientifically cautious.

Temporally based models of pitch focus on the phase locking of neural firing along auditory pathways. They assume that pitch is related to the rate of neural firing because the two are clearly correlated, at least up to a limit of a few thousand hertz (Verschooten et al. 2019). Research into these issues often employs artificial or unnatural sounds, such as complex tones in which the phase relations between the partials are independently manipulated and all variables are carefully controlled. But we can also learn about pitch perception in experiments using real, ecologically valid environmental sounds in all of their complexity. We need to take a broad perspective that includes many different empirical observations. Plack and Oxenham (2005) pointed out that "when developing a model of pitch perception, or when identifying a cell type or brain region that may be involved in pitch perception, it is important to ensure that the results are consistent with the wide range of psychophysical observations, and not to focus on a single property of pitch that may provide an easy solution" (7).

Physiologically and temporally based models of pitch perception involve different physical processes, including autocorrelation in the timing of neural impulses in the auditory nerve (Cariani 2001), cancelation (Cheveigné 1998), and cochlear filtering and coincidence detection (Shamma and Klein, 2000). Neural pulse codes may be channel-based, temporal-pattern, or time-of-arrival codes (Cariani 1999). It is tempting to explain aspects of musical harmony on the basis of such models (e.g., Cariani 2001; Tramo et al. 2001), but the approach is misleading if it assumes Just tuning and intonation (ratios of small integers; see chapter 7). Mark Tramo and colleagues (2001) noted that "for the consonant intervals (the fifth and fourth), the pattern of major and minor peaks in the autocorrelation is perfectly periodic, with a period related to the fundamental bass . . . suggest[ing] that the consonance of harmonic intervals reflects regularities in their temporal fine structure" (100–101). Apart from assuming that musical intonation is ideally Just, this claim skims over three kinds of complexity: that of music (which includes both deliberate and accidental variations in intonation), that of the brain (which comprises enormous numbers of neurons that could potentially be involved in a single perceptual act such as perceiving a musical interval), and that of relevant philosophy and metaphysics (in which the mind-body problem remains unsolved).

Pitch shifts and octave stretch pose additional problems. Perceived pitches often depart significantly from pitches predicted on the basis of neural-temporal patterns. There is also research showing that tonotopic representations (frequency–place mappings) are essential for pitch perception (Oxenham et al. 2004). That does not mean the neural-temporal patterns are unrelated to pitch. On the contrary, it suggests that pitch perception generally depends on a complex mixture of, and possibly interaction between, temporal and spectral (or tonotopic) factors.

Creators of temporal models of musical interval perception are in good company. Beliefs in music of the spheres and musical intervals as number ratios have been seriously held by scientists as eminent as Nicolaus Copernicus in the fifteenth century; Leonardo da Vinci and Gioseffo Zarlino in the sixteenth; Galileo Galilei, René Descartes, Johannes Kepler, and Marin Mersenne in the seventeenth; Isaac Newton, Leonhard Euler, Jean-Philippe Rameau, and Immanuel Kant in the eighteenth; and even (to a lesser extent) Hermann von Helmholtz in the nineteenth (Pesic 2014; Proust 2009). Besides, neural periodicities can indeed correspond to pitch. When a tone is perceived, temporal behaviors of neurons are often phase locked to stimulus waveforms along auditory pathways (Cariani 1999; Langner 1997), and their periods correspond inversely to perceived pitch. But there is a danger of jumping to conclusions. Correlation does not imply causality. Just because something that we experience is correlated with something happening in the physical world does not mean that the physical states or changes are causing the experience.

Misunderstandings

Differences between schools of thought in psychoacoustics or pitch perception may involve simple operational definitions of pitch or audibility. I am assuming that pitch is measured by the frequency of a pure tone that is judged subjectively to have the same pitch as a successively presented test sound. That is true regardless of what spectral or temporal aspect of the signal is causing the pitch. For example, if pitch is perceived in a fast train of clicks, the pitch is still measured by the same psychoacoustic procedure. I am also assuming that a partial within a complex sound is audible if a listener notices a change when the partial is turned on and off. Others may define pitch and audibility differently.

Misunderstandings can also involve differences in unstated assumptions about the nature of reality (metaphysics). The auditory scene is often treated as if it were physical or physiological in nature; I have argued that it is not. Of course, the auditory scene has a physical or physiological foundation, but the measurements upon which it is based are psychoacoustic—mediated by human consciousness. We should not pretend that we are making physical measurements when in fact we are not.

For the music-theoretical purposes of this book, I am sidestepping the physiological complexity of complex tone perception and corresponding unresolved empirical and philosophical issues, and focusing instead on the psychoacoustic relationship between physical parameters on the one hand (the frequencies, amplitudes, and phase relations of the partials of a complex sound in air) and experiential parameters on the other (pitch, loudness, timbre, perceived duration). This psychoacoustic approach is appropriate for investigating the physical and psychological foundations of MmT as an experiential, subjective, and cultural phenomenon, and for addressing questions such as: Why is (Western) musical structure based on hierarchically structured pitch-time patterns? Why is conventional music notation, like representations of the auditory scene in psychoacoustics, similar to a graph of pitch against time? Both these questions refer to the ear's extraordinary sensitivity to pitch by comparison to loudness, timbre, or phase (see chapter 6).

What we know about the auditory scene (what it contains, what it does not, what parts of it attract attention, and so on) comes from psychoacoustic experiments. Data of that kind are always mediated by the consciousness of the listener. In that sense, the auditory scene is a fundamentally subjective entity. It is a representation of subjective experience. While it could not exist without the underlying physiology, the underlying physiology is not what is measured in experiments of that kind. If a measurement is based on a subjective judgment, which is mediated by consciousness, it should be labeled "subjective"—if only for scientific clarity and consistency.

In the three-worlds concept of Karl Popper and John Eccles (1977), the pitches that we perceive in the auditory scene belong to World 2 (subjective)—not World 1 (physical). The pitches that we perceive certainly *depend on* World 1, but since our perception is inherently subjective, we should not jump to conclusions about that dependency. We need to keep our feet on solid scientific ground.

Many psychologists, including many music psychologists, disagree with the previous arguments. The past century has seen a tendency toward physicalism—the belief that only the physical world exists, denying the existence of subjectivity (Papineau 2013). Within that broader academic and social context, many colleagues in music psychology are studying aspects of musical subjectivity, and some are devoting their lives to that task. Of them, a large proportion may paradoxically believe that the object of their research does not exist. They are right, from a physical perspective, that only the physical world exists. They are also right that most of the sciences operate most successfully on a physicalist-monist assumption. The trouble is that they are studying art, and art is inherently subjective. Without subjectivity, music would be meaningless.

This book is about the pitch structures that people experience when listening to music. For every such experience, highly complex physical correlates in the brain can be identified. These correlates can and should be studied in detail in order to understand the anatomic structures and their biological functions better, both within and between neurons. But the observations of neuroscientists are limited to the physical world. Regardless of unpredictable future technological developments, and despite the fascinating fantasies of science fiction, neuroscientists will never directly observe the subjective experience of live subjects within their brains.

14 Consonance and Dissonance

In a psychoacoustic approach, simultaneous consonance/dissonance (C/D) depends on smoothness (lack of roughness), harmonicity, and familiarity. Roughness is produced by fast beating (amplitude modulation) when partials in a complex sound are close in both frequency and amplitude; its physiological origin is the inner ear. Harmonicity is the degree to which a spectral-pitch pattern (the audible partials of a complex sound) resembles a harmonic series. Familiarity can be either musical or nonmusical; it is the familiarity of a sound in musical or everyday environments. The theory explains why the most prevalent trichords in MmT are major (in semitones: [047]), minor [037], suspended [057], diminished [036], unnamed [025], and unnamed [035], in that order. Musical chords are tone simultaneities whose internal intervals are culturally learned pitch distances and whose sonorous or consonant qualities and voice-leading possibilities emerged gradually in musical psychohistory. The theory solves an old music-theoretical problem by parsimoniously explaining the origin of the minor triad and its importance for MmT. The theory also generalizes to explain the relative prevalence in tonal musical repertoires of all possible trichords and tetrachords (chords of three or four pitch classes).

History of C/D

During the second millennium of the Christian era (AD 1,000–2,000), Western music was increasingly dominated by multivoiced textures. Early melodies were often played or sung in parallel octaves or 5ths. Toward the late twelfth century, polyphony emerged in which voices moved quasi independently of each other.

There are well-known examples in extant manuscripts from the Notre-Dame school in Paris and the eleventh-century Winchester Troper (van der Werf 1992). These were largely oral traditions in which the music was memorized, improvised, or both (aural paradigms; Treitler 1991, 83). Polyphony or counterpoint was created by contrary motion between pairs of melodies, often departing from and then moving back toward a perfect unison, 5th, or octave interval as a point of rest. For example, an octave sonority might have been preceded or followed by a dissonant major 6th (or vice versa), the two voices moving in stepwise contrary motion. As the number of voices increased to three or four, new vertical pitch combinations emerged.

These developments were influenced by the history of ideas. The ancient idea of the harmony of the spheres connected mathematics, the cosmos, the human soul, and music. Such beliefs can be seen as part of a broader philosophical and religious search for truth and beauty (J. A. Martin 2014)—motivated, in part, by preferences for consonant sounds and passages in both religious and secular music traditions.

In that context, consonant sonorities tended to happen more often than dissonant ones, and dissonant ones resolved in specific ways to consonant ones. This principle may even be applied to single sung tones in (Gregorian) chant, the C/D of which may have depended on the relationship between their audible partials and the prevailing scale (Parncutt 2019a). The compositional rules that governed two-part counterpoint can be seen retrospectively as strategies to avoid or control horizontal and vertical dissonance and to promote horizontal and vertical consonance—if we understand consonance in the general sense of any simultaneous or successive sounds that go together musically.

Rules of two-part writing were later applied to music in three parts, causing new pitch structures to emerge (Eberlein 1994). In the fourteenth century, "A systematic study of sonority usage in Machaut's motets . . . reveals his preference for particular combinations of intervals over others" (Hartt 2010, 179). In that context, Sarah Fuller (1986) explained that "perfect sonorities are the most stable, doubly imperfect the most unstable, and imperfect sonorities lie somewhere between the other two" (Hartt 2010, 197).

A standard recurring pattern in early and common-practice Western polyphony is consonance-dissonance-consonance. It can happen at different hierarchical levels: the structure of a melody, longer chord progression, or whole pieces (e.g., exposition, development, recapitulation). Dissonant tone combinations must be suitably prepared by consonant combinations and then resolved by consonant combinations. The resolution should follow voice-leading rules and lead to a perceptible reduction in musical tension.

Components of C/D

James Tenney (1988) surveyed changing consonance–dissonance concepts (CDCs) in the history of Western music theory. In the pre-polyphonic era, the prevailing CDC was melodic, involving relationships between successive tones (tonal affinity). CDCs in early polyphonic music were diaphonic, focusing on the sonorous character of harmonic dyads. Renaissance music developed more complex contrapuntal CDCs that depended on the perceptual salience of the lower voice. In the classical period, CDCs became more triadic, focusing on individual tones within chords. In the nineteenth century, the acoustic theory of roughness contributed to musical CDCs, linking them to timbre.

Music perception is a mixture of nature and nurture. On the one hand, we perceive the stimulus directly, and our response is influenced by its perceptible structure (the psychoacoustic aspect). On the other hand, our response is influenced by our personal history of perceiving similar stimuli, which in turn depends on the cultural history that influenced the styles of music with which we are familiar (the psychohistoric aspect). A comprehensive theory of C/D in MmT should bring together these two aspects. The psychohistoric aspect reminds us that C/D is culture specific; the psychoacoustic aspect reminds us that it is not arbitrary.

Today, the C/D of a sonority in MmT depends primarily on its roughness, harmonicity, and familiarity (Harrison and Pearce 2020; Parncutt, Reisinger et al. 2019). Something similar may have applied to medieval polyphony as actually perceived at the time—even if each component was very different from today. Said another way, the C/D of a given tone simultaneity changed from one period to the next as its harmonicity, smoothness, and familiarity changed—even if the sound was physically unchanged. As MmT emerged from sixteenth-century polyphony, the most consonant chords (major and minor triads) began to function as psychological references.

Successive and simultaneous aspects of C/D can be called horizontal and vertical, respectively, by analogy to the musical score. They may also be called melodic and harmonic; I will use these terms interchangeably. Let us now look at vertical (harmonic, simultaneous) aspects of C/D in more detail. They involve:

- *Smoothness*: lack of acoustic roughness originating from fast beating or amplitude modulation in the inner ear—also called sensory smoothness,

- *Harmonicity*: spectral similarity with the harmonic series, also called spectral harmonicity; and
- *Familiarity*: how often the pitch pattern in question happens in music that the listener has heard, considering the relationship between the pattern and its context.

Roughness and harmonicity are the main two psychoacoustic aspects of vertical C/D (Terhardt 1977). Both depend on relationships among the audible partials of a complex sound. Roughness was investigated in detail by Hermann von Helmholtz, and harmonicity (perceptual fusion of harmonic spectra) by Carl Stumpf—both in the late nineteenth century. Much later, quantitative models of roughness were developed by William Hutchinson and Leon Knopoff (1978) and by Wilhelm Aures (1985). Peter Harrison and Marcus Pearce (2020) developed a model for the C/D of musical simultaneities that combined roughness (interference), harmonicity, and (cultural) familiarity.

Roughness is how rough a sound is perceived to be due to fast modulations in the temporal amplitude envelope. As such, it is purely subjective and belongs to World 2. It is an important aspect of subjective musical experience. If neighboring tones on the piano keyboard (creating a harmonic major or minor 2nd) are played simultaneously, the sound is rough; smaller intervals tend to be rougher than larger intervals, and the same interval (e.g., a major 3rd) becomes rougher if played in a lower register. Roughness is associated with nearby partials in the sound's overall spectrum (lying within a critical bandwidth; Plomp and Levelt 1965).

A sound's harmonicity is the degree to which its spectrum is harmonic, or the similarity of its audible spectrum to a harmonic series, as it occurs in voiced speech sounds, and focusing on the audible lower harmonics (Bowling et al. 2018; DeWitt and Crowder 1987; McDermott, Lehr et al. 2010). If the evaluation of harmonicity is subjective, harmonicity, like roughness, belongs to World 2.

When discussing harmonicity, I will assume that the number of perceptually relevant (audible) harmonics in isolated harmonic complex tones is typically approximately ten. While this number can vary considerably from one sound to the next, the literature suggests that a limit of ten is both psychoacoustically realistic and a useful rule of thumb for music-theoretical purposes. The intervals among the first ten harmonics are therefore familiar to the auditory system (see chapter 13).

Harmonicity is part of a more general concept of musical familiarity. It is about the similarity of musical sounds and familiar voiced speech sounds. It suggests that simultaneous musical intervals such as octaves, 5ths, 4ths, major 3rds, and minor 3rds correspond to intervals between audible partials in voiced speech sounds. Other foundations for the C/D of such intervals are smoothness and musical familiarity.

Harmonicity has been discussed at length in many studies as if it were a quantifiable variable (Bregman 1990; Bowling and Purves 2015; Harrison and Pearce 2020). But there is little consensus on how to quantify harmonicity. Virtual-pitch salience (Terhardt et al. 1982b) is a possible solution. A virtual pitch is predicted to be more salient if it has a greater number of spectral pitches above it at approximately harmonic positions; if the matching spectral pitches are better tuned to the expected harmonic pitches; and if the spectral pitches are more perceptually salient.

The physiological foundations of harmonicity and familiarity are located centrally in the brain (in neural networks), whereas roughness originates in the peripheral auditory system (the basilar membrane in the inner ear). The brain areas can be identified more specifically: the processing of deviations from expected musical patterns, based on musical experience, involves the superior temporal gyrus, deep frontal operculum, insula, rostral cingulate zone, middle cingulate gyrus, pre-supplementary motor area, and putamen (Tsogli et al. 2022). The physiological separation of roughness and familiarity is additional evidence that C/D is not a unitary concept but rather depends on partly independent factors.

C/D also applies to relationships perceived between successive sounds. Horizontal (melodic) aspects of C/D in MmT include pitch commonality and pitch proximity—the inverse of pitch distance (see chapter 18). In short, pitch commonality is the degree to which two successive complex sounds have perceptible pitches in common. A pitch that is perceived in both sounds is considered common if it is perceived to belong to the same chromatic scale step. This definition relies on the idea of categorical pitch perception in music that is based on something close to the standard chromatic scale. In a progression of complex sounds such as musical chords, we may consider the pitch commonality of each successive or non-successive pair.

Pitch distance is the perceived distance between two successive complex sounds. If two sounds are perceived to comprise the same number of pitches or voices, overall pitch distance depends on the pitch distances traversed within each voice. In good voice leading, pitch distance is kept to a minimum—a simple case of promoting horizontal consonance. In the simplest case, each sound is a harmonic complex tone with one main pitch, and pitch distance is the interval between successive tones. The holistically perceived pitch distance between two musical chords depends on the probability of perceiving different voice-leading connections. To model that, we first need to estimate the probability of perceiving each pitch in the chord (its perceptual salience). On that basis, we might estimate the probability of perceiving melodic connections between specific successive pitches. To solve this problem mathematically, we need to consider all possible voice-leading or streaming possibilities that might link the two chords (Parncutt 1989).

Voice Leading

The relative prevalence of different trichords and tetrachords (harmonic vocabulary: which chords appear, and how often they appear) could depend on voice leading: how easily they can be approached and quitted. Dmitri Tymoczko (2010) argued that chords that are almost equally spaced around the pitch-class circle may be preferred for reasons of voice leading. Eytan Agmon (1991) argued similarly that "triads and seventh chords are the only cyclic chords within the diatonic system which form progressions for which a linear transformation satisfying a certain 'efficiency constraint' always exists" (15). These ideas are compelling but difficult to confirm quantitatively. If the chord prevalence profiles shown in figure 14.1 can be accounted for by considerations of vertical consonance alone, a theory that focuses on vertical consonance and ignores voice leading is more parsimonious and, in that sense, preferable.

To demonstrate quantitatively that a theory of trichord and tetrachord prevalence based on voice leading has explanatory value beyond that of a psychoacoustic theory of vertical C/D, one would need to demonstrate statistically that the new model significantly improved existing predictions. Given the practical advantages of parsimonious theories and the success of C/D theory in predicting the harmonic vocabulary of MmT, I will assume for the purpose of argument that the prevalence of specific chords in MmT depends mainly on their C/D in isolation.

Harmonic Intervals

The harmonic vocabulary of common sonorities in MmT can be predicted by evaluating the harmonicity and roughness of all possible pitch-class sets within given constraints using simple psychological models. The predictions can then be compared with statistical analyses of music. The procedure is parsimonious and empirically based, and it readily and directly explains why some chords are more common than others.

Starting with intervals: perfect octaves, 5ths, and 4ths are consonant because they are both smooth and harmonic. The tritone is dissonant because it is neither smooth nor harmonic, whereas 2nds and 7ths are dissonant mainly because of their roughness. Lying between these extremes are 3rds and 6ths, which are therefore preferred in two-part writing in MmT.

Favoring harmonicity means favoring not only intervals from the harmonic series, but also sounds for which the "root" (the tone corresponding to the fundamental of the harmonic series) is in the bass. Hindemith (1937) assigned a root to every interval in the chromatic scale. According to him, the root is the lower tone of the perfect 5th, the major and minor 3rds, and the major and minor 7ths; it is the upper tone of the perfect 4th, the major and minor 6ths, and the major and minor 2nds. The simple algorithm of Parncutt (1988)—revised such that the minor 3rd is not a root-support interval—also allows the roots of individual intervals to be predicted. It predicts that the root of an interval is the lower tone for the perfect 5th and the major 3rd, and the upper tone for the inversions of those intervals (the perfect 4th and minor 6th)—which therefore sound inverted. For the major 2nd and minor 7th, the theory makes no prediction. Those intervals are both weak root-support intervals and inversions of each other, so the root-support effects cancel each other out.

Since the harmonic series is familiar from complex tones in speech, listeners expect the fundamental or its octave equivalent to be lower in pitch than the other tones. The tone in the bass of a harmonic interval tends to be perceived as the root simply by virtue of being in the bass (Hurwitz 1970; Parncutt 1997). That can explain why chords in root position sound more consonant and therefore appear more often in music than chords in inversion: putting the root in the bass clarifies the root. The same applies to individual intervals.

Putting these ideas together, we might predict that all intervals inside an octave are perceived to be in root position, with two exceptions: the perfect 4th and minor 6th. In those two cases, the root (or the fundamental of an implied harmonic series) is perceived to be in the upper voice. The upper voice is more salient according to virtual pitch theory, provided the two tones would have been equally loud if heard alone. That can explain why the perfect-4th and minor-6th intervals were regarded as more dissonant that the perfect 5th and major 6th, respectively, in the history of music theory (Tenney 1988).

Another possible reason for the dissonance of the minor-6th interval in medieval theory is the rough semitone intervals in its combined spectrum. If the fundamentals are A3 and F4, the spectrum is A3 F4 A4 E5 F5 A5 C6 C#6 E6 F6 G6 <u>A6</u>, where underlining denotes that a partial is common to both tones. Here, we see dissonant semitones between E5 and F5, C6 and C#6, and E6 and F6. For the perfect 4th interval, with fundamentals A3 and D4, the spectrum is A3 D4 A4 D5 E5 <u>A5</u> C#6 D6 E6 F#6 G6 <u>A6</u>, with dissonant semitones between C#6 and D6 and between F#6 and G6. This simple approach allows the roughness of different intervals to be compared.

Measuring C/D

The C/D of simultaneous musical intervals can be measured in a simple psychological experiment. Present the intervals or chords to listeners (e.g., on the piano) and ask them to rate their C/D. How well do the tones go with each other, in a musical sense? Answers can be given on a rating scale (Likert scale); an eleven-point scale might have the end points "very poorly" = 0 and "very well" = 10. If different people offer similar ratings for the same intervals, they can be averaged.

Constantine Malmberg (1918) played harmonic intervals on the piano or pipe organ, or with tuning forks, to a group of experts that included professional musicians, psychologists, and teachers. They listened repeatedly to each interval, separately assessed four aspects (blending, smoothness, fusion, and purity), and discussed their opinions before presenting their collective conclusions. The results, which were presumably strongly influenced by the music-theoretical knowledge of the participants, were then used as a benchmark against which to measure the musical ability of nonexperts: those who produced similar results were held to have more musical talent.

Several of my students in Graz have informally carried out a simpler experiment, in which nonmusicians hear a harmonic interval and are asked to rate its consonance. The results can be surprising: 3rds and 6ths may be preferred over octaves, 5ths, and 4ths. A possible reason is musical familiarity—the cultural component of C/D, alongside the psychoacoustic components harmonicity and roughness. Intervals of a 3rd or 6th are more common than other intervals in two-part writing, either because their relative lack of fusion allows the two voices to be heard more clearly (Huron 2001; DeWitt and Crowder 1987) or because listeners prefer an intermediate degree of C/D—not too much and not too little (cf. Berlyne 1971).

Some psychological and neuroscientific studies of C/D have separated cultural from sensory aspects. In an electrophysiological (ERP) approach, in which participants heard musical intervals and rated the feelings they evoked, Daniele Schön and colleagues (2005) showed that ERP latencies depended on musical expertise, implying a stronger role of musical experience for musicians than for nonmusicians. Paula Virtala and Mari Tervaniemi (2017) presented evidence that both the major-minor and the consonance–dissonance distinction have "partly innate, biologically hard-wired properties" that can be "drastically modified" by "musical enculturation and active music training" (387). Another study looked for, but failed to find, an innate preference for consonance over dissonance (Plantinga and Trehub 2013), suggesting an important role for learning. In any case, the C/D of a chord depends on the musical context in which it is presented (Johnson-Laird et al. 2012), which can be regarded as an aspect of familiarity: more familiar contexts lead to more consonant impressions.

Cultural Influences

Imagine a set of empirical data on the perceived C/D of different sounds—intervals, chords, and perhaps also short passages, limited to certain styles or periods—that is both reliable (empirically reproducible) and valid (reflecting what we mean or intend by C/D). Imagine also that we have reliable data on the frequency of occurrence of corresponding sonorities in selected musical repertoires, again limited to styles and periods.

Both sets of data can be explained in various ways. They may depend on physical factors (nature: temporal/spectral envelopes, octave register, tuning), personal factors (what pleases me might not please you but depends on my personal experience), or cultural factors (more general background of listeners). Personal and cultural factors can depend on both musical familiarity (e.g., familiarity with pitch intervals in music) and nonmusical familiarity (e.g., familiarity with pitch intervals within the spectra of everyday nonmusical sounds).

The frequency of occurrence of a pitch structure in music may depend on all of these factors. Listeners may develop an arbitrary taste for certain sounds, which are then favored—by analogy to fashions in the clothing industry. In more familiar tone combinations, listeners hear the pitches more accurately (McLachlan, Marco, Light et al. 2013). Familiarity may be acquired either individually (e.g., M. W. Weiss et al. 2020) or culturally (e.g., Harrison and Pearce 2020; Lahdelma and Eerola 2016). Tudor Popescu and colleagues (2019) explained: "the place a sonority occupies on the consonance/dissonance spectrum is by no means the sole factor predicting the pleasure it will elicit . . . the 'positive' end of this spectrum does not, by itself, automatically imply pleasantness in any given musical style . . . new models are needed that are capable of explaining the ecological experience of listening" (9).

The harmonic octave interval is not necessary considered more consonant than other intervals when presented to modern nonmusician listeners in a simple experiment. But octave equivalence clearly plays an important role in MmT, and in a parsimonious account of C/D, octave equivalence and octave consonance are almost the same thing. Let us therefore suppose for a moment that mainstream music theory is correct and that the octave really is the most consonant interval. If so, its consonance can be explained in different ways:

- A well-known theory says that its consonance comes from its frequency ratio. As explained in chapter 10, we should not rely on abstract mathematical ideas of this kind, given empirical evidence that octaves are usually stretched in music and that listeners tend to prefer stretched octaves.
- Another explanation refers to the lowest two partials of a harmonic complex tone. The octave interval between them is familiar to the auditory system from exposure to speech. That can explain the octave stretch phenomenon, given what we know about the dependence of spectral pitch on SPL and masking (pitch shifts; chapter 13).
- Then there is roughness. If two harmonic complex tones lie an octave apart, all the partials of the higher tone line up with partials of the lower tone, minimizing roughness. In this case, octave stretch is not predicted, but roughness may be imperceptible for small deviations from a 2:1 ratio.

- Finally, there is musical experience and familiarity. We constantly hear octave relationships in music that reinforce our perception of octave similarity and octave equivalence.

The interval of a perfect 5th is considered the second most consonant interval because it lies between harmonics 2 and 3. For that reason, it is presumably the second most common simultaneous interval in music. Its C/D can be explained in the same ways as that of the octave.

The C/D of less consonant intervals depends increasingly upon how they are used in music and their roles in musical scales. For example, a minor-3rd interval may be perceived as consonant for three reasons:

- The interval may somehow remind us of the interval between partials 5 and 6, audible in most harmonic complex tones in speech.
- When two harmonic complex tones sound a minor 3rd apart, the 6th harmonic of the lower tone aligns with the 5th harmonic of the higher. Mistuning can produce beats or roughness.
- The widespread use of standard pentatonic and diatonic scales in Western and other musics suggests that scales with many 4th and 5th intervals between nonadjacent scale degrees are favored. That principle could explain how pentatonic and diatonic scales emerged in the prehistory of Western music. The minor-3rd interval could be perceived as consonant simply because it is frequently heard in such scales.

Summarizing, while the perceptual foundations of Western C/D (roughness and harmonicity) may be perceptual universals, their aesthetic evaluation is culture specific. There is no particular reason why roughness and harmonicity should be important in non-Western or nonclassical music, although they sometimes are (e.g., Ambrazevičius 2017).

Trichords and Tetrachords

One approach to quantifying C/D is to count how often each interval or chord happens in a representative database of musical scores. The assumption is that the more often an interval or chord is notated or perceived, the more consonant it is.

Music theory texts traditionally regard the main trichords in Western music as major [047], minor [037], diminished [036], and augmented [048], implying that these simultaneities are more consonant than others. But musical trichords are more diverse than theory texts suggest, and the relationship between basic and derived chords is less clear. Chords are not necessarily constructed by stacking 3rds. In a systematic, psychologically founded approach, we might assume initially that all pitch combinations in the chromatic scale are possible, but some are preferred according to some criterion (which we here assume to be consonance). To test that, we need a body of data ("ground truth") with which to compare model predictions. That means carefully counting how often each chord happens in the musical literature. To test models, we need to compare their predictions with such counts as well as with the results of listening experiments.

Figure 14.1 shows the relative prevalence of vertical trichords and tetrachords in a representative database of vocal polyphony. The profiles for before and after 1600 differ in interesting ways, but they are also remarkably similar. Using modern (anachronistic!) terminology, the most common trichords in both periods are major [047], minor [037], suspended [057], diminished [036], unnamed [025] (e.g., CDF), and unnamed [035] (e.g., DFG)—more or less in that order. That list is not much different from the traditional four chords listed at the start of music theory texts (major, minor, diminished, augmented), but it is interesting that augmented triads happen so seldom and that chords with suspended 4ths are so common. The analysis also suggests that sus4 chords can act as independent sonorities in their own right rather than as derivatives of major and minor triads, which happened increasingly often in twentieth- and twenty-first-century music.

The lower part of the figure shows that the most common tetrad in music of the seventeenth to the nineteenth centuries was {0368}, the Mm7 chord. The chord is notated here as a prime form (in curly brackets), which includes all four inversions; it may be more easily recognized in root position as [047T]. The next few tetrachords, in approximate rank order, are {0358} = [037T] (minor 7th = m7 chord; in inversion, added 6th = add6), followed by {0257} = [057T] (7sus), {0158} = [047L] (maj7), {0247} (add9), and {0258} = [036T] (half-diminished 7th or ø7).

This list seems to agree with music theory with the exception of the 7sus chord, which—like the sus4 triad—is not regarded as a chord in its own right in conventional theory because it often resolves to a regular 7th chord. But since 7sus happens so often in music of the past century and so often fails to resolve to a Mm7 or m7 chord—suggesting it became a chord in its own right—and since the chord plays such an important role in jazz/pop harmony

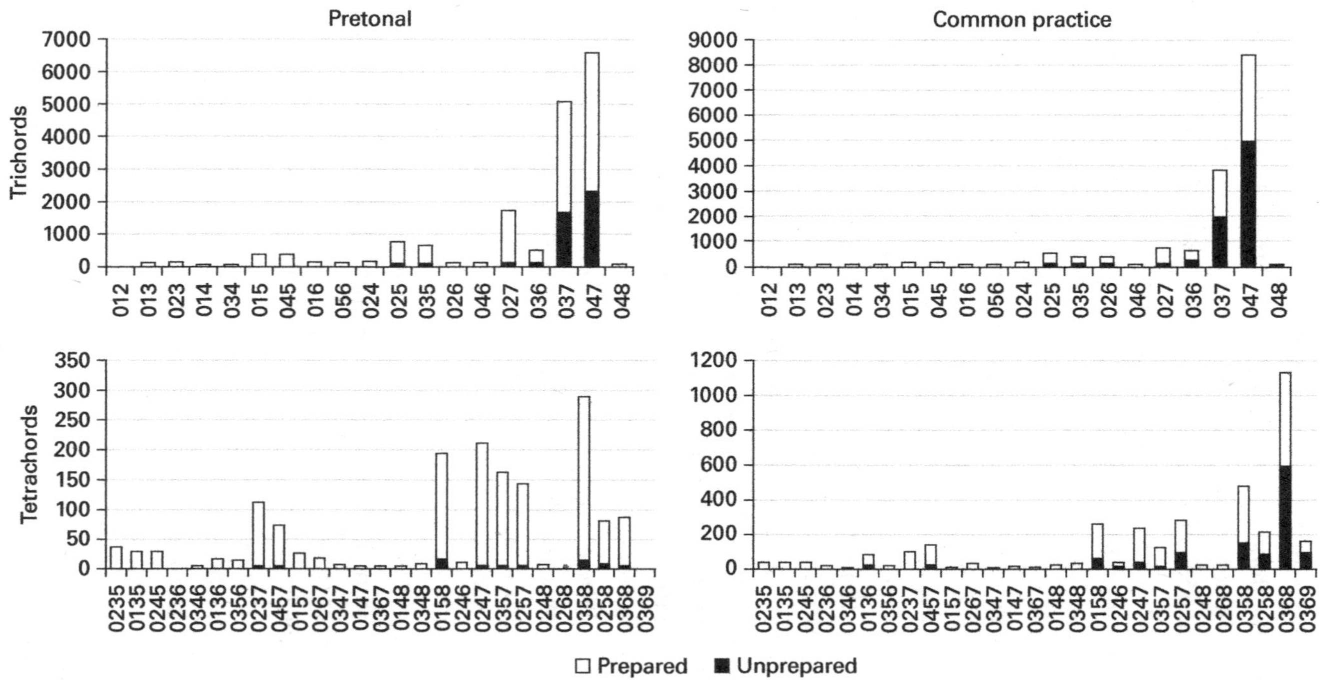

Figure 14.1

Prevalence of trichords (upper part of figure) and tetrachords (lower part of figure) in a database of Western notated vocal polyphony from the thirteenth to the sixteenth centuries ("pretonal") and from the seventeenth to the nineteenth centuries ("common practice"). The data for trichords also appear in Parncutt, Reisinger, et al. (2019, figure 3).

(being similar to a diatonic dominant 11th chord, V11), perhaps the music-theoretical idea of suspension needs revisiting.

The absence of the diminished-7th chord [0369] from this list may come as a surprise, given that any diatonic triad can be preceded by a secondary dominant that can, and often does, morph into a diminished-7th chord when the root is raised a semitone. Many readers will be familiar with the dramatic prominence of diminished-7th chords in the music of Beethoven, or their use in spooky silent movie soundtracks. Perhaps the relative rarity of this chord explains why it is so dramatically important.

Lists of this kind do not regard more complex triads (or tetrads) as variants of simpler triads (or tetrads). But the idea is interesting, and music theorists since Rameau have entertained it. The most important triads can be derived from the major triad, and the most important tetrads from the Mm7 chord, by shifting selected tones through semitone intervals. Tetrads can also be regarded as triads with added tones.

Still, such theoretical procedures do not necessarily reflect what actually happened in music history. For centuries, composers and performers have been trying out different tone combinations while tending, on average, to prefer the more over the less consonant. Sometimes, a composer consciously added a new tone to an existing chord; other times, a new chord emerged from voice leading. When a "new" chord becomes prevalent in a given style, we can regard it either as a collection of equally important pitch classes or as an existing chord with an added or shifted tone. An added 9th (add9) chord, for example, can be regarded either as a major triad plus a dissonant 9th (as it has been for most of music theory's history) or as four equally important tones that happen to sound together. In popular music and jazz, the latter interpretation may sometimes feel more correct, with no expectation that the 9th will resolve or should be resolved.

If we compare the frequency of occurrence of different possible triads and tetrads in music written before versus after 1600, we see a bigger difference for tetrads than for triads. Unprepared tetrads did not exist before 1600, but an automatic database analysis reveals various prepared tetrads—created more or less accidentally by voice leading. Another difference before 1600 was the relative absence of tritone intervals. Consequently, the Mm7 chord, with its tritone interval between the major 3rd and minor 7th—the most common tetrachord after 1600—was only number 6 in rank order of prevalence before 1600.

The most common unprepared tetrachords before 1600 were major or minor triads with extra tones created by voice leading. The most common tetrachord, in modern terminology, was the minor 7th [037T] (or equivalently added 6th [0479]). The chord could be created in various ways such as adding a passing 7th to a minor triad or a passing 6th to a major triad. Next in order of prevalence was the added 9th chord [0247], often created by adding a passing 9th to a major triad. Then came the major-7th chord [0158] = [047L], created by adding a major 7th above a major triad. Then came [0357], created by adding a passing 4th to a minor triad. After that was the suspended chord [0257], created by adding a 9th to a suspended triad. These combinations are music-theoretically unsurprising; the interesting claim is that the relative prevalence of these sonorities was mainly determined by their consonance, a combination of smoothness and harmonicity (Parncutt, Reisinger et al. 2019)—not necessarily by rules or principles of voice leading.

Having considered explanations for the prevalence of different tone combinations in MmT, we can return to a basic question: Why is MmT dominated by trichords? If the aim was to optimize consonance, we would expect simultaneities to be avoided and music to be dominated by unaccompanied melody—as in many non-Western traditions. The Western love for relatively consonant trichords can be explained if we like a moderate degree of dissonance—not too much and not too little—but such an explanation has ad hoc character: it does not predict how much dissonance is the preferred amount. Another reason for favoring trichords is that scales comprise seven tones per octave and that major and minor triads imply those scales via their pitch-class salience profiles (see figure 1.1), but the idea seems overly complex.

Major and Minor Triads

Having analyzed in detail the prevalence of different sonorities in Western polyphony and MmT, we can return to a basic question that has bothered music theorists for centuries: Why is MmT based on major and minor triads? If MmT is about prolongation of the tonic triad, why are there just two modes—major and minor—whereas there had previously been many more church modes?

Of the many theorists who tried to solve this problem in terms of number ratios (including Rameau [1722] and Riemann [1882]), all faced a basic problem. The ratio assigned to the minor triad (10:12:15) is more complex than the ratio assigned to other less common or more dissonant chords (e.g., the diminished triad, which is sometimes regarded as 5:6:7).

Another approach is to consider the harmonics of each tone. Rameau proposed that a tone can generate a chord: the 3rd and 5th harmonics of a harmonic complex tone generate the 5th and 3rd of the major triad, respectively. If there had been subharmonics, they might have generated the minor similarly. The trouble was, there were no subharmonics—at least not in the physical world.

Rameau's ideas were not always convincing (Christensen 1993). For example, if we imagine a major triad in root position and close position (e.g., CEG), the major 3rd (CE) lies below the minor 3rd (EG) within the framing 5th (CG); in the minor, the reverse is the case. Perhaps, thought Rameau, the order of the intervals does not matter, explaining why the minor triad is almost as important as the major for MmT. But that idea is as arbitrary as the idea that a vibrating body generates both harmonics and subharmonics or undertones, which at different times was held by both Rameau and Riemann and later abandoned (Christensen 1987; Mickelsen 1977; Riemann 1905).

In fact, all chords, and all musical sounds, including major and minor triads, depend on both nature and culture. Each musical chord that we now regard as a normal part of musical vocabulary emerged at some time in history. Before that time, it was not used intentionally, although it may have happened accidentally.

When we (Parncutt, Reisinger et al., 2019) analyzed a corpus that included twelfth- and thirteenth-century scores, we assumed that every tone in the score could be mapped onto a tone in the chromatic scale. The composers (or musicians to whom the scores are attributed) included Adam de la Halle, Alfonso el Sabio, Pérotin, and anonymous (Montpellier Codex). The analysis suggested how new chords originated and later became accepted as part of musical vocabulary in two steps:

- New tone combinations were first heard quasi by accident, although usually obeying existing rules of voice leading. Voice leading generated many sonorities that were not consciously created or intended by the involved composers or musicians. Prepared trichords such as [015] (e.g., BCE), [045] (e.g., CEF), [025] (e.g., CDF), and [035] (e.g., DFG) happened repeatedly in early polyphony, although the composers or musicians of the time had no vocabulary to describe them.

- After hearing many such "accidental" simultaneities, musicians started to get a feeling for them. Something clicked, and they started to use them deliberately. From then on, we may say that a chord exists as a sonic-cultural-psychological entity, even if it does not yet exist in the world of music theory or musical ideas.

There is nothing especially generative about this process. It may be more appropriate to regard the origin of specific chords as a multigenerational, cultural, experimental process. Each chord in today's familiar tonal harmonic vocabulary was created gradually by generations of musicians and composers as they experimented with different tone combinations in musical contexts. Much later, the chord in question was recognized and labeled by music theorists. In that sense, we may claim that major and minor triads existed for centuries before being labeled as such (Parncutt, Reisinger et al. 2019).

Return now to the question of why MmT is dominated by major and minor triads. Why were these two sounds preferred over other trichords? The simplest and most direct answer does not involve abstract physical, mathematical, philosophical, or mystical ideas. Rather, historic Western listeners simply liked these two sonorities. The tones were perceived to go together. So, the chords were perceived to be consonant. People also like dissonances of course, but in a first pass at the problem, we cannot fail but to be impressed by the long-term stability of preferences for major and minor triads, despite the enormous diversity of Western tonal styles. Of all possible trichords, the major triad (in semitones relative to the root: [047]) and the minor triad ([037]) are the most consonant for two reasons: only they include both a perfect 4th (or 5th) interval ([5] or [7] semitones, promoting harmonicity) and no 2nds ([1] or [2] semitones, avoiding roughness; Parncutt 1988).

An intriguing additional possibility involves prenatal perception of audible harmonics within the maternal voice (Parncutt & Chuckrow 2017). Martine Van Puyvelde and colleagues (2010) observed that, in early mother–infant singing interactions, the mother often sings tones from the major triad or harmonic series. That is unremarkable if the mothers are Western (in this case Belgian) and clearly learned the music from their culture. But what if we are trying to understand where central aspects of that culture itself came from?

> Cross-culturally, music has one common nature-related aspect: namely, sound. And almost every sound in our environment is a multi-tone complex consisting of a number of simultaneous pure tones called harmonics, nature tones or overtones.

As we will explain in more detail below, together with the pentatonic series—a five-tone series system that is deduced from the harmonic series—harmonics can be considered universal tonal aspects. The human voice also produces tones rich in harmonics. Already in the womb, an infant is surrounded by these natural ratios of the harmonic series produced by its mother's voice. A foetus is able to identify the maternal voice (Kisilevsky et al. 2003, 2009) that is internally transmitted via body tissues and bones (Lecanuet 1996). (Van Puyvelde et al., 2010, 388)

Whereas the augmented triad [048] is traditionally considered one of four basic triads, two other trichords without traditional names occur more often in tonal music. In semitones relative to a reference pitch, they are [025] and [035] (e.g., CDF and DFG; Parncutt, Reisinger et al. 2019). In pitch-class set theory, they belong to the same set, as intervallic inversions of each other. Joel Lester (1992) called them "minor-3rd-plus-whole-step" and presented a familiar example of their use as passing or neighboring sonorities: the opening bars of Mozart's Piano Sonata K. 331 in A major. Although they may be perceived as incomplete tetrachords (7th chords or added 6th chords)—being relatively unequally spaced on the pitch-class circle—they can also function as sonorities in their own right and emerge from common voice-leading patterns. They deserve a more prominent place in music theory.

Trichord [026] is also quite common and often functions as an incomplete Mm7 chord [0269]. The example clarifies that the prevalence of a sonority in music depends not only on its psychoacoustic smoothness and harmonicity when presented in isolation, but also on the smoothness and harmonicity (and hence prevalence) of more complex sonorities to which it belongs (or of which its pitch classes are a subset).

Avoiding Anachronism

Margaret Bent (1998) has argued cogently and forcefully against applying modern ideas (whether from music theory or music cognition) to early music:

Presentist anachronism and ahistorical formalism converge. Formalism cannot achieve the neutrality to which it aspires because, as far as early music is concerned, it is created out of an alien set of expectations. Both formalist approaches and those that seek the origins of tonal harmonic in "modal" early music set out to demonstrate qualities assumed a prior to be there, such as tonal coherence, goal-directed motion, prolongation, and elaboration, but without showing either from within that music or from relevant theory that these are appropriate categories to be looking for, without telling us why we should expect to find the particular of coherence

> that the analyst seeks to demonstrate in that particular repertory, and even worse, without setting out clearly the criteria of tonal schemes, consonance and dissonance, closure, tension and resolution by which such qualities, and deviations from them, are being defined. It assumes the outcomes of an evolutionary process of which medieval harmony was an early or primitive step, not a different world. (24)

Those are serious objections, and they warrant serious discussion. My psychohistorical approach applies principles based on perceptual quasi universals to early music. The intention is to be objective, but the choice of universals is inevitably biased by my experience of MmT. The principles are applied in a quasi-objective formalist way (based on pitch-class sets) to music of different historical periods on the assumption that the chosen universals could influence the structure of any music.

Despite these problems, the approach is promising. When the methods and data are clearly presented, the reader can evaluate them. The question of what perceptual factors influenced early music, and in what way, becomes a matter of interpretation. In any case, it should be clear that early music was and is primarily valuable for its own sake and not as something that was leading to something else. Early musicians heard music differently based on their music listening experience and could not guess how music would develop in the future.

In retrospect, it seems likely that both roughness and harmonicity influenced the perception of C/D in medieval music. Indeed, it would be surprising if they had not. In fact, C/D perception in early music may have depended more on psychoacoustical aspects of C/D than it depended on music-theoretical knowledge or compositional convention. The tables may have turned later as music became more complex and C/D perception depended increasingly on musical familiarity.

There may be a misunderstanding here about the difference between prescriptive and descriptive music theory. Much music theory in the humanities is prescriptive, implying how music *should* be written, or how we *should* hear it. The present approach is intended to be descriptive. The aim is to describe musical structure in a relatively systematic and objective way as a step toward understanding why the music is like it is.

There may also be a misunderstanding about determinism—the idea that history is predetermined by previously existing causes. The theory I have presented seems to imply that MmT was going to emerge anyway because it was inherent in the musical materials. But the emergence of MmT also depended on music's social (religious and secular) functions. Was it inevitable that Western music would be based on progressions of trichords and tetrachords, in which trichords predominated because they were optimally consonant or similar to a prenatally audible harmonic series? Questions of that nature are hardly answerable, given the complexity of society and its history. But it is difficult to avoid falling into a determinist trap when trying to explain a historical development reductively. Reductionist explanations can nevertheless be valuable. Clearly, history (including music history) is partly a result of preexisting causes and partly a result of unpredictable or pseudo-random events, including the free will of influential people. The extent to which free will is actually free is the subject of continuing debate in neuroscience and philosophy. Newtonian mechanics implies determinism, but it fails when distances are very small or velocities are very fast. Quantum mechanics is intrinsically nondeterministic.

Mathematical Approaches

Gerald Balzano (1980) conceived of musical intervals as:

> members of a group of transformations generating a pitch system C_n. Three isomorphic representations of C_{12} (our current system) bear striking parallels to the melodic, harmonic, and key relations as exhibited in tonal music, and the diatonic scale is revealed as a special subset of C_{12} that is in fact the simplest embodiment of the abstract relations given in two of the three representations . . . The three groups are conceived of as controlling pitch-set structure at nested time levels, with the space of "semitones" the most local, the space of "thirds" next, and the space of "fifths" the most global. (83)

Another mathematical approach to modeling tonality was presented by Elaine Chew (2000), inspired by the *Tonnetz* of Euler and Riemann:

> The Spiral Array generates representations for pitches, intervals, chords and keys within a single spatial framework, allowing comparisons among elements from different hierarchical levels. Structurally, this spatial representation is a helical realization of the harmonic network (tonnetz). The basic idea behind the Spiral Array is the representation of higher level tonal elements as composites of their lower level parts. The Spiral Array assigns greatest prominence to perfect fifth and major/minor third interval relations, placing elements related by these intervals in proximity to each other. As a result, distances between tonal entities as represented spatially in the model correspond to perceived distances among sounding entities. The parameter values that affect proximity relations are prescribed based on a few perceived relations among pitches, intervals, chords and keys. (abstract)

The explanatory power of such approaches is difficult to evaluate if they ignore not only the physical structure of the stimulus and empirically founded psychoacoustic foundations such as roughness and harmonicity, but also the historical development of tonal musical structures.

To illustrate the two dimensions of musical pitch (pitch register and pitch class) and their psychological reality, Révész (1913) visualized a helix with one rotation per octave. Shepard (1982) explained: "Rectilinear scales of pitch can account for the similarity of tones close together in frequency but not for the heightened relations at special intervals, such as the octave and perfect fifth, that arise when the tones are interpreted musically. Increasingly adequate accounts of musical pitch are provided by increasingly generalized, geometrically regular helical structures: a simple helix, a double helix, and a double helix would around a torus in four dimensions or around a higher order helical cylinder in five dimensions" (305).

In this approach, musical scales have underlying cognitive structures that can be discovered by multidimensional scaling of empirical data. But like other cognitive structures, the ontology of the musical pitch helix is questionable. Does it really exist in the musical mind? The helix model suggests that a major-7th or minor-9th interval is similar to an octave interval, which is clearly not the case.

The "heightened relations" that are observed at octave and 5th intervals can be explained more directly in terms of acoustics and psychoacoustics. The ear distinguishes the lower harmonics of a harmonic complex tone. Successive tones at octave and perfect 5th have audible harmonics in common. Simultaneous harmonic complex tones at octave and perfect 5th intervals create sonorities that are high in harmonicity and low in roughness. Shepard's finding that "increasingly adequate accounts of musical pitch" are provided by increasingly complex structures is not surprising: the more degrees of freedom a model has, the more data it can account for.

Shepard (1982) also commented that "the structural properties of pitch seem to be manifested even in purely melodic sequences of purely sinusoidal tones . . . In the absence of physical overlap of upper harmonics . . . wherein does this structure reside?" (306). The answer to this question is psychohistoric. Our perception depends not only on the structure of the stimulus that we hear in real time, but also on our experience of similar stimuli in the past and the contexts in which they were heard. People learn interval relations from music comprising harmonic complex tones and then apply that sound knowledge to patterns of pitch created by pure tones. That is an example of the ear's ability to learn large numbers of pitch patterns. We effortlessly recognize human voices (including their foreign accents) on the basis of complex patterns of pitch, loudness, timbre, and temporal structure. We can similarly remember and recognize thousands of melodies, regardless of whether they are played with pure or complex tones—independent of timbre, tempo, and even interval size—provided rhythm and contour are not affected (Dowling 1978).

In the ecological approach of James Gibson (1979), perception is direct, without cognitive processing. It is more realistic to regard pitch (whether musical or not) as a simple one-dimensional psychological scale from low to high, corresponding to the frequency of a pure tone. Certain intervals are privileged if they happen more often in both musical and nonmusical sounds; the octave is one of them. The term "octave equivalence" means that pitches an octave apart are perceived to be equivalent or strongly related; it does not imply that we have a pitch helix in our mind.

It is possible to explain aspects of a typical person's extensive sensory knowledge of music by invoking cognitive structures or representations, but I find it more parsimonious to imagine a library of remembered and recognized patterns. Because the patterns overlap, each pattern being similar to many other patterns, the structure of memory is a mixture of network and hierarchy (Fuster 1997). The patterns can be musical or nonmusical (e.g., the pitch patterns of the partials of complex tones). In both cases, the auditory system is capable of storing enormous amounts of information and recalling it effortlessly and instantaneously. Relative to that, the cognitive structures proposed by Shepard and Krumhansl are oversimplified.

It is sometimes unclear what mathematical approaches to musical structure can achieve that has not already been achieved by more direct, parsimonious theory. Take for example the problem of tracking changes in tonality as a passage modulates. Given the inherent difficulty of modeling this process, it is hard to demonstrate the quantitative or qualitative superiority of one model over another. Piet Vos (2000) explained: "The fuzziness of the key notion 'tonality' itself . . . leads to divergent conceptualizations of tonality induction, such as 'key finding,' 'tonal induction,' 'tonal center perception,' and 'tonal feeling.' I argue that the fuzziness is largely due to the lack of well-defined historical or otherwise categorical boundaries of 'Western tonal music.' The second impediment arises from various dualities in music, notably tone versus note representations, tonal versus intervallic representations, horizontal ('leading voice') cues for tonality induction versus vertical ones ('fundamentals'), and tonal versus temporal tonality induction cues" (abstract).

Neuroscientific approaches can also be problematic if not grounded in musical reality. Toso Pankovski and Ana Pankovska (2022) proposed that patterns of consonant and dissonant intervals emerge from a complex neural network. That raises some interesting questions. Do we need a new theory of the origin of intervals? Probably, scales and tuning emerged gradually as musicians became sensitive to pitch commonality of successive tones; changes in scale structures were passed from one generation to the next in oral-aural tradition. In a similar way, Western C/D came to depend on smoothness and harmonicity. Tuning was a compromise among partially contradicting criteria. But the predictions of Pankovski and Pankovska point toward Just intonation, which real music does not; evidently, the predictions arose from the Just intervals in the harmonic series with which the authors' Hebbian-learning network was trained. The theory, like other Just intonation theories, belongs to an ancient Pythagorean mystical tradition in which the elegance of mathematics is preferred over empirical observation.

A psychoacoustic theory of C/D that is based on what listeners directly perceive in real music (harmonicity, smoothness, familiarity) predicts the relative prevalence of different possible trichords and tetrachords without making assumptions about abstract cognitive structures. How one chord progresses to another is explained in terms of voice-leading principles, pitch commonality, and the relationship to the prolonged tonic triad—phenomena that are also perceived relatively directly.

15 The Root of a Chord

Chord roots are omnipresent in the theory and practice of MmT. Jazz standards and pop songs are notated as melodies with lyrics and chord symbols; the first character in a chord symbol (e.g., the C in C7) is the root. Schenker's Ursatz *is a prolongation of the tonic triad, the root of which is the tonic. How can the root be explained scientifically? Following Rameau, a musical chord can be compared with a harmonic complex tone that is perceived as one sound although it comprises many audible partials. The root of a chord is then analogous to the fundamental. In a musical context, the root is a pitch class that acts as a psychological pitch reference. Chords with relatively clear roots (major triads, Mm7 chords) and chords in root position tend to be more consonant and are therefore preferred and happen more often. Theories of the root that invoke simple frequency ratios or periodicity perception ignore the flexibility of musical intonation and underestimate the complexity of the underlying neurophysiology or the intractability of the mind-body problem. An ecological-psychological approach to musical intervals as intrinsically subjective, approximate pitch distances is more realistic. Musical intervals are first heard and learned within harmonic complex tones in voiced speech sounds and, in that sense, are intrinsically linked to virtual-pitch perception and chord roots.*

Early Chords

Chord roots have a long history in musical practice. A statistical analysis of a database of representative polyphonic scores (Parncutt, Reisinger et al. 2019) suggested that as early as the fourteenth century, simultaneities that were later recognized as chords were already appearing more often in root position than (chordal) inversion. Two contrasting explanations can be formulated. One refers to the history of music theory—in particular, the theory of intervallic C/D in the Renaissance (e.g., P. Schubert 1999, 19), according to which it was normal to voice 3rds and 5ths above the bass. Consonances were divided into perfect (unison, octave, 5th) and imperfect (3rd, 6th). If there was already a 5th above the bass, the only possible additional tone was a 3rd, because a 6th would create a dissonant 2nd between the 5th and 6th; the 4th or 2nd above the bass would also have been dissonant. Therefore, the most common voicings of three different pitch classes featured a 5th and a 3rd above the bass. Another common voicing with three pitch classes was a major 6th and a major 3rd above the bass for similar reasons; that was the penultimate chord of the fourteenth-century double-leading note or Landini cadence, and in later music, it was a first-inversion minor triad. The other explanation refers to modern perceptual theory. Voicing the root in the bass made sonorities sound more similar to a single harmonic complex tone and hence more consonant according to the principle of harmonicity. In the Middle Ages and Renaissance, composers and performers created their music not only according to existing rules of composition, but also following their musical intuition, which can today be modeled, at least in part, by theory of pitch perception.

Whatever the reason, the chord voicings that emerged over the centuries tended toward later theories of chord roots and chord voicings. That being the case, the psychological foundations of chord roots may be much older than the corresponding writings of music theorists. Said another way, chord roots may have been psychologically real for at least two centuries before they were discovered in Renaissance polyphony, after which theorists started to write about them. For example, French motet composer Jean Mouton seemed aware of triadic harmony, especially in his later works (in the early sixteenth century), but not of tonality in the sense of one triad acting as a reference. That suggests in turn that it takes a lot of nurture (cultural history) for listeners to start to perceive chords as unified sonorities in musical contexts, in their diverse voicings and arpeggiations, and subsequently to perceive those sonorities relative to reference pitches or roots.

History of Theory

One of the earliest music theorists to discover and verbalize chord roots was Johannes Lippius, who "undoubtedly stands foremost because of his clear and insistent reference to the three-note chord as the basis of all harmony. It was he who coined the term "trias harmonica" (Rivera 1978, 43). Here is what Lippius (1612/1977) had to say about the triad: "The harmonic, simple, and direct triad is the true and unitrisonic root of all the most perfect and most complete harmonies that can exist in the world. . . . The triad is the image of that great mystery, the divine and solely adorable Unitrinity (I cannot think of a semblance more lucid)" (41). This religiously colored statement is consistent with a prenatal origin of triads: the basilar membrane of the fetus starts to generate a running spectrum in which the lower harmonics of the maternal voice are separated (Parncutt 2019b).

Lippius also explained aspects of voice leading in triadic chord progressions on the implicit understanding that most triads would be in root position and that intervals between successive roots would often be 4ths and 5ths: "The bass, because it must always lay the foundation for the triads, is allowed to proceed slowly by such leaps. In this way various kinds of triads will more elegantly, more easily, and more wonderfully be mixed, combined, and arranged in proper order. Thus also, the upper melodies will be able to flow forth, as it were, with a more natural and more even gait" (Lippius 1612; as cited in Rivera 1984, 71).

Before Lippius, Johannes Avianius (1581) had used the term *basis harmoniae* for the bass and noted that the *basis* can sometimes appear in other voices and even the highest voice, anticipating the later concept of chordal inversion (which was also absent in Zarlino). Avianus also suggested (somewhat unclearly or inconsistently) that chords are constructed by building a 3rd (*media*) and a 5th (*summa*) on the *basis* to form a "perfect consonance" (Rivera 1978).

Psychological Approaches

There is surprisingly little music-psychological literature on chord roots. A possible reason is the Schenkerian turn in North American music theory in the 1970s and its influence on North American music psychology. In Schenker's linear approach to analysis, chord roots play a subsidiary role; perhaps for that reason, it became unfashionable to talk about them. Today, we can adopt a more balanced perspective. Political problems aside, Schenker's linear approach was no more or less important than the more vertical approach of Rameau or Riemann. The vertical and the horizontal always interact; to understand MmT, we need both.

Another possible reason is the failure of psychological studies around the middle of the twentieth century to verify the psychological reality of chord roots—perhaps due to a combination of music-theoretical misunderstandings and methodological issues. Chord roots are normally understood in music theory to be pitch classes—not specific pitches in specific octave registers: a C-major triad has root C and not C2 or C7. Chord roots are also assumed, tacitly or intuitively, to exist in lower rather than higher pitch registers—but not in specific registers at the expense of others. Attempts to investigate the perception of chord roots that used harmonic complex or pure tones in specific octave registers (e.g., Hurwitz 1970; Pritschet 1992; Reichweger 2010; J. R. Smith 1967) were unable to verify the root's existence clearly and tended to conclude that chord roots are culturally arbitrary with no psychoacoustic foundation.

Another problem is the experimental task, which should correspond to the expertise of the participants. Robert Hurwitz (1970) played simultaneities of two regular musical tones and asked listeners to "decide which one is the more important, that is, the one which seems to be the more fundamental of the two, or seems to include the other within itself" (17). One can only guess how participants responded to this instruction. Musicians may have tried to recognize the interval and respond on the basis of music theory. Nonmusicians may have responded to the most salient pitch, which in this case depended in a rather complex way on the amplitude spectra of the two tones, created using a Conn electronic organ. As we have seen, there is an important difference between salience and stability in music theory.

An additional problem is the duration of the sounds. In Hurwitz's experiment, they lasted for five seconds—much longer than a musically typical tone and inappropriately encouraging analytic listening (the hearing out of spectral pitches). Shorter tones (less than one second) would have been more appropriate, given the spontaneity of root perception in musical contexts. Moreover, spectral listeners may have responded differently from fundamental listeners.

In an experiment by John Platt and Ronald Racine (1990), a major triad of pure tones, in close position and one of three different inversions, was presented in each trial followed by two of the three chord tones, one after the other. Participants were asked which of the two tones was most similar to the chord. All possible tone combinations were presented. Some listeners, called root trackers, were more likely to find a tone corresponding to the conventional root to sound more similar to the chord; others, called melody trackers, were more likely to focus on the highest tone (the difference is reminiscent of fundamental versus overtone listeners; Seither-Preisler et al. 2007). The authors concluded that their results offered "considerable support for root tracking based on a missing fundamental for harmonic triad types, at least when the dominant melody tracking strategy is suppressed and the subjects have some musical training" (427). However, the experiments "did not reveal variations in the strength of root tracking with variations in triad position" (428).

If chord roots are pitch classes whose octave register is not defined, the chords and test tones used in an experiment should be created from octave-complex tones, with a pure tone in every octave register, ideally spread across the audible spectrum. Octave-complex tones, sounding like a pipe organ in a church with all the stops pulled out, were first synthesized by Shepard (1964); much later, Krumhansl and Kessler (1982) reported an experiment on the perception of MmT, in which chord progressions and probe tones had been created from Shepard tones. That experiment revolutionized psychological understanding of MmT. The empirical investigations into chord roots by Parncutt (1993) and Parncutt, Sattmann et al. (2019) similarly used octave-complex tones.

Stacked 3rds

Arguing that the root of a chord depends on the harmonic series, Rameau (1737) regarded the major triad and the Mm7 chord as basic chords from which other chords deviate by shifting one or more tones by a semitone. He also developed the idea of chord construction by stacked 3rds, the root being the lower note of the lowest 3rd. This idea, sometimes called tertian harmony (as opposed to quartal harmony), can still be found in today's music theory texts. It corresponds nicely (and perhaps coincidentally) to the lines and spaces of conventional music notation.

A psychological explanation might run something like this. Intervals of a 2nd are avoided in chords due to their roughness. At the same time, chords are preferred if they are similar to a harmonic series (harmonicity). Harmonics 4–7, which resemble a stack of 3rds, play an important role because they are clearly audible in typical voiced speech sounds and relatively close to each other (cf. the tones of a pentatonic scale). Beyond that, we can consider each stacked 3rd separately, but problems arise at every stage:

- Imagine a chord that comprises only two tones a 3rd apart: the first 3rd in the theoretical stack. We normally think that a perfect 5th interval is more consonant than a 3rd. Why start with a 3rd? A possible answer might refer to the 3rds and 6ths that predominate in two-part tonal counterpoint. Huron (2001) explained that in his three-part Sinfonias, J. S. Bach avoided octaves and 5ths because they tend to fuse, making it more difficult to hear the separate parts. The same idea can explain preferences for 3rds and 6ths in two-part writing.
- Two stacked 3rds create a triad. But major and minor triads are the most common tetrachords due to their consonance (presence of a perfect 5th or lack of major or minor 2nds)—not because of their stacked 3rds.
- Three 3rds create a 7th chord. But 7th chords can be explained differently. In the late sixteenth century, when the sound of major and minor triads was familiar (although theorists did not have a terminology or clear theory for them), composers such as Monteverdi (G. Chew 1989), exploring expressive tonal possibilities, tried adding a fourth pitch class to a major or minor triad. Guided by their ears and conventions of voice leading, seventeenth-century composers continued to favor tone combinations that sounded more consonant or less dissonant. If we start with the major triad CEG, the two most consonant possibilities involve adding A (creating an added 6th chord, a chordal inversion of a minor-7th chord) and B♭ (creating a regular Mm7 chord). Of the two, B♭ is not inherently superior.
- Similar principles can explain the origin of 9th, 11th, and 13th chords—but again there are caveats. If we wish to add a fifth pitch class to a Mm7 chord, the tone that minimizes the dissonance of the new chord will be a major 9th above the root. If we wish to add a sixth pitch class, cumulatively and according to the same criterion, the most consonant additional tone will be an augmented 11th; after that, a major 13th. Stacked 3rds do not *in themselves* explain why chords are constructed in this way. In a scientific approach, an argument based on C/D is preferable because C/D can be measured and experienced, whereas stacked 3rds is an abstract idea.

The idea of stacked 3rds can help us imagine and list various harmonic possibilities. It also helps us label and notate them. But it does not itself explain why chords in MmT are like they are. A psychological theory of C/D can do that more convincingly. The tones that are usually added to existing, familiar chords are simply those that maximize consonance (harmonicity) and minimize dissonance (roughness). The underlying principle is C/D, not stacked 3rds.

Stacked 3rds do not consistently explain which chord tone is the root. The root is often ambiguous, and stacked 3rds can explain only one option. A diminished triad BDF has three possible roots—G, B, and D—but stacked 3rds account for only one (B). A second-inversion major triad that proceeds to a dominant–tonic cadence (e.g., GCE-GBD-CEG) can be perceived as a 6/4 to 5/3 resolution, implying that the bass tone of the first chord (the conventional 5th of the chord) functions as the root (some argue that GCE is not even a chord here), whereas stacked 3rds suggest the root is C. Stacked 3rds cannot account for this familiar voice-leading pattern, nor can they account for the ambiguity of the minor-7th chord: a first-inversion minor triad or minor-7th chord on the supertonic (ii^6, $ii^{6/5}$) can sound like a root-position added-6th chord on the subdominant (iv^6, $iv^{6/5}$).

Chromatic Chord Symbols

One might conclude from this that the stacked 3rds idea should be abandoned. Instead, we should label chords relative to 12-EDO, which also avoids the problem of ambiguous enharmonic spellings. I am not advocating that because the stacked 3rds approach also has clear advantages. At the very least, it is well known.

It is nevertheless interesting to imagine what a musical world without stacked 3rds would be like. I once experienced that world as a piano accompanist for pop singers. When my singer asked me to learn a new song, I wrote down the chords relative to the chromatic scale, inspired by pitch-class set theory.

Here's an example. Let's say I didn't know the jazz standard *All of Me* (written by Seymour Simons and Gerald Marks in 1931) and needed to learn it in a hurry. One option would be to write down the chords in C major, like this:

C _ E7_ A7 _ Dm7 _
E7 _ A-7 _ D7 _ Dm7 G7
C _ E7_ A7 _ Dm7 _
F F♯^{o7} C/G A7 D7 G7 C Dm7-G7

Here, I am using the underscore symbol (_) to mean "repeat the previous chord" (halving the harmonic tempo) and the hyphen (-) to mean the opposite: chords joined by the hyphen are played in half the time (with twice the usual harmonic tempo). The note following a slash symbol (/) is the bass note if the chord is inverted. Note also that I am not including melody notes in these symbols. For example, the melody above the 4th chord Dm7 includes a suspended 4th (G) which is not included in the chord symbol.

Another option would be to use scale-step symbols from music theory and think in terms of intervals relative to the tonic. In the following, lower case means minor:

I _ III7_ VI7 _ ii7 _
III7 _ vi7 _ II7 _ ii7 V7
I _ III7_ VI7 _ ii7 _
IV ♯iv^{o7} I/V VI7 II7 V7 I ii7-V7

Yet another option would be to write down the progression chromatically, with the chord roots in semitones relative to the tonic and the other chord tones as superscripts relative to the root, like this:

0^{47} _ 4^{47T} _ 9^{47T}_ 2^{37T} _
4^{47T} _ 9^{37T} _ 2^{47T} _ 2^{37T} 7^{47T}
0^{47} _ 4^{47T} _ 9^{47T}_ 2^{37T}
5^{47} 6^{369} $0^{4}/7$ 9^{47T} 2^{47T} 7^{47T} 0^{47} 2^{37T}-7^{47T}

It is also possible to abbreviate these symbols. There is no need to write "47" for a major triad if it is clear that a chord without a chord symbol is a major triad:

0 _ 4^{T} _ 9^{T}_ 2^{3T} _
4^{T} _ 9^{3T} _ 2^{T} _ 2^{3T} 7^{T}
0 _ 4^{T} _ 9^{T}_ 2^{3T}
5 6^{369} 0/7 9^{T} 2^{T} 7^{T} 0 2^{3T}-7^{T}

As a performing musician, I found this system convenient and clear, especially in cases of enharmonic ambiguity. The approach trained me to think relative to the tonic when playing the same piece in different keys; I could also have done that with Roman numerals but was attracted by this elegant solution to the problem of enharmonic equivalence. The idea that the note a major 3rd above $\hat{3}$ (part of III) is the same as the note a minor 3rd above $\hat{4}$ (within iv) seemed not only practical, but also theoretically appropriate.

Origin of the Root

Hindemith (1937) argued that the root of a chord is a combination tone, drawing on Guiseppe Tartini's (1754) observation that two simultaneous tones can produce an audible third tone that violinists can use for tuning purposes. Later psychoacoustic research demonstrated that Tartini's *terzo suono* arises from distortion and nonlinearity on the basilar membrane in the inner ear (J. L. Hall 1974; Goldstein 1967).

Combination tones can be heard by a violinist whose left ear is close to the vibrating body of the violin, and by performers in recorder (Blockflöte) ensembles. But they are seldom audible for audiences in concerts of classical music, given their low intensity at typical listening levels, and psychoacoustic masking by other tones (Plomp 1965; Smoorenburg 1972). For that reason, they can hardly explain the root of a chord. Almost by definition, a sound that is not audible cannot contribute to auditory perception. Other reasons to doubt a theory of the root that is based on combination tones include their dissonance when heard in musical contexts, their tuning (Tartini to the contrary, in practice they usually do not sound in tune, even when the tones that produced them sound in tune, due to familiarity with 12-EDO and octave stretch), and the failure (so far) of a theorist to predict the roots of diverse chords on that basis.

Terhardt's (1974a) theory of pitch perception offered an alternative that was consistent with psychoacoustic data about the pitch of complex sounds. Chord roots depend on both nature (perceptual universals) and nurture (cultural learning). From the natural viewpoint, chord roots are analogous to the pitch of a complex tone; on average, their perceptual salience is greater than that of other chord tones. But complex tone perception also depends on a learning process. The auditory system is constantly exposed to pitch patterns, including patterns of musical pitches. As the auditory system learns to recognize invariances in those patterns, it uses them to recognize sound sources.

The harmonic series is perceptible among the audible harmonics of harmonic complex tones in voiced speech sounds and other environmental sounds, but it is often incomplete. First, some partials are masked by other sounds. Second, the spectra of some natural sounds, such as bells, are only partly harmonic: some partials correspond to an incomplete harmonic series, while others do not. Consequently, the pitch of a complex sound often corresponds approximately to the fundamental of an incomplete harmonic series of audible partials that does not include all the sound's partials. In a musical chord, the harmonic pattern is incomplete if not all tones in a chord correspond to harmonics of some fundamental. In the minor triad, for example, the root and 5th correspond to harmonics 2 and 3 (respectively) of the conventional root, but the minor 3rd does not correspond to an audible harmonic.

Terhardt (1982) presented a simple algorithm that predicted the root of most chords. Since the root is defined as a pitch class (not a specific pitch), the algorithm was entirely octave generalized. But the model still had trouble with the minor triad. My solution (Parncutt 1988) was to weight the five root-support intervals relative to each other (see table 15.1). Those intervals are the octave/unison, 5th, major 3rd, minor 7th, and major 2nd/9th.

These models were computer implementations of the following train of thought. There are twelve intervals in the chromatic scale. Five of them (called root supports) determine chord roots, as follows:

- The root is often determined by the perfect 5th interval (in semitones: [7]) between harmonics 2 and 3 (e.g., the tones C and G in the major triad CEG or [047]). If a chord contains two tones a perfect 5th (or 4th) apart, the lower tone of the 5th (or the upper tone of the 4th) is usually the root or at least a strong root candidate.
- The major-3rd interval [4] between harmonics 4 and 5 can also support the root. It can explain why the root of the major triad [047] (and hence the tonic of a passage in a major key) is clearer (less ambiguous) than the root of the minor triad [037] (and the tonic of a minor key). But the contribution of the major 3rd [4] to the root is weaker than that of the 5th [7].
- The minor 7th [T] can support the root, but to an even smaller extent. That can explain why the root of a Mm7 chord [047T] is less ambiguous than that of a major-7th chord [047L]. The Mm7 can appear in any inversion; its root is always clear. The root is also clear when the chord sounds (in any inversion) just before another chord, acting as a secondary dominant or augmented-6th chord. Other 7th chords do not have this property.

- The major-2nd (or major-9th) interval [2] can slightly strengthen the root but even less so than the minor 7th [T]. For example, the root of a regular dominant 9th chord (G9 in the key of C major) becomes less clear if the 9th is flattened to create a dominant minor-9th chord [0147T] (e.g., G7♭9).
- Usually, the root is present as one of the tones of the chord. In other words, the perfect unison [0] is a root-support interval. Because the presence of the root has such a big effect, the unison is the strongest root-support interval. If the root is not physically present, at least two of the other tones must point to it, otherwise it would be too ambiguous to perceive. If, for example, the diminished triad BDF [036]) is perceived as having the root G [8] (as in Rameau's theory of fundamental bass; Christensen 1993), it is because all three chord tones lie at root-support intervals above G.

In jazz, a chord can be created from a tritone interval, heard as the 3rd and 7th of a Mm7 chord (sometimes called guide tones and equivalent to a Mm7 with missing root and 5th). Usually, there are also one or more tension tones. Depending on context, the tritone interval alone can be enough to determine the root, but the root becomes clearer if it is physically added.

Terhardt's theory of pitch perception takes Rameau's original idea and makes it more concrete, allowing more aspects of harmony, progression, and tonality to be explained. The root-support intervals are derived from the harmonic series that is audible in voice speech sounds. The theory is not limited to an analysis of conventional triads but makes reasonable predictions for any simultaneity in the chromatic scale. Nor is its relevance confined to MmT: "the recognition of non-traditional options such as the use of 10 (minor 7th) and 2 (major 9th) as functional consonance would appear crucial for the analysis of music composers such as Scriabin and Debussy" (Väisälä 2002, 271).

Modeling the Root

The root-support weights, as used in Parncutt (1988) and this book, are shown in table 15.1. The weights are only rough estimates, but they suffice for music-theoretical purposes. They cannot be measured directly in an experiment, but it may be possible to adjust them incrementally by trial and error to improve the correlation between empirical data and predictions on the basis of this

Table 15.1
Root-support intervals and their relative importance ("weight") in the chord-root algorithm of Parncutt (1988) and in the calculations in this book.

Root-support interval	Unison (octave)	Perfect 5th	Major 3rd	Minor 7th	Major 2nd	Minor 3rd	Other intervals
Weight (1988)	1	1/2	1/3	1/4	1/5	1/10	0
Weight (here)	10	5	3	2	1	0	0

model (as attempted in Parncutt 1989, 1993). The auditory system is assumed familiar with the harmonic series in voiced speech sounds in which typically about ten harmonics are aurally relevant.

We can create an appropriate octave-generalized model by first transposing the harmonics down by octaves until they lie in the same octave register. The result is the list of root-support intervals in the table. The perfect unison is the most important for two reasons: the corresponding partials are most clearly audible in typical speech sounds, and several partials are octave equivalent with the fundamental (harmonic numbers 1, 2, 4, and 8). A high value for this weight (10 in the table) is consistent with the music-theoretical observation that the root of a chord is usually one of the chord tones—chords with missing roots being the exception rather than the rule. The perfect 5th interval is assumed to support the root (or the fundamental of a harmonic complex tone) to about half the extent that the perfect unison does (with weight 5 rather than 10); it may be less clearly audible in typical voiced speech sounds and is represented only twice among the first ten harmonics (at harmonic numbers 3 and 6). The remaining intervals are understood to be progressively less important for complex tone perception and hence for the perception of chord roots.

Table 15.2 shows how Terhardt's concept can be used to predict the root of a major triad. Each row in the table represents the subharmonics of one of the chord tones, expressed as pitch classes. The subharmonics do not exist physically in the sense implied by harmonic dualists, but they do exist in the world of subjective experience as potential virtual pitches on the assumption that a virtual pitch is generally a subharmonic of a salient spectral pitch.

The first row of the table shows that subharmonics of C are C, F, A♭, D, and B♭. Each subharmonic is weighted according to table 15.1. The next two rows are obtained by transposing the first row to get the subharmonics of E and G. The last row, labeled CEG, is the sum of the previous three rows.

Table 15.2
Calculation of the pitch-class salience profile of a C-major triad according to Parncutt (1988, with revised root-support weights).

	C		D		E	F		G		A		B
C	10	0	2	0	0	5	0	0	3	0	1	0
E	3	0	1	0	10	0	2	0	0	5	0	0
G	5	0	0	3	0	1	0	10	0	2	0	0
CEG	**18**	**1**	**3**	**3**	**10**	**6**	**2**	**10**	**3**	**7**	**1**	**0**

The procedure is a simple mathematical realization of the idea that a pitch in a chord is more likely to be salient, and to function as the root, if it corresponds to a subharmonic of one or more of the chord tones.

The last row of the table suggests that the root is C—not E or G. The root can nevertheless be perceived as E or G if that tone is in the bass and the voice leading points to that interpretation. For example, in the chord progression EGC-EGB, the root of the first chord can be heard as E if the C above it is a dissonance that resolves to B. Similarly, the progression GCE-GBD can be interpreted as a double suspension on G (6/4 to 5/3), G being the effective root of both chords.

Apart from the chord tones, the most salient pitch classes in a C-major triad are predicted in table 15.2 to be A and F. If A is added in the bass, the chord becomes Am7. If F is added in the bass, the chord becomes Fmaj9 with missing 3rd. The other pitch classes that are implied by the chord are predicted to be relatively weak; little would be gained by trying to interpret them musically.

Table 15.2 can be used as a model for predicting the pitch-class salience profile of any chord. Readers are encouraged to try that out and think about how the results might be interpreted in each case. Those who are wondering whether this procedure generalizes to any combination of pitch classes are asked to check the experimental tests of Parncutt (1993) and Parncutt, Sattmann et al. (2019).

Table 15.3 shows how the theory predicts the pitch-salience profile of a C-minor triad. The last row suggests that the chord has two possible roots, C and E♭, consistent with music theory. For example, the root-position minor triad often has the function ii, which is often realized in first inversion as ii^6. In classical MmT, chords ii and IV have a similar function as diatonic

Table 15.3
Calculation of the pitch-class salience profile of a C-minor triad according to Parncutt (1988).

	C		D		E	F		G		A		B
C	10	0	2	0	0	5	0	0	3	0	1	0
E♭	0	1	0	10	0	2	0	0	5	0	0	3
G	5	0	0	3	0	1	0	10	0	2	0	0
CE♭G	**15**	**1**	**2**	**13**	**0**	**8**	**0**	**10**	**8**	**2**	**1**	**3**

predominants. The first-inversion minor triad (ii^6) can have the same root as a root-position major triad (IV), namely $\hat{4}$.

After C and E♭, the next pitch class in order of salience in table 15.3 is G, the third chord tone. After that come the tones F and A♭, both of which can be implied as roots of a C-minor triad. If F is added in the bass, the chord becomes an F9 chord with missing 3rd. If A♭ is added, the chord becomes A♭maj7. The remaining pitch classes are relatively unimportant.

In summary, the model predicts that the root of a C-minor triad is usually C, but it can also be E♭. The chord also implies roots F and A♭, and the scale C E♭ F G A♭ B—similar to the C harmonic minor scale. That is consistent with the idea that a passage of music in a minor key is a prolongation of its tonic triad.

The low predicted salience of D is surprising because this tone often goes with the chord as a scale step or added tone (added 9th), whereas D♭ does not. But the model only predicts virtual pitches. There may also be audible spectral pitches in a typical C-minor chord. They include pitch-class D, which is the 3rd harmonic of G and the 9th harmonic of C (see chapter 16).

The important role of the minor triad in MmT can now be explained. The minor triad is one of many possible combinations of pitch classes. It is the second most consonant triad after the major because, like the major, it includes a perfect 5th interval but no major or minor 2nds. That gives it high harmonicity and low roughness.

The minor-triad problem bothered music theorists for centuries. The solution that I am proposing is parsimonious and grounded in musical and psychological reality: the approximately tuned chromatic scale and psychological research on consonance and dissonance perception.

Table 15.4 analyzes the C-diminished triad CE♭G♭ using the same procedure as before. The analysis suggests that the chord has four possible roots: the three chord tones and the missing root at A♭, consistent with music theory. Next in line is B, of which E♭ and G♭ correspond to the major 3rd and perfect 5th, respectively (enharmonic differences playing no role). Adding that tone creates a B (or C♭) major triad. Then there is F, which when voiced in the bass creates a m9 chord with missing 3rd. Finally, adding D in the bass makes the chord a D7♭9 chord with missing 5th.

Table 15.4
The pitch-class salience profile of a C-diminished triad (after Parncutt 1988).

	C		D		E	F		G		A		B
C	10	0	2	0	0	5	0	0	3	0	1	0
E♭	0	1	0	10	0	2	0	0	5	0	0	3
G♭	0	0	3	0	1	0	10	0	2	0	0	5
CE♭G♭	**10**	**1**	**5**	**10**	**1**	**7**	**10**	**0**	**10**	**0**	**1**	**8**

Table 15.5 shows the pitch-salience profile of a C7 (Mm7) chord, CEGB♭. The striking thing about this chord is the clarity of its root, that is, its lack of root ambiguity. The last row of the table suggests that no pitch other than C can reasonably compete for the status of root, but there are several possible secondary roots. From strongest to weakest, they are E♭, A, F, G♭, A♭, and D, which can be compared with the blues scale on C. One well-known version of that scale is C E♭ F F♯ G B♭ (cf. figure 17.1). The list includes all tones of the C7 chord except E, which if added would produce adjacent half steps (Pressing 1978). The blues scale is also related to the standard pentatonic scale C

Table 15.5
The pitch-class salience profile of a C7 chord (after Parncutt 1988).

	C		D		E	F		G		A		B
C	10	0	2	0	0	5	0	0	3	0	1	0
E	3	0	1	0	10	0	2	0	0	5	0	0
G	5	0	0	3	0	1	0	10	0	2	0	0
B♭	2	0	0	5	0	0	3	0	1	0	10	0
CE♭GB♭	**20**	**0**	**3**	**8**	**10**	**6**	**5**	**10**	**4**	**7**	**11**	**0**

E♭ F G B♭. The tones E♭, F♯, and A in the blues scale may function as rising leading tones to chord tones in a tonic Mm7 chord (see chapter 8).

Figure 15.1 shows calculated pitch-class salience profiles for the more common trichords in MmT. Although the predicted profiles are intrinsically approximate (given the approximations inherent in the modeling and ignored effects of specific musical experience), they are nevertheless consistent with empirical data (Parncutt and Radovanovic, 2023; Parncutt, Sattmann et al. 2019).

The profiles of the major and minor triads resemble the key profiles of Krumhansl and Kessler (1982). The model accounts for the high salience of the minor 3rd in the minor triad and hence the high stability of the minor 3rd in the minor scale. The role of the leading tone (eleven semitones above the nominal root) is not accounted for: in the major triad, predicted salience at eleven semitones is lower than at three, six, eight, or ten semitones. For the minor triad, predicted salience at eleven semitones is relatively high but for the wrong reason: the leading tone is enharmonically equivalent to a major 3rd below the minor 3rd. A better model of major and minor scales might separate the missing-fundamental or virtual-pitch effect from the leading-tone effect, regarding these as two separate components with different psychohistorical origins.

Returning to figure 15.1, the most salient predicted pitch in the suspended-4th triad is the 4th itself: the 5th of the chord lies a major 2nd above the 4th, and the major 2nd is a weak root-support interval. But this chord usually appears with the [0] of [057] in the bass. The reason may be due to voice leading: the resolution onto a major triad [047] involves both relaxation (falling) and a small interval (a semitone). Another salient pitch in [057] is [10], the minor 7th, which goes well with the chord, creating a sus7, often on $\hat{5}$.

The diminished triad [036] is an interesting case. The predicted profile has four peaks, all about equally salient: the three chord tones and the missing root [8], which when added to the chord creates a Mm7 chord. This implication is clearest when the chord appears on the leading tone in a major or minor key; it disappears if the implied root contradicts the prevailing scale, such as when the chord is the diatonic supertonic triad in a harmonic minor key (ii°). The other salient tones are [11], which turns the chord into a major triad with added minor 9th, and [5], which often acts as a tone of resolution.

The chords [025] and [035] happen remarkably frequently in MmT, although they have no theoretical names. In both cases, the implied root is

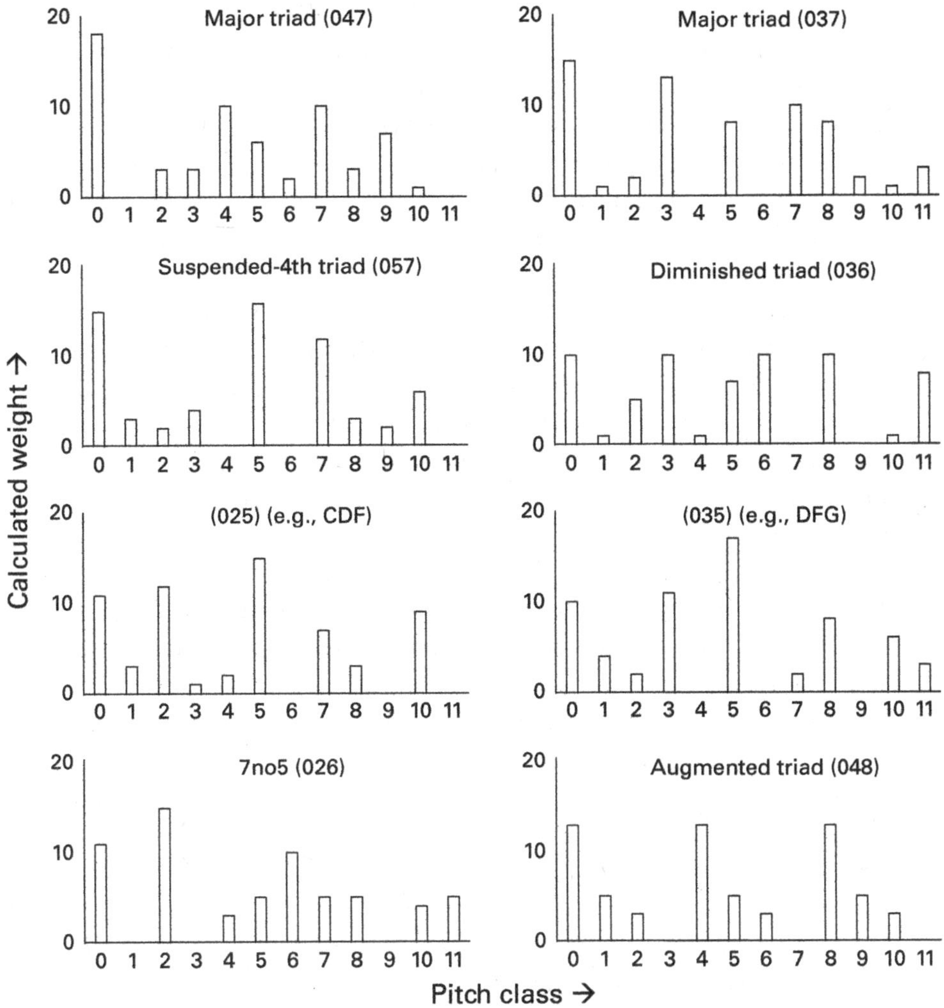

Figure 15.1
Calculated pitch-class salience ("weight") profiles of common trichords in MmT according to Parncutt (1988), with root-support weights 10 for unison, 5 for perfect 5th, 3 for major 3rd, 2 for minor 7th, and 1 for major 9th.

[5]. Beyond that, [025] implies [7] and [10], and [035] implies [8] and [10]. Adding these tones creates a standard pentatonic scale. The reason why these chords have no names in music theory may involve spacing around the pitch-class circle (Tymoczko 2010): the chords are perceived as incomplete because they cover the circle so unevenly.

A pertinent example of [025] can be found in the first few bars of the original recording of *(They Long to Be) Close to You* by the Carpenters. It is the first sonority we hear, and it resolves immediately to [047] as the lowest tone falls through a whole tone, implying that the [025] had been an added 9th chord with missing root. As the music progresses, the [025] increasingly takes on a life of its own, inviting us to hear it as a sonority in its own right. The chord even ends the short instrumental introduction as if it were a consonance, alongside the other conventional dissonances that are treated as consonances in this style: Mm7, major 7th, minor 7th, added 9th, and dominant 11th.

The augmented triad [048] is symmetrical relative to the pitch-class circle. A possible root is a semitone above one of the chord tones, which when added creates a minor triad. The chord 7no5 [026] is an incomplete Mm7 chord (here, [9] is missing) with root [2].

As shown in figure 15.2, the Mm7 chord [047T] has a clear root, consistent with its frequent appearance in all chordal inversions; none of the chord's four inversions threatens the supremacy of the conventional root. The pitch-class salience profile of the Mm7 chord is reminiscent of the blues scale, just as the pitch-class salience profile of the major and minor triads is similar to the major and minor scales. The root of the m7 chord [037T] is ambiguous ([0] or [3]), consistent with its frequent appearance as an added 6th chord; a possible missing root is [8], creating a major-9th chord. Similarly, the maj7 chord [047L] has a possible missing root at [9], creating a minor-9th chord. The profile of the added-9th chord [0247] is similar to that of a major triad [047]. The root of the 7sus chord [0257] is ambiguous, with three main root candidates [0], [5], and [T]—consistent with the chord's floating or static quality. The main root of the half-diminished chord [036T] is [3], consistent with its frequent appearance in inversion as a m6 chord (minor triad with added major 6th); a common missing root is [8], which turns it into a dominant 9th chord. The 7♯5 chord [048T] conforms to the whole-tone scale; the absence of a perfect 4th or 5th interval means there is no clear root. In a musical context, the root is determined by context and voicing. Finally, the

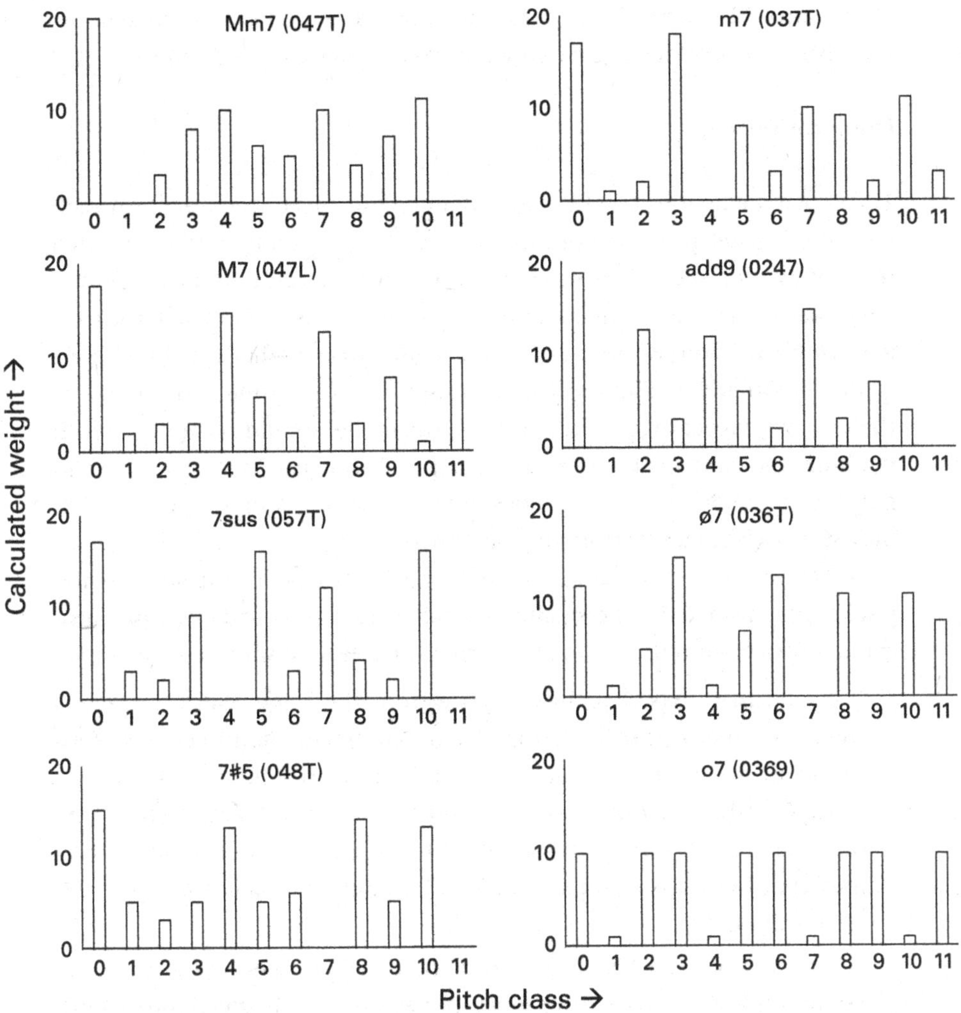

Figure 15.2
Calculated pitch-class salience profiles of common tetrachords in MmT according to Parncutt (1988), with root-support weights as before.

diminished-7th chord o7 [0369] is ambiguous, as expected from its symmetry, with a possible missing root a semitone below each chord tone.

Empirical Testing

The above predictions can be tested in a psychological experiment. In each trial, the listener hears a chord followed by a single tone; both are created from octave-complex tones and randomly transposed around the pitch-class circle to avoid effects of pitch register. The listener is asked how well the tone goes with the chord or (in a different experiment) whether the tone is part of the chord, and responds using a rating scale. If the tone corresponds to the root, we expect high ratings. Ideally, the model should account for both masking and harmonic pitch-pattern recognition (Parncutt 1993), but in practice, a simpler model that does not consider masking (Parncutt 1988) may suffice (Parncutt, Sattmann et al. 2019).

But the tonal expectations following a chord depend not only on the pitches perceived in the chords and their saliences. They also depend on musical experience—and these two approaches often lead to similar predictions.

- Listeners may expect to hear a tone from the same diatonic scale as the chord, diatonic scales being so familiar, having dominated Western music for millennia. A C-major triad is diatonic in the major keys of C, F and G. After hearing it, we may expect tones that belong to the same scale, such as D, F, A, or B—but also F♯ (if the chord is heard as IV in G major) or B♭ (if it is heard as V in F major). But we would not expect C♯/D♭, D♯/E♭, or G♯/A♭.
- Listeners may expect to hear chords that include tones lying a perfect 5th or 4th interval away from tones in previous chords, given the overriding importance of these intervals (after the octave) in tonal music. After a C-major chord, for example, we may expect to hear the tones D, F, A, or B, but not C♯D♭, D♯/E♭, F♯/G♭, G♯/A♭, or A♯/B♭. Expectations based on this second principle differ little from those based on the previous diatonic principle because diatonic scales are based on perfect 5th and 4th intervals.
- Listeners may expect to hear tones that are implied by the chord as missing fundamentals. For example, the C-major triad implies D, F, and A as missing fundamentals: C and E correspond to harmonics of D, C and G

correspond to harmonics of F, and E and G correspond to harmonics of A. According to this principle, a C-major triad does not imply C♯/D♭, D♯/E♭, F♯/G♭, G♯/A♭, A♯/B♭, or B. There is evidence for a weak but significant effect of this kind in both perceptual experiments (Parncutt, Sattmann et al. 2019) and statistical database analyses (Parncutt and Reisinger 2021).

- Listeners may expect to hear a completion tone—a tone that completes a familiar chord of greater cardinality (the number of pitch classes it contains). If we hear a C-major triad, we may expect to hear one of the following: B♭, which if added to the chord would create a familiar C7 chord; A, which would create a C6 chord; B, to create Cmaj7; or D, to create Cadd9. According to this principle, we are less likely to expect F, C♯/D♭, D♯/E♭, F♯/G♭, or G♯/A♭.

These predictions overlap, but they can be separated to a limited extent. The results of a perceptual study in which listeners heard a chord followed by a tone and were asked how well the tone went with the chord (Parncutt, Sattmann et al. 2019) suggested that for isolated triads, all four theories influenced expectations. If, however, listeners were asked whether the following tone was in the chord, two of the four principles stood out: listeners were more likely to respond in the affirmative if the tone stood in a 5th relation to a chord tone or corresponded to a missing fundamental of the chord.

Results of a statistical analysis of tonal music from the thirteenth to the nineteenth century (Parncutt and Reisinger 2021) produced similar results. Tones that immediately preceded or followed chords in tonal music belonged most often to the same diatonic scale or stood in 5th relations to the chord. Those were the most important two of the four principles. In addition, missing fundamentals and completion tones tended to happen more often than other possibilities.

These results are consistent with the assumption that musical chords imply pitch-salience profiles, and that scales are perceived as compatible with chords if they correlate with those profiles. But chord-scale compatibility is more complex than that. There is also a tendency for a compatible scale to include tones that are a semitone lower than chord tones, functioning as lower chromatic neighbors (rising leading tones). If that is true, compatible scales are determined by a mixture of two effects, one harmonic and one melodic (see chapter 17).

Reflection

The results of Parncutt, Sattmann et al. (2019) were consistent with the prediction that CEG implies F and A, ACE implies the tones D and F, and so on for selected chords. No data contradicted these predictions, despite overlapping predictions based on familiarity with perfect-5th intervals, diatonic scales, and common tetrachords (chords of four pitch classes). But the effects were small and on the verge of significance. Could something else have caused them? If so, do the hypothesized virtual pitches really exist? The following arguments suggest that they do:

- *Simple logic*: When we hear a voice on the telephone, we only hear higher harmonics. The telephone doesn't transmit the fundamental of a typical speaking voice. We nevertheless hear the pitch at the missing fundamental. There is no reason why we would not also hear pitches at missing fundamentals in musical chords.
- *Music theory*: The predicted implied pitches correspond to missing roots in mainstream music theory and provide a parsimonious explanation.
- *Generalizability*: The theory that predicts these pitches also predicts diverse nonmusical psychoacoustic phenomena (Terhardt et al. 1982a).

16 Spectral Pitch in Chords

The timbre of a musical sound depends on the pitch pattern created by its audible partials—its spectral-pitch pattern. *In a musical chord, that pattern comprises audible harmonics of chord tones (spectral pitches). A simple back-of-an-envelope analysis shows that typical musical chords evoke spectral pitches that are not octave equivalent with chord tones. For example, a C-major triad CEG [047] evokes spectral pitches at B [L] (3rd harmonic of E; 5th of G) and D [2] (3rd harmonic of G; 7th of E; 9th of C). In composition or performance, these tones may be added to the chord to create a new chord of higher cardinality (more pitch classes). The new chords are Cmaj7 or Cadd9, and the change in timbre is relatively small because the new tone corresponds to a salient spectral pitch that is already present in the triad. Any chord can be analyzed and extended in this way; for example, CDF [025] has audible partials at G and A, which can be added without changing the chord's basic character.*

The Audible Spectrum of a Chord

The amplitude spectrum of a musical chord typically comprises a large number of partials (pure-tone components). Some are inaudible because they mask each other. The audible partials are heard as spectral pitches.

These pitches are interesting for harmony theory if some of them represent pitch classes that differ from the notes. A C-major triad includes audible spectral pitches at B (3rd harmonic of E, 5th harmonic of G) and D (3rd harmonic of G, 7th of E, 9th of C). A C-minor triad includes audible spectral pitches at B♭ (3rd harmonic of E♭, 7th of C) and D (3rd harmonic of G, 9th of C). These spectral pitches can explain why chords such as maj7, add9, and maj9 in jazz are common substitutes for major triads and collectively described as "major." Adding B and D to a C-major triad (to create a major 7th and 9th) does not change the "major" sound very much, because the chord already includes audible tones at B and D. The same idea can explain why the chords m7, m-add9 and m9 can act as sustitutes for minor triads.

Some pop styles from the 1970s and 1980s featured diatonic 7th chords including Δ7 (I7, IV7), m7 (ii7, iii7, vi7), and diatonic varieties of 7sus such as V11 (IV/V) or V13 (IV7/V, the first chord of "Y.M.C.A.," released in 1968 by Village People). Proponents of this style include Burt Bacharach, Carly Simon, and the Carpenters, and relevant genres include Motown (e.g., Marvin Gaye) and disco (e.g., the Bee Gees). The music sounds consonant, despite the frequent harmonic 2nd and 7th intervals, including psychoacoustically rough major 7th intervals (and their inversions, minor 2nds) in Δ7 chords. The consonant feeling (despite the roughness) may be due to constancy of scalar context if most tones are confined to a simple major diatonic scale.

Ear-training courses train musicians to recognize the notes that were played to create a given chord—which harmonic complex tones the sound physically contains. But a musical chord evokes pitches that are not notated, which is one reason why ear training can be difficult. To find out which tones music listeners (from concertgoers to music students struggling with ear training) actually hear in musical chords, we first need to check which partials (pure-tone components) are present and audible (as spectral pitches) in the chord's spectrum. On that basis, we can ask which virtual pitches are heard. In a first step, this chapter focuses on spectral pitches, although virtual pitches tend to be more salient in both music and everyday life.

Major Triad in Close Root Position

We start with the familiar C-major triad in root position and close position in the middle register, C4E4G4. The first ten harmonics of each tone ($n = 1$–10) are listed in table 16.1.

Table 16.1
Spectrum of a C-major triad in close position and middle register.

n	1	2	3	4	5	6	7	8	9	10
C4	C4	C5	G5	C6	E6	G6	B♭6	C7	D7	E7
E4	E4	E5	B5	E6	G♯6	B6	D7	E7	F♯7	G♯7
G4	G4	G5	D6	G6	B6	D7	F7	G7	A7	B7

The top row (n) is the harmonic number. The left column lists the note names. In the body of the table are the first ten partials of each tone.

Note that some harmonics are mistuned relative to 12-EDO. In particular, the 7th harmonic is about 1/3 semitone flatter than an equally tempered minor-7th interval. This discrepancy lies within the auditory system's range of tolerance when recognizing harmonic patterns (B. C. J. Moore et al. 1986).

Combining the three rows of the table into one, we obtain the following overall spectrum:

C4 E4 G4 C5 E5 G5 B5 C6 D6 E6 G6 G♯6 B♭6 B6 C7 D7 E7 F7 F♯7 G7 G♯7 A7 B7

The underlined partials are common to two of the three tones. For example, G5 is both the 3rd harmonic of C4 and the 2nd harmonic of G4. The double-underlined partial (D7) is common to all three tones. In musical renditions of a chord, we might expect doubled partials to be more audible on average.

The aural relevance of these partials is limited by psychoacoustic masking. Whereas most or all of the listed partials are probably separately audible when the chord tones are sounded alone, some become inaudible when the chord tones sound together. A pure tone (such as any one of the partials) becomes inaudible if it is close in frequency to a pure tone of higher amplitude. As a rough rule of thumb, if two partials lie a semitone apart, the weaker of the two is masked by the stronger, making the weaker inaudible. For that reason, partials with higher harmonic numbers are more likely to be inaudible: the intervals between them are smaller.

Following these principles, we can delete some of the partials in the chord C4E4G4 because they probably mask each other, crossing them out like this:

C4 E4 G4 C5 E5 G5 B5 ~~C6 D6~~ E6 G6 ~~G♯6 B♭6~~ B6 ~~C7~~ D7 E7 ~~F7 F♯7 G7 G♯7 A7 B7~~

Here, I have assumed that doubled (underlined) tones have higher amplitude than tones that are not doubled. For example, since B5 is doubled, it is likely to mask C6—but not vice versa. Therefore, C6 is likely to be inaudible. Note also that the highest partials in the list have been deleted. That is because, in a real musical chord, the spectrum typically continues toward higher frequencies, with smaller intervals between the partials. At some point, no more partials are audible.

We might reasonably expect the following partials to be audible in a typical musical example of the chord C4E4G4. We could call it the audible spectrum; Terhardt called it the spectral-pitch pattern:

C4 E4 G4 C5 E5 G5 B5 E6 G6 B6 D7 E7

The described procedure is inherently approximate. Each tone in a musical chord can be played by a different instrument and have a different amplitude envelope. Many other aspects of the stimulus can vary in complex ways. There can be subtle variations in attack or release that affect the relative salience of pitches within the chord according to principles of auditory scene analysis. The tones can be more or less loud, and the partials within each tone can be more or less audible.

Within these limitations, we can nevertheless approach the question systematically. The following procedure, inspired by the psychoacoustics of masking, can be applied to any chord:

1. List ten harmonics of each tone, relative to 12-EDO.
2. Create a single ordered series of partials. If a partial is doubled (a harmonic of two different fundamentals), underline it. If three fundamentals, double-underline it.
3. Delete every tone that is not doubled and lies a semitone from an upper or lower neighbor.
4. Delete the highest pitch in the list unless it is doubled.

At the end of this procedure, the lower partials should be generally audible, and the upper partials should only be audible if doubled. Uncertainty can arise if a tone is not doubled and lies midway between two tones four semitones apart, such that the interval is two semitones on both sides. In a typical complex tone, harmonics up to about the 10th are audible (the interval between the 9th and 10th is two semitones).

The audible spectrum of the chord C4E4G4, obtained according to this procedure, includes audible partials at B5, B6, and D7. That is consistent with the timbral similarity of major triads and Cmaj7, Cadd9, and Cmaj9 chords.

Preliminary Experiment

The predicted tones have been shown to be audible (aurally relevant) in a preliminary experiment (Parncutt, Engel, et al., 2023 submitted). In each trial, listeners heard a chord (duration 300 ms) followed by 300 ms of silence and then the same chord with one partial removed, and they were asked to rate similarity (cf. Terhardt 1998, sound example 17). Here are the tested chords and tentative results:

- [047] (major triad) in root position and both inversions: in all three cases, the chord CEG evoked B and D.
- [037] (minor triad) in root position and both inversions: in all three cases, CE♭G evoked B♭ (and perhaps D or F).
- [036] (diminished triad): CE♭G♭ evoked B♭ and D♭.
- [048] (augmented triad): CEG♯ evoked F♯.
- [025]: CDF evoked G and A.
- [035]: CE♭F evoked G and B♭.
- [047T] (Mm7): CEGB♭ evoked D and B♮.

The extra pitch classes that were evoked by these trichords were often tones that can be added to create tetrachords. But given other psychoacoustic reasons for adding particular tones (minimizing roughness, maximizing harmonicity), the listed tones do not necessarily predict the most common tetrachords.

The findings for the major triad can explain why major-7th and major-9th intervals are often added. The general idea is this: if the sound changes relatively little when a given tone is added, that addition is more likely to be made. Musicians are not aware that these partials are present in a major triad (the effect is not a result of conscious theorizing). Instead, they are sensitive to how the sound of a chord changes when a tone is added. The principle is not confined to jazz; the same process applied in the eighteenth and nineteenth centuries as different passing tones were heard against major triads.

Virtual pitches may also predict added tones. It is common to add a major 6th to a major triad, coinciding with the virtual pitch at A (see chapter 15). There is also a virtual pitch at F, but that tone is added less often, being a semitone above chord tone E (EF sounds quite rough). Despite the semitone, triadically or diatonically oriented clusters that include [0457] (an add4 chord) have become popular in modernist choral arrangements. In the choral works of Eric Whitacre, multiple traditional dissonances and their resolutions are often sung simultaneously to create new cluster sonorities (e.g., *When David Heard*; A. Larson 2006). The present theory offers a promising new approach for the analysis of such sonorities.

Pitch classes B and D are not consciously perceived within a C-major triad—except perhaps by some analytic-hearing individuals (cf. Schneider and Wengenroth 2009; Seither-Preisler et al. 2007). Nor are the other spectral and virtual pitches, including those corresponding to chord tones, necessarily consciously perceived in typical musical contexts. The point is that pitch classes

B and D are present in the chord, even if they are not noticed. They therefore contribute to the chord's overall character. Adding harmonic complex tones at those pitch classes changes the overall sound of the chord relatively little.

This informal spectral analysis of a simple musical chord illustrates how a theory of chord perception can be applied in music theory pedagogy. Whereas music theory is often speculative, there is little speculation here. There can be no doubt that the predicted audible partials (spectral pitches) of a given chord really are audible and hence psychologically real.

Revisiting Virtual Pitch

Presented with a complex spectrum such as the chord C4E4G4—created with harmonic complex tones—the brain looks for incomplete harmonic series within it, as it does for any sound. The fundamental of the best-fitting spectrum corresponds to the most salient perceived pitch. A harmonic series can be matched to the above audible spectrum at various transpositions with various degrees of success. In this way, Terhardt's model predicts several possible pitches, and they do not always correspond to chord tones.

Take the pitch C3, an octave below the bass. It is not physically present in the chord as a pure tone, but we might expect the chord to imply C3 as a missing fundamental because most lower harmonics of C3 correspond to audible partials in the audible spectrum. The first ten harmonics of C3 are C3 C4 G4 C5 E5 G5 B♭5 C6 D6 E6, and the spectrum of the chord is C4 E4 G4 C5 E5 G5 B5 E6 G6 B6 D7 E7. Comparing these, we see that six harmonics of C3 (C4, G4, C5, E5, G5, and E6) correspond to audible partials in the chord, and four do not (C3, B♭5, C6, and D6). Therefore, C3 is implied. (I am assuming here that partials above the 10th harmonic of C3—i.e., above E6—are not relevant for the perception of C3.)

Also implied is A2, whose first ten harmonics are A2 A3 E4 A4 C♯5 E5 G5 A5 B5 C♯6. Four of these—E4, E5, G5, and B5, with harmonic numbers relative to A2 of 3, 6, 7, and 9—are audible within the spectrum of C4E4G4, making A2 a possible root. Thus, the chord implies a minor-7th chord on A (Am7), which is a common harmonic substitution (e.g., ii^7 instead of IV in the key of G major).

The idea that CEG implies A can also be explained another way. Musicians are free to add any tone to any chord to produce a chord of higher cardinality, but they tend to add tones that minimize dissonance. Adding A to CEG produces a tetrachord with relatively low dissonance. That can explain why the 6 or m7 chord is used so often by Western composers and performers. An analysis of this kind cannot distinguish between two possibilities: the "nature" idea that we perceive A2 directly as a missing fundamental of C4E4G4, or the "nurture" idea that we perceive A2 in this situation because Am7 is so familiar (which in turn is because it is relatively consonant).

The chord C4E4G4 also implies F2. The first ten harmonics of F2 are F2 F3 C4 F4 A4 C5 E♭5 F5 G5 A5. Comparing this list with the chord's audible spectrum, three harmonics of F2—C4, C5, and G5 (harmonic numbers 3, 6, and 9)—are present within the spectrum. If we then add F2 to the chord, we get a more dissonant chord than before. F2C4E4G4 (or C/F) is more dissonant than A2C4E4G4 (or C/A) due to the major 7th between F and E, which can explain why it is used less often.

Table 16.2
Spectrum of a C-minor triad in close position and middle register.

n	1	2	3	4	5	6	7	8	9	10
C4	C4	C5	G5	C6	E6	G6	B♭6	C7	D7	E7
E♭4	E♭4	E♭5	B♭5	E♭6	G6	B♭6	D♭7	E♭7	F7	G7
G4	G4	G5	D6	G6	B6	D7	F7	G7	A7	B7

Minor Triad in Close Root Position

Now consider the C-minor triad shown in table 16.2. Following the same procedure as before, the spectrum is:

C4 E♭4 G4 C5 E♭5 G5 B♭5 C6 D6 E♭6 E6 G6 B♭6 B6 C7 D♭7 D7 E♭7 E7 F7 G7 A7 B7.

Some of these partials can be deleted according to the above rules, as follows:

C4 E♭4 G4 C5 E♭5 G5 B♭5 C6 ~~D6 E♭6 E6~~ G6 B♭6 ~~B6 C7 D♭7~~ D7 ~~E♭7 E7~~ F7 G7 ~~A7 B7~~.

The predicted audible spectrum or spectral-pitch pattern is then:

C4 E♭4 G4 C5 E♭5 G5 B♭5 C6 G6 B♭6 D7 F7 G7.

Four spectral pitches in this pattern are not octave equivalent with chord tones. They are B♭5, B♭6, D7, and F7. If we want to create a tetrachord by adding a pitch class to this C-minor trichord while changing the sound as little as possible, we have various options:

- add a minor 7th to make a Cm7 chord,
- add a major 9th to make a Cm-add9,
- add a perfect 4th to make Cm-add4, or
- add two or all three of these tones.

Intuitively, none of these options affects the chord's "minor" character.

The virtual pitches evoked by this chord include A♭ and F. F can be added as a kind of color tone; in the bass, it creates an F9 chord with a missing 3rd. Adding A♭ in the bass changes the root from C to A♭, and the chord becomes major.

The analysis of a minor triad can also be done in an octave-generalized fashion, ignoring the octave registers of the partials. Starting with the A-minor triad ACE, the tone A usually has audible harmonics at chromas A, E, C♯, G, and B; the tone C, at C, G, E, B♭, and D; and the tone E, at E, B, G♯, D, and F♯. Of those partials, the following have pitch classes that are not included among the chord tones: C♯, <u>D</u>, F♯, <u>G</u>, G♯, <u>B</u>, and B♭ (here, underlining denotes pitch classes that are harmonics of two chord tones). Applying the above masking rules, this list can be reduced to D, G, and B. These are tones that can be added to an A-minor triad without changing the sound very much, and the chords that result are Am-add4, Am7, and Am-add9.

What virtual pitches are implied by an A-minor triad in an octave-generalized approach? The chord first evokes virtual pitches at the chord tones: A, C, and E. Of those, A is the most salient, but C is not far behind. The chord also evokes virtual pitches at three missing fundamentals. Starting with the most salient, they are D (because all three tones are harmonics of D), F (because A and C are harmonics of F), and B (because A is a harmonic of B). If we ignore the masking procedure in the previous paragraph and consider all harmonics of all tones up to the 10th, the missing fundamental at D could be associated with partials at D, A, F♯, C and E; the missing fundamental at F, with C, A and G; and the missing fundamental at B, with B, F♯, A, and C♯.

Table 16.3
Spectrum of chord CDF in close position and middle register.

n	1	2	3	4	5	6	7	8	9	10
C4	C4	C5	G5	C6	E6	G6	B♭6	C7	D7	E7
D4	D4	D5	A5	D6	F♯6	A6	C7	D7	E7	F♯7
F4	F4	F5	C6	F6	A6	C7	E♭7	F7	G7	A7

[025] in Close Root Position

As a third example, consider the chord CDF (table 16.3)—a familiar sound that has no name in traditional music theory. The partials of C4D4F4 are:

C4 D4 F4 C5 D5 F5 G5 A5 C6 D6 E6 F6 F♯6 G6 A6 B♭6 C7 D7 E♭7 E7 F7 F♯7 G7 A7.

Applying the above rules, several partials can be deleted:

C4 D4 F4 C5 D5 F5 G5 A5 C6 ~~D6 E6 F6 F♯6 G6~~ A6 B♭~~6~~ C7 D7 ~~E♭7~~ E7 ~~F7~~ F♯~~7~~ ~~G7 A7.~~

The predicted audible spectrum is then:

C4 D4 F4 C5 D5 F5 G5 A5 C6 A6 C7 D7 E7.

The spectral pitches that do not correspond to chord tones are G5 A5 A6 E7. The chord implies A, which, if added, makes it an F6 or Dm7 chord, to which an additional E may be added without changing the chord's character. The chord also implies G, making it a Csus-add9 or G7sus chord.

Reflection

We have seen that the audible partials of a musical chord not only affect its timbre and musical function—they also influence what tones can be added to the chord to create a chord of higher cardinality. That is particularly true for trichords, in which a few audible partials can usually be identified whose pitch classes do not correspond to chord tones. It is less true for tetrachords, whose audible partials less often have pitch classes that differ from chord tones because the partials are closer together and mask each other more.

Among pitch classes that do not correspond to chord tones, virtual pitch theory predicts that virtual pitches tend to be more salient than spectral pitches—at least when the chords are perceived spontaneously, without focusing attention on specific pitches (whereas in analytic listening, virtual pitches tend to disappear). For example, virtual pitches at F and A, when implied by a C-major triad, are predicted to be more salient in a musical context (and hence musically more important) than spectral pitches at B and D. There are also effects of voicing. A tone that immediately precedes a chord in a musical passage may attract attention to a nearby audible partial according to the principle of pitch proximity in auditory scene analysis.

17 Jazz Harmony

As it developed during the twentieth century, jazz harmony significantly expanded the range of sonorities in Western music. Previous tonal music had been characterized by preferences for consonant over dissonant sonorities, C/D depending on harmonicity, roughness, familiarity, and diatonicity. (Bebop) jazz continued along the same path. In mainstream jazz theory, pitch is confined to approximate 12-EDO. Any chromatic tone can be added to any chord (often as a tension). The classical principle of harmonicity is reflected by tertian chord labels and upper-structure chords. Consistent with classical C/D theory, minor-9th intervals are avoided between inner voices (e.g., between a major 3rd and a perfect 4th above the root). There are interesting similarities between classical and jazz theory in areas such as scale and chord construction, intonation, enharmonic spelling, and tonality as prolongation.

Chord Roots

As in Western classical music, chord roots in jazz are determined primarily by root-support intervals (see chapter 15). Usually, a perfect 5th, major 3rd, minor 7th, and/or major 9th are voiced above the root, which may be missing. Other tones, if perceived to be dissonant, are called tensions. These principles are consistent with the central role of five standard tetrads, whose chord symbols are 7, -7, Δ7, ø7, and °7 (or Mm7, min7, maj7, half-dim7, dim7). Common variants include 7sus (Mm7 with suspended 4th, creating a quartal feel) and 7+(Mm7 with augmented 5th, often with tensions like 9 and ♯4, creating a whole-tone feel).

The regular 7 chord (Mm7) includes all root-support intervals except the major 2nd (or 9th). If a major 9th is added to create a 9 chord, the chord includes all root supports, making its root particularly clear. Starting with the Mm7 chord, we can create other 7th chords by shifting individual tones up or down by a semitone. The Δ7 (major-7th chord) can be created by shifting the minor-7th interval of a regular 7 chord up a semitone; the m7 chord, by shifting the major 3rd down a semitone. The ø7 is created by shifting two root-support intervals down a semitone (major 3rd and perfect 5th); the °7 chord, by pushing down three root-support intervals or pushing up the root. Every change to a root-support interval makes the chord root more ambiguous.

Scale Construction

Western musical scales, including the usual jazz modes, are subsets of the chromatic scale. A jazz musician who is fluent in different scales, each of which can be transposed to twelve different starting pitches, is effectively treating all twelve chromatic tones as equal. The scales are part of an oral-aural tradition that includes sound recordings.

The idea of chord-scale compatibility in jazz theory is consistent with the idea of a scale as a prolongation of a chord—more precisely, of a scale that is abstracted from a musical passage that is a prolongation of a chord. It is also broadly consistent with pitch-salience profiles of chords as predicted by Terhardt's pitch model (see chapter 15), whereby scale tones that lie between chord tones may be either virtual pitches or rising leading tones (a semitone below a chord tone) or both.

Familiar examples of chord-scale relations are shown in table 17.1. Starting with the right-hand column, we see that:

- The maj9 chord (CEGBD) implies a major scale (CDEFGAB).
- The 9 chord (CEGB♭D) implies a Mixolydian scale (CDEFGAB♭).
- The m9 chord (CE♭GB♭D) implies a Dorian scale (CDE♭FGAB♭).
- The ø7 chord (CE♭G♭B♭) implies a Locrian scale (CD♭E♭FG♭A♭B♭).
- The o7 chord (CE♭G♭A) implies a diminished scale (CDE♭FG♭A♭AB).

Whether these relationships are consistent with musical practice, on the one hand, and the predicted pitch-salience profiles of the chords (see chapter 15), on the other, is an interesting question. Another question concerns the extent to which the tonicizing function of leading tones plays a role. To answer this, we would need statistical data. One could take a corpus of transcribed jazz solos and count how often each pitch class is played over each kind of chord and compare the profile for each chord with theoretical predictions.

Table 17.1
Examples of chord-scale compatibility, adapted from Jamey Aebersold (2000, 14).

Chord symbol	Chord/scale name	Whole- and half-step construction	Scale	Basic chord
C	Major	W W H W W W H	C D E F G A B	C G D B D
C7	Mm7/ Mixolydian	W W H W W H W	C D E F G A B♭	C E G B♭ D
Cm	Minor/Dorian	W H W W W H W	C D E♭ F G A B♭	C E♭ G B♭ D
Cø7	Half-dim./ Locrian	H W W H W W W	C D♭ E♭ F G♭ A♭ B♭	C E♭ G♭ B♭
C°7	Diminished/ Octatonic	W H W H W H W H	C D E♭ F G♭ A♭ B	C E♭ G♭ A

Jazz Tonality

Whereas a passage of common-practice music can be regarded as a prolongation of a major or minor triad (classical tonality), a passage of jazz, rock, or blues may seem like a prolongation of a Mm7 chord (blues tonality). It is not unusual to start or end a passage in jazz or blues with a tonic Mm7 chord and for that chord as a whole to act as a consonant tonic. The familiar twelve-bar blues pattern (e.g., I7 IV7 I7 I7, IV7 IV7 I7 I7, V7 IV7 I7 V7) can be heard relative to, and as a prolongation of, the first chord I7.

The Mm7 chord can act as a tonic chord in jazz because its spectrum is similar to a harmonic series, giving it high harmonicity. Its roughness is also relatively high, but roughness is not necessarily avoided or carefully prepared and resolved in jazz—at least not to the same extent as Western classical music.

Whereas music theorists and musicians often call the Mm7 a dominant 7th, the word "dominant" is inappropriate if the chord functions as a tonic. The most stable pitches in jazz and blues are often the chord tones of the tonic Mm7 chord; relative to a major scale, we can label them $\hat{3}$, $\hat{5}$, and ♭$\hat{7}$. The scale degree ♭$\hat{7}$ is diatonic (in the sense of belonging to the prevailing scale) in blues, although notated as non-diatonic (♭ rather than ♮). The major triad built on ♭$\hat{7}$ (♭VII)—part of the familiar double-plagal (♭VII-IV-I) and Aeolian (♭VI-♭VII-I) progressions (Biamonte 2010)—has a similarly diatonic feel. The classical leading tone is conspicuous by its absence.

Allan Moore (1995) has argued that in rock (which emerged from blues), "it is *intrinsic* to what rock music has been, that the use of the 'flattened' diatonic seventh scale degree (and sixth, third, second and occasionally fifth, and also 'sharpened' fourth"), far from being aberrant, should not even be viewed as *departures*. Such a view leads inevitably to the model of rock as a deformed offspring of classical tonality" (186, italics in original).

Seventh chords can also be prolonged in classical and romantic music, but over more limited time periods (Goldenberg 2008). Besides, not all music of the eighteenth and nineteenth centuries ended with major or minor triads. The first song in Schumann's *Dichterliebe* cycle "Im wunderschönen Monat Mai" and Chopin's F-major Prelude Op. 28 No. 23 ended on Mm7 chords. Another chord that can potentially act as a tonic sonority is the suspended triad (sus4; cf. Herbie Hancock's composition "Maiden Voyage"), but the chord is dissonant relative to the major and minor triads onto which it can easily resolve without departing from a standard diatonic scale. In medieval and Renaissance polyphony, closure was often created by an open 5th sonority, but that sonority (like the open 5th at the end of the "Kyrie" in Mozart's *Requiem*) did not by itself imply a scale of about seven tones per octave—at least not in the way that major and minor triads do via their pitch-class salience profiles.

Blues Scale

The standard of tuning and intonation in jazz is tacitly assumed to be 12-EDO. Noticeable deviations are called blue notes, which are also found in the blues scale. Which tones belong to the blues scale and which do not is a matter of opinion (scale membership is sometimes ambiguous). One well-known version is shown in figure 17.1.

As the figure shows, blue notes tend to be flat relative to the tones of the tonic Mm7 chord. The resultant pitches often lie between $\hat{3}$ and $\flat\hat{3}$, $\flat\hat{7}$ and $\hat{6}$, and $\hat{5}$ and $\sharp\hat{4}$ / $\flat\hat{5}$. Another common microtonal pitch lies between $\hat{6}$ and $\flat\hat{6}$

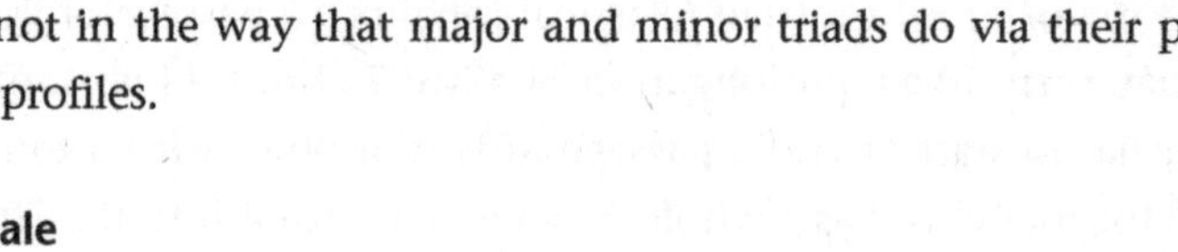

Figure 17.1
The blues scale ("negro scale") according to Leonard Bernstein (1939). From Asher Chodos (2018, 147).

or between the major and minor 3rd of IV. That being the case, we might ask why blue notes are flat rather than sharp relative to the tonic Mm7 sonority? Here are three possible answers:

- An interval that lies midway between major and minor (usually a 3rd or 6th) is sometimes called neutral. It can also be seen as a clash of simultaneous major and minor tonality or a compromise between the two. If the background chords in a blues progression (e.g., I7, IV7, V7) imply a major or Mixolydian scale, and the melody implies a pentatonic scale (e.g., $\hat{1}, \flat\hat{3}, \hat{4}, \hat{5}, \flat\hat{7}$), the optimal melodic solution may be a compromise or intermediate quarter tone. Trevor De Clercq and David Temperley (2011) observed that "it is often difficult to categorise rock songs in the common-practice tonal systems of 'major' or 'minor'; many songs adhere to other modes (such as Mixolydian), freely mix triads from the major and minor modes, or even combine elements of major and minor simultaneously (e.g. a minor third in the melody over a major triad)" (55).
- Blue notes may act as rising leading tones (without actually rising), tonicizing the chord tones in the tonic sonority, and thereby reinforcing the tonal structure. Blue notes nevertheless have a different character from classical leading tones, with their different intonation and the frequent use of portamento or slide (Wicks 1989).
- Scale degrees $\flat\hat{3}$ and $\flat\hat{5}$ may be perceived as missing fundamentals or implied roots of the Mm7 chord (see chapter 15). Like scale tones in classical tonality, they are implied by the tonic sonority.

Claims that ratios of small integers or forms of Just intonation represent the foundation of jazz or any other music, whether Indian (Danielou 1943) or African (Kubik 2010), are misleading. The tuning of any scale degree in any style of music can be, and often is, changed for expressive purposes. The blues 7th is simply a tone that lies somewhere between $\flat\hat{7}$ and $\hat{6}$. Court Cutting (2018) to the contrary, there is no statistically significant empirical evidence that musicians playing a blues 7th are trying to create a ratio of 7/4 (969 cents) as opposed to 9/5 (1018 cents), just as there is no convincing evidence in measurements of intonation in Western classical music that major-3rd intervals sometimes have the ratio 5/4 and other times 81/64.

The existence of the blues scale used to be denied—perhaps driven by racist or colonial prejudice (Chodos 2018). Hopefully, we are wiser today. A scale is a set of tones that is often (not always) used to make music and whose tuning is generally approximate. These two considerations apply to the blues scale in the same way as they apply to the major and minor scales of classical European music and indeed any other melodic music.

Quartal Harmony

The idea of stacked 3rds or tertian harmony is a theoretical abstraction. It may be a useful and practical pedagogical tool, but it lacks a clear scientific foundation. The idea of stacked 4ths or quartal harmony is similarly abstract.

The third most prevalent trichord in music of the Renaissance was the quartal 4/5 sonority, later called the suspended triad (Parncutt, Reisinger et al. 2019). It was less prevalent than major and minor triads, and the 4th resolved to a 3rd. In both cases, the reason was roughness—not theoretical considerations of tertian versus quartal harmony. The same applies to the 7th chords with suspended 4ths (7sus) that were commonplace in classical and romantic music as well as jazz. Both cases are sometimes considered "quartal" because they can be created by stacking 4ths.

Jazz harmony, like classical harmony, depends on three psychological components of C/D: harmonicity, roughness, and familiarity. The role of harmonicity is clear from tertian chord symbols and upper structures, but many jazz voicings seem to disregard roughness completely or even delight in producing a lot of it—reminiscent of some styles of Balkan polyphony (Brandl 2008) or Lithuanian traditional music (Ambrazevičius 2017).

Sonic familiarity seems to have been as important for the development of jazz harmony as it had been for Western polyphony and tonality. New harmonic ideas were introduced gradually, and the vocabulary of aurally familiar chords grew step by step as players and listeners got used to new kinds of roughness and harmonicity.

Stacked 3rds have dominated Western harmonic theory since Rameau. They remain popular in jazz because many chords (including the most complex chords) can be notated that way; jazz harmony is still based on familiar triads and 7th chords. If the Mm7 chord is regarded as jazz harmony's harmonic foundation, and it corresponds to harmonic numbers 4, 5, 6, and 7 (although its intonation deviates from 4:5:6:7), the link between harmonicity and stacked 3rds is clear. Tones that do not fit the stacked-3rds pattern can be notated as 9ths, 11ths, and 13ths.

Chords constructed from stacked 4ths can make musical sense in jazz insofar as the individual tones can be interpreted traditionally as 3rds, 5ths, 7ths, and so on. The chord CFB♭E♭, for example, is a C chord if the m7 interval between C and B♭ determines the root, but it can also be an F chord, with C and E♭ acting as root supports. In other words, its root is ambiguous, which is part of its appeal. The spacing by 4ths gives the chord a recognizable texture and harmonic ambiguity. Also relevant is the link to standard pentatonic scales, which are constructed from adjacent tones on the circle of 5ths such that all tones are related to other tones by 4th or 5th intervals.

Upper Structures

Music theorists have often wondered about the relevance for chord roots of the 11th harmonic, which lies almost midway between a perfect and an augmented 4th above the 8th harmonic (e.g., between F and F♯ above C). I will assume here that the 11th harmonic plays a negligible role in determining chord roots, first because it is usually inaudible in everyday sounds such as voiced speech sounds, and second because it is noticeably mistuned relative to the chromatic scale. Olli Väisälä (2002), after considering empirical findings on the perception of mistuned harmonics (Moore et al. 1985), offered a third reason: "The interval of a [perfect] fourth strongly points to the pitch class of its upper tone as the root, a feature probably linked with its conventional status as a dissonance and likely to override any slight root-supporting effect deriving from its connection with the eleventh harmonic" (210).

Augmented 11th chords may instead be perceived as having a separate upper structure and hence two roots. An example is the chord D/C7: a D-major triad over a C7 chord (CEB♭DF♯A). The C7 chord (CEB♭) is the lower structure, and the D-major chord (DF♯A) is the upper structure.

An upper-structure chord, also called a hybrid, is a simultaneity of two chords. As such, it is an example of an extended chord. Each of the two chords within it is relatively simple, such as a major or minor triad or Mm7 chord. For example, A♭/C7 is an A♭-major triad above a C7 chord.

The characteristic jazz sound of upper-structure chords depends on a juxtaposition of, and interaction between, two harmonic patterns with different roots. Whereas polychordal structures originated in Western classical music (e.g., Ravel or Stravinsky), the style of harmony that we now recognize as quintessentially (bebop) jazz emerged in the musical practice of twentieth-century jazz performers, guided by performance and perception.

A note about terminology: When an upper-structure chord is notated as two chords, the upper chord is written first and the lower second, for example E/C7. But if we list all the tones in order, we normally start with the lowest tone (here, CGB♭EG♯B♮). Often, the two chords have a common tone (here, E), but there are exceptions (e.g., D♭/C7).

The combined chord can be heard either as two independent chords or as individual tensions relative to the lower root (Mulholland and Hojnacki 2013; cf. Bigand et al. 1996). In the first case, the root of the upper triad of E/C7 is a major 3rd above the root of the lower 7 chord. In the second case, the tensions are G♯ (an augmented 5th above C) and B♮ (a major 7th).

A psychohistoric approach invites us to understand an upper-structure chord as an elaboration of a simultaneous interval or bichord—just as a triad can be imagined as an elaboration of a single harmonic complex tone. Both a single tone and a chord have various virtual pitches. In both cases, the intervals between the pitches correspond to intervals in the harmonic series. There is a three-way relationship between the tonic as tone, the tonic as triad, and passages of tonal music within which tones and triads function as tonics: the tone is prolonged to create a chord, which in turn is prolonged to create a passage. In this way, the gradual introduction of upper-structure chords into jazz may be compared with the emergence of two-part polyphony in twelfth century Europe, both in theory and in practice.

Voicings of upper structures tend to be preferred if the upper and the lower structures do not overlap: "extensions (ninths, elevenths, and thirteenths) usually appear at least an octave above the root, and elevenths and thirteenths are rarely placed below any members of the core triad" (McGowan 2011, 168). Voicings may also be preferred if they avoid minor-9th intervals, which are dissonant from the viewpoint of both harmonicity and roughness: "Because the tones of the triad—the core of all tonal jazz dialects—are inherently stable, the pitches a minor ninth above them are not permitted as possible chord tones, and are thus inherently unstable" (McGowan 2011, 168). Jeff Pressing (1978) listed examples:

(1) 3rd and 11th do not normally occur simultaneously in dominant chords
(2) A minor 9th is a very rare addition to a minor 7th chord
(3) Dominant 7th chords typically do not contain both a 13 and a +5. (29)

Diatonicity

A complete theory of C/D in jazz harmony involves not only smoothness, harmonicity, and familiarity, but also diatonicity or belongingness to a familiar diatonic scale. Diatonicity offers an alternative explanation of why major triads with added major 6ths, 7ths, and/or 9ths are called "major" in jazz theory: the added notes sound well together, despite their varying degrees of roughness, because they belong to same familiar diatonic scale. A similar argument can explain the consonance of suspended chords such as Dm7/G (V11 in C major).

For another example, consider the 7♭9 chord (Mm7 with added minor 9th). The chord is familiar from classical music, where it often appears as a diminished-7th chord with added root in the bass. The minor 9th undermines the root because it is a semitone higher (whereas a semitone lower might reinforce the root). At the same time, the minor 9th, like all tones in the chord, is diatonic in a harmonic minor scale. If C7♭9 appears in the key of F (harmonic) minor, all tones in the chord—CEGB♭D♭—are diatonic. The chord emerged in classical music history when the familiar melodic progression $\flat\hat{6} - \hat{5}$ was heard over a dominant 7th chord.

Enharmonic Spelling

Enharmonic spelling tends to be less important in jazz than in classical theory. Whereas jazz arrangers may sometimes be sloppy about spellings—at least from a classical viewpoint—the musicians reading the scores are seldom

Figure 17.2
Extract from Beethoven's Piano Sonata Op. 27. No. 2 "Moonlight," 3rd movement. The upper tone of the trill (B4) combined with the dominant 7th chord in the left hand (G♯B♯F♯) creates what today's jazz musicians call a ♯9 chord. Source: musescore .com.

concerned. If correct enharmonic spelling is seen as a classical thing, jazz theorists who insist on getting it right seem pedantic. Originality, expression, and improvisation are more important than notation.

An example of enharmonic ambiguity in jazz is the ♯9 chord—a Mm7 (usually with a missing 5th) with added minor 3rd or 10th (called ♯9), usually in the top voice. The chord has a long history, as suggested by figure 17.2. In this case, the ♯9 sound is created by the upper tone of the trill. That is an example of how more complex chords gradually make their way into harmonic language. At the start, the chord is carefully prepared by its context, as in this case, or it emerges as a by-product of voice leading. Later, as listeners get used to the sound, the chord is played without preparation. In that way, the 7♭9 chord was also a springboard for upper structures in jazz such as F♯m/C7 (or C13♭9♯11).

In classical music and theory, the tension note in the ♯9 chord would normally be notated ♭3 relative to the root—like the B♯ in the Beethoven example—and not as ♯2 or ♯9. In jazz, the ♯9 chord label is convenient, concise, and unambiguous. At a stretch, it also allows the chord to be treated as a stack of 3rds in the tradition of Rameau.

V Perception of Progression

18 Pitch Commonality

The pitch commonality of two chords is the degree to which they have perceived pitches in common. It depends on the salience of coinciding pitches—those that lie in the same psychological categories corresponding to chromatic scale steps. Relevant pitches include both audible partials (spectral pitches) and (missing) fundamentals (virtual pitches). Pitch distance is the overall perceived distance between successive sonorities; it depends on pitches in different perceptual (chromatic) categories and voice leading. Pitch commonality and distance may be the simplest conceivable psychoacoustic foundations for perceived relationships between successive musical sounds, but they also involve a mixture of nature and nurture. The nurture aspect depends on familiarity with frequently occurring relationships in music. When musical polyphony emerged in the European Middle Ages, and early chord progressions appeared in the Renaissance, we may assume that C/D was perceived both vertically and horizontally, the horizontal aspect involving pitch commonality. Later, an additional source of consonance was the relationship between each chord and the tonic triad (a psychological reference). Passages of MmT became prolongations of the tonic triad—enabled by pitch commonality.

Examples

For a simple example of pitch commonality, take the chord progression from C major to D minor (CEG-DFA). It may help to imagine the first two chords of either Mozart's *Ave Verum Corpus* or the introduction to the 1954 popular song *Mr. Sandman* by Pat Ballard (ignoring the 7ths and transposition into different keys). These two triads have no notes in common, but they sound strongly related. There are two possible reasons for that, which loosely can be called "nature" and "nurture".

In the nature explanation, the chord CEG implies pitch classes D, F, and A as missing fundamentals. The idea comes from pitch theory and is consistent with both psychoacoustic experiments (Parncutt, Sattmann et al. 2019) and database analyses (Parncutt and Radovanovic, 2023; Parncutt and Reisinger 2021). Thus, chords CEG and DFA have three perceptible pitch classes in common. The nurture reason involves familiarity with scales. The major or standard diatonic scale is highly familiar. Since both chords belong to the same scale, they sound related. Moreover, since the two chords often follow each other in music (as I-ii or V-vi), they are perceived as related.

When we hear the relationship between these two chords in a musical context, it is not immediately clear what is more important: nature or nurture. Nor can an experiment answer that question directly. But we can infer from the various empirical sources that the listener's experience of this chord relationship depends on both.

A more complex example is the Tristan chord. At its first appearance in Wagner's opera, it is notated F3B3D♯4G♯4. The tonal context in which it is presented is unclear until after it resolves. That context turns out to be the key of A minor, but in the ensuing music, Wagner repeatedly avoids sounding the tonic triad, which contributes to the music's mysterious, magical, suspended feel.

If the context was clearly A minor, we could talk about the perceived relationship between the Tristan chord and the A-minor triad. One point of contact is the tone F—a tone in the Tristan chord that is implied by the A-minor triad as a missing fundamental (recall that ACE implies F and D). In addition, the tones B and G♯ in the Tristan chord belong to the scale of A harmonic minor. The remaining tone, D♯4, is the leading tone of E ($\hat{5}$ in A minor) and the 3rd of a double dominant harmony (II7) in A minor—a familiar harmonic function. The individual tones of the Tristan chord are strongly related to the key of A minor, but the simultaneous combination of pitches is unusual.

The Tristan chord functions as a kind of sub- or predominant—similar to an augmented-6th, diminished-7th, or minor-7th chord, but with one tone shifted by a semitone. The chord is familiar in some ways but not others. Its mysterious sound arises when the familiar half-diminished-7th sound is heard in an unfamiliar context. The chord is more familiar in a context of E♭ minor, in which it is ii7—a diatonic 7th on $\hat{2}$. Another familiar context is G♭ major, in which it is a diatonic V9 with missing root. But Wagner has the chord resolve to an E7 chord, the dominant 7th of A minor—very distant from both E♭ minor and G♭ major. This surprising move was made more digestable by careful adherence to familiar rules of voice leading, with non-diatonic tones moving by semitone steps to diatonic neighbors. In short, the Tristan chord's C/D depends in part on the simultaneous C/D of the half-diminished-7th chord in isolation and in part on the successive relationship between the chord and its context.

The Tristan chord is also an example of the irrelevance of enharmonic spellings for a psychologically oriented music analysis. None of the above arguments depend on whether the chord's non-diatonic tones are notated as D♯ or E♭, or as G♯ or A♭. Chord-tone F could have been written E♯, and chord-tone B could have been C♭, without changing our conclusions about the chord's function or perceived consonance/dissonance (C/D).

Nature or Nurture?

Imagine a simple chord progression of the kind C-X-C, where C is a C-major triad and X is some other chord. How strong is the perceived relationship between the second chord and the first and last? Emmanuel Bigand and colleagues (1996) played such progressions to musicians and nonmusicians and asked them to rate the musical tension of the second chord. Findings were compared with predictions of various models, which again can be roughly classified into nature and nurture.

A nature or sensory model depends on the stimulus as directly perceived, whereas a nurture or cognitive model depends on musical experience. Possible nature models in this case are the C/D of chord X (roughness amd harmonicity), the pitch commonality of chords C and X, and the perceived pitch distance between them.

Sensory aspects depend on short-term or echoic memory (Bigand et al. 2014; Leman 2000). Regardless of the listener's attentional focus, each sound is retained in short-term memory for a few or several seconds. During that time, it can be compared with other sounds. That does not mean that successive tones are perceived as if they are simultaneous, so arguments based on beating or roughness are not relevant. Instead, the pattern of pitches in the first sound is compared with the pattern in the second.

Nurture is about musical experience. If we hear a certain kind of musical relationship many times, we start to like it, and it starts to sound consonant—even if our initial impression was negative. Possible nurture models in this case include simple statistics (how often a certain progression happens in music to which the listeners in an experiment had probably been exposed) and cognitive structures (e.g., whether the second chord is perceived to belong to the same or a different tonality, and if different, how different). Bigand et al. (1996) did not compare empirical data with statistics from database analyses, but instead tested two cognitive models: Krumhansl (1990) and Lerdahl (1988).

Moment-to-moment expectations in music depend on our familiarity with music in similar styles. Imagine an experiment in which a piece of music suddenly stops, and the listener is asked whether a given sound (e.g., a single tone) would have been a good continuation (Narmour 1990). If we could answer that question in general, we would know a lot about the perception and cognition of musical structure. We might also have a better understanding of musical emotion (L. B. Meyer 1956).

In the case of unaccompanied melody, the expected continuation depends mainly on the previous few tones and the listener's familiarity with similar melodies in music. The expected continuation is usually a small interval (a tone or semitone) away from the last tone. The expected direction of motion may be the same following a step and opposite following a leap (Narmour 1990). Familiar diatonic scales also play a role. If the melodic fragment is played within a familiar scale, the next tone is expected to belong to that scale. If the scale is familiar but the last tone of the fragment is not in the scale, the listener expects resolution to a neighboring scale degree.

Bigand et al. (1996) found that both nature and nurture contribute to the perception of musical tension in short chord progressions. Since the study was correlational, and many different intercorrelated models could conceiveably account for the data—remembering that "Lerdahl's (1988) cognitive model and Parncutt's (1989) psychoacoustic model produced highly correlated predictions" (137)—the conclusions of the study (like those of other empirical studies cited in this book) were necessarily tentative. In addition, "nonmusicians appeared to have based their responses on the most easily perceivable features (i.e., melodic arrangement between the tones of successive chords). Musicians probably had greater knowledge of the tonal hierarchies and greater sensitivity to roughness, developed through musical tuition and practice" (Bigand et al. 1996, 139).

In another study, in which successive pairs of musical elements (triads, dyads, and octave-complex tones) were compared for goodness of fit, W. F. (Bill) Thompson and Parncutt (1997) provided evidence that nature (in the form of pitch commonality) influenced the goodness of fit of single tones that followed major triads, major-3rd dyads, and perfect-5th dyads. Pitch commonality also influenced how well a triad followed another triad or dyad in all twelve transpositions. All sounds were built from octave-complex tones to avoid pitch-distance effects. We also found an effect of the circle of 5ths and preferences for trials ending with a triad, both of which were probably learned from music (nurture).

Given these findings, it is interesting to ask whether the nature models might represent the ultimate psychohistoric origin of the nurture models. To answer such a question, we need first to consider what criteria a scientific theory of harmonic relationship between successive musical chords, and hence of musical pitch structure, should fulfill:

- There should be valid, reliable, quantitative empirical data against which the theory can be tested. The above studies collected data of that kind.
- The theory should be generalizable: the ideas should be drawn from, and to some extent also apply, outside the domain for which they are claimed to be true. This criterion is satisfied by Terhardt's more general theory of pitch perception. Similarly, Krumhansl's theory of tonality can be seen as a special case of more general theories of cognitive representation (e.g., in language).
- The theory should be parsimonious and falsifiable.

Regarding parsimony, the idea of pitch commonality is relatively simple and uncontroversial. It refers to the degree to which two successive musical sounds or patterns of sound are perceived to have pitches in common—that is, pitches that correspond to the same chromatic scale-degree categories. The perception of pitch commonality depends on the perceptual salience of each coinciding pitch (a possible mathematical formulation was presented in Parncutt 1989). Pitch commonality may also depend on salient pitches that do *not* coincide or are perceived to belong to adjacent rather than identical chromatic scale steps. A correlation coefficient between two tone profiles (such as those calculated by Krumhansl 1990) encapsulates both aspects.

Pitch commonality is comparable with Helmholtz's (1877) *Tonverwandtschaft* or tonal affinity, but there is an important difference. Helmholtz was thinking of individual partials (as were Dean et al. 2019), whereas pitch commonality involves all (salient) spectral and virtual pitches (Stoll and Parncutt 1987; Terhardt 1976; Terhardt et al. 1986). Since virtual pitches tend to be more salient than spectral pitches in everyday sounds, including music, pitch commonality depends in a first approximation on virtual pitches weighted by salience. Adding spectral pitches to the model may improve predictions.

Bigand et al. (1996) were unsure which was more important for their experiment: nature based on perception of specific pitches in the musical surface, or nurture based on musical experience. Other research projects in music psychology have tried to pit these two alternatives against each other in diverse ways (e.g., Huron 2001; Lahdelma and Eerola 2016; Lerdahl 1996; Linnavalli et al. 2020; McDermott, Lehr et al. 2010; Trainor and Trehub 1994). Given how many such studies are consistent with either nature or nurture or both, it is safe to assume that both play a role in general.

In a psychohistoric approach, there is a historical relationship between nature and nurture. Musical sound combinations may initially find their way into music for nature reasons, after which the relationships are perceived for nurture reasons. If we hear two musical sounds as related, it could be due to our musical experience if those two sounds often happen together in music. If we then ask why the composers or performers of the past often put those sounds together in music, or what it was about the sounds that made them go together or facilitated the perception of relationships between them, the answer inevitably involves nature.

If the above argument is correct, the nature side of the argument is an important route toward understanding musical structure. Even if modern listeners do not directly experience the posited natural relationships between pitches in the musical surface, those relationships may have influenced the historical development of music.

Riemann's Approach

The theory of pitch commonality was inspired by existing theories of chord relationships in tonal music, including the neo-Riemannian operations for pairs of triads with two tones in common (see chapter 10). Table 18.1 shows the pitch-class salience profiles of triads involved in the neo-Riemannian transformations P, R, and L and the calculated pitch commonality in each case.

All three Riemannian operations produce relatively high pitch commonality due to the two common tones. But pitch commonality also depends

Table 18.1
Pitch-salience profiles of five triads according to Parncutt (1988), with root-support weights as shown in table 15.1.

Pitch class	C-major triad	C-minor triad	A-minor triad	E-minor triad	E-major triad
C	18	15	13	8	3
	0	1	0	2	7
D	3	2	8	1	1
	3	13	0	3	0
E	10	0	10	15	18
F	6	8	8	1	0
	2	0	2	2	3
G	10	10	1	13	3
	3	8	3	0	10
A	7	2	15	8	6
	1	1	1	0	2
B	0	3	2	10	10
Comparison	C-C	C-c (P)	C-a (R)	C-e (L)	C-E (LP)
r	1.00	0.55	0.66	0.55	0.52

The last two rows ("comparison") evaluate the P, R, and L operations of neo-Riemannian theory and the combination LP. The final row, *r*, is Pearson's correlation coefficient between column 2 (C-major triad) and the column in question—a measure of pitch commonality.

on pitch-class salience profiles. The table suggests that of Riemann's three cited operations, R represents the closest relationship. That is consistent with compositional practice: the falling 3rd progression (e.g., R = C-a) happens more often than the rising third (L = C-e) or switching from major to minor on the same root (P = C-c). The correlations for operations P and L are about the same. We also need to consider pitch distance, which is greater for R, where one of the three tones moves through a whole tone rather than a semitone. Thus, R represents a harmonically closer relationship, whereas P and L represent melodically closer relationships.

In Riemann's approach to harmonic function, any triad in any context can be classified as tonic, dominant, or subdominant. Table 18.1 suggests how this classification may be made. Calculate the pitch-class salience profile of the chord in question. Compare it with the pitch-class salience profiles of stereotypical tonic, dominant, and subdominant chords. The higher the correlation coefficient, the more likely the chord has the corresponding has the corresponding function.

Riemann argued that other harmonic progressions or relationships can be regarded as combinations of R, P, and L. Take for example a chord progression from C major to E major. Riemann imagined it as a combination of two operations: C → e → E, that is, L followed by P. That is an interesting way to think of the C-E progression, but the pitch-commonality approach is more conceptually parsimonious. It also corresponds more closely to the everyday experience of hearing chord relations directly, without conscious awareness of intervening steps. In a first approximation, the perceived strength of the relationship is the number of common notes (in this case, one). In a second approximation, the perceived strength of the relationship is the correlation coefficient between the two pitch-class salience profiles.

The analysis suggests that Riemann's combined relationship LP is harmonically weaker than L, as we might expect. We should also consider pitch distance. In the LP progression CEG-BEG♯, pitch distance is relatively high because two parts move (C-B, G-G♯) rather than one. At the same time, overall perceived distance is reduced by contrary motion (the average pitch of the three tones stays the same)—an aspect of good voice leading.

Melodic Octaves and 5ths

When two melodic tones span a rising octave (for examples, see chapter 18), they have several pitches in common. In particular, the pitch at the 2nd harmonic of the first tone (see figure 13.1) lines up with the pitch at the fundamental of the second tone. Both pitches are usually virtual; if the pitch at the 2nd harmonic is virtual, the listener is hearing is a series of even-numbered harmonics—a tone with a tone. Both pitches can also be spectral if the listener focuses attention on one partial. We know that subsidiary pitches are audible in harmonic complex tones from two kinds of experiment: those in which listeners hear a complex tone followed by a pure tone and are asked if the pure tone goes with, or was part of, the preceding tone; and those in which listeners rate the similarity of successive tones spanning octave and other intervals (Parncutt 1989; Terhardt et al. 1986).

Now, imagine it's 1961 and Elvis Presley is recording the first three tones ("Wise men say . . .") of *Can't Help Falling in Love* by Hugo Peretti, Luigi

Creatore, and George David Weiss. The first two tones lie a perfect 5th apart, and the third returns to the first. The 5th relationship that we perceive depends on our prior experience of major scales and perfect 5th intervals. But those two tones also have pitches in common—to a lesser extent than tones an octave apart, because fewer spectral and virtual pitches line up, or those pitches are less salient (Parncutt 1989; Stoll and Parncutt 1987). The virtual pitch at the 2nd harmonic of the higher tone (the fundamental of harmonics 2, 4, 6, 8, and 10) lines up with the virtual pitch at the 3rd harmonic of the lower tone (the fundamental of harmonics 3, 6, and 9). Regarding spectral pitches, the 3rd harmonic of the lower tone lines up with the 2nd harmonic of the upper one, and some other audible harmonics line up similarly. In different octave registers, there may be different contributions to pitch commonality.

The pitch relationship that we perceive between successive tones an octave or 5th apart depends on a mixture of pitch commonality and experience of similar patterns in music. Comparisons of experimental results for musicians and nonmusicians, and for pure tones and harmonic complex tones, have suggested that despite the obvious effect of musical experience, pitch commonality remains a significant factor (Parncutt 1989).

In prehistoric music (or any world of unaccompanied melody with no polyphony), pitch commonality may have played a role. In the West, it may gradually have driven the emergence of pentatonic, diatonic, and eventually chromatic scales. It may explain why, in traditions where consonance was intuitively valued, scales were preferred if they corresponded to continuous segments of the circle or line of 5ths.

19 Tonality

A chord progression's subjective coherence depends on perceived relationships between successive pairs of chords and between each chord and the hierarchy of tonic, dominant, and subdominant. A passage of music in MmT can be heard as a prolongation of its tonic triad, suggesting that emotional connotations of major versus minor triads are also prolonged. The idea seems intuitively correct when diatonic melodies are compared with the three pitch classes of the tonic triad: each tone in the melody lies within one scale step of a chord tone in the tonic triad. The idea is also consistent with correlations among three profiles: pitch-class prevalence profiles in tonal music, Krumhansl's key profiles, and calculated pitch-salience profiles of tonic triads. If we combine that with the tonicizing role of rising leading tones, we can account for the origin of major and minor scales. MmT emerged historically as major and minor triads increasingly served as tonic sonorities in modal contrapuntal contexts. Today, MmT is popular for a combination of political and psychological reasons. Politically, it was promoted during a long and bloody history of colonialism. The psychological aspect may involve the role of the voice in human sociality and the audibility of the harmonic series within it. MmT may facilitate cognitive processing of pitch structures, and there may be neural rewards for finding structure in complex signals.

Key Profiles

Krumhansl and Kessler (1982) had participants listen to a chord progression followed by a tone and asked them how well the tone followed the progression. For each chord progression, the experiment generated a profile of twelve numbers—one for each chromatic scale step. This key profile (figure 1.1) was different for major and minor keys and surprisingly independent of the chord progression used to instantiate a key.

From a music-theoretical perspective, Krumhansl's key profiles are profiles of tonal stability. They may depend on several factors, including the acoustic structure of major and minor triads, the effect of harmonic pitch pattern recognition within those triads (virtual pitch), the diatonic scales to which the chords belong, and the tonicizing effect of rising leading tones. These, one might argue, are the psychoacoustical and psychohistorical foundations of major-minor tonal structure. A hypothesis of that kind is difficult to test for statistical reasons: each profile comprises only twelve elements, so there may be insufficient degrees of freedom. Besides, correlation does not imply causation. But the ideas can be evaluated both quantitatively and qualitatively.

One way to model Krumhansl's profiles is to compare them with predicted pitch-class salience profiles of the chords used in her experiment (Parncutt 1989), weighting the chords to account for memory decay (Huron and Parncutt 1993). Another is to compare the key profiles with the tonic triad only (Parncutt 2011), consistent with the Schenkerian idea that passages of MmT are prolongations of their tonic triads.

Consistent with an account based on virtual pitch, Krumhansl found scale degree $\hat{1}$ to be considerably more stable than scale degree $\hat{5}$ in both major and minor keys. In a minor key, she found scale degree $\hat{3}$ to be more stable than scale degree $\hat{5}$, contrary to what music theorists might have predicted. In both cases, Terhardt's model provides an explanation (cf. Parncutt 1988). The 5th interval between the root and the 5th of a minor triad ensures that the root is perceived as such. The 3rd of a minor triad is a root candidate because it lies a M3 interval below the 5th, but (contrary to Riemann's dualist theorizing) the 5th of a minor triad is not a root candidate (it is not supported as a root by any other tone). Virtual pitch theory can also account for differences between tone profiles following the individual chord progressions of Krumhansl and Kessler (1982): II-V-I, IV-V-I, IV-V-I, and I alone, each in both major and minor keys (Parncutt 1994; see also Parncutt and Bregman 2000).

Origin of Scales

A theory of MmT as prolongation of the tonic triad can explain the origin of major and minor scales. I propose two main principles: one harmonic and the other melodic. The harmonic principle says that scale tones in major and minor keys should include not only the chord tones of the tonic triad, but also tones implied by the tonic triad as virtual pitches or missing fundamentals. Thus, the scale of C major should include not only C, E, and G, but also D, F, and A; and the scale of C minor should include not only C, E♭, and G, but also F and A♭. In addition, the melodic principle says that the scale should include a rising leading tone a semitone below the tonic (i.e., a semitone below the root of the tonic triad; cf. chapter 11).

Taken together, these principles can explain why the major scale comprises the seven tones C-D-E-F-G-A-B. They can also explain why the minor contains the six tones C-E♭-F-G-A♭-B, but they don't clearly explain why it contains D. A possible reason is that the dominant–tonic progression is important for other reasons, and D is a perfect 5th above the dominant G. There is nevertheless some ambiguity about $\hat{2}$ in a minor key, consistent with the more frequent use of Neapolitan harmonies in minor keys (e.g., D♭/F relative to C) than in major keys.

MmT and Emotion

The association between major keys and positive affect, and between minor keys and negative affect, is well known and frequently researched. From a psychological viewpoint, the effect is robust, although the effect of tempo on emotion is usually stronger, faster tempi being associated with more positive affect as well as higher arousal. Also, children learn the emotional cues of tempo earlier (at around five years of age) than those of major versus minor modes (Dalla Bella et al. 2001).

The idea of tonality as prolongation of the tonic triad is relevant here. The emotional difference between major and minor modes may stem from a difference in the clarity (or its opposite, ambiguity) of the roots of major and minor triads, and consequently of reference pitches or tonics when passages of music are perceived as prolongations of major or minor triads (Parncutt 2014). The emotional implications of the tonic triad may be amplified when the chord is prolonged.

Chords that divide the octave equally (augmented triad, diminished 7th) do not have clear roots (are harmonically ambiguous) and are perceived as musically tense (Lahdelma and Eerola 2016). Prolongations of such chords (to the extent they can be psychologically prolonged) would amplify the tension. Passages based on diminished triads may evoke fear or despair, while augmented or whole-tone passages may evoke a feeling of amazement, mystery, or weightlessness (cf. Willimek and Willimek 2017).

Scales in minor tonalities are linked to negative emotion. There is no widely accepted reason for the connection, but the question is perhaps simpler than we think. Positive emotions tend to mean that things are OK or normal, whereas negative emotions mean the opposite: "The happy person is energetic, optimistic, assertive, socially outgoing, eager to start new projects, and makes substantial investments in people and projects with confidence that they will payoff. The sad person is lethargic, pessimistic, submissive, socially withdrawn, excessively realistic about personal abilities, and has little initiative for new relationships or projects" (Nesse 1990, 273).

In Western music, happiness is unsurprisingly linked to consonance in general and harmonicity in particular. The minor triad is more dissonant than the major because it is less harmonic, that is, less similar to a harmonic series. Any departure from harmonicity may be experienced as a departure from the social normality represented by the harmonic series, which is audible within every voiced speech sound. In that regard, minor triads and scales may be perceived as less OK or less normal than major. The harmonic minor scale is more dissonant than the major because it contains more semitone and tritone intervals, and fewer whole-tone intervals and perfect 4ths/5ths. The harmonic minor scale does not correspond to a closed segment on the circle of 5ths as the major scale does.

Apart from the major-minor distinction, there is no psychological foundation for the oft-asserted emotional character of specific musical keys. Johann Mattheson (1713) regarded the key of D minor as "rather devotional, calm . . . agreeable and satisfying," G minor as "tender as well as enlivening, yearning as well as contenting," and so on for other keys (Powers 1998: 278). Whereas such traditions played an important role in the history of musical thought, they are hardly relevant in a scientific approach. The purported emotional implications cannot be perceived by listeners unless they have absolute pitch and extensive experience of repertoires in which such traditions are manifested (Powell and Dibben 2005).

There is evidently much more to MmT and emotion than this short introduction implies. Millions of people from all over the world perceive a rich and intense palette of emotions in response to their favorite examples of MmT. It is remarkable that we know so little about the origin of that emotion. The term "tonality" does not appear in the index of the currently leading summary of research on music and emotion (Juslin and Sloboda 2010). Instead, we find the phrase "tonal expection." It leads to a discussion of the implication-realization theory of Leonard Meyer (1956), as elaborated by

Narmour (1990) and Huron (2006)—a theory that despite its promise and popularity is remarkably unspecific and undifferentiated about predicted emotional responses. We also find the phrase "tonal stability" in the index, which leads to a discussion of research on that topic that focuses on tension and release (i.e., arousal rather than valence). The index entry "harmony/harmonic" initially seems more promising, but the cited passages do not reveal new insights about the relationship between harmony and emotion beyond well-known connections between major-minor and consonant–dissonant on the one hand and happy–sad or positive–negative on the other (Crowder 1984; Parncutt 2014).

Whereas we know a lot about many relevant aspects of musical emotion—basic emotions in music across cultures (Argstatter 2016), the remarkable sensitivity of infants to musical emotion (Swaminathan and Schellenberg 2015), the emotional implications of expressive timing and dynamics in Western performance (Gabrielsson and Juslin 1996), the paradoxical enjoyment of sad music (Vuoskoski et al. 2011)—little of that research is specifically tied to the pitch structures of MmT. The role of musical humour (Nagy et al. 2022) seems pertinent, but the research again falls short of linking diverse specific musical structures to diverse specific emotional responses. There are many references to melody in the index of Patrik Juslin and John Sloboda (2010), but that topic is less directly linked to MmT. This quick literature search leaves us none the wiser about the ultimate origin of the specific emotional connotations of MmT. The problem is perhaps that we lack an accepted theory of the origin of music in general. If someone asks me why they like a certain chord progression so much, I'm afraid the honest answer is that we don't really know.

Origins of MmT

Whereas the origins of emotion in MmT are hard to explain, it is easier to explain the origin of the pitch-time structures. Previous chapters have laid the foundation for a parsimonious, psychohistoric account of the nature and origin of those structures. In a broad-brush approach, we might hypothesize that MmT emerged in four stages:

(1) Prehistoric and ancient melody
The emergence of pentatonic and diatonic scales as (European) musicians became sensitive to the pitch commonality of melodic octave and 5th intervals and came to prefer them.

(2) Medieval polyphony

The emergence of leading tones as musicians became sensitive to variations in the consonance of individual tones in diatonic chant and early polyphony and hence to the consonance and prevalence of *fa* relative to *mi.*

(3) Renaissance polyphony

(a) Simultaneous C/D: The growing sensitivity of listeners to the roughness and harmonicity of tonal sonorities, leading to preferences for major and minor triads (as they were later known).

(b) Chord profiles: Growing sensitivity to variations in pitch salience within musical chords as represented by pitch-class salience profiles and leading to the emergence of chord roots—in particular, for major and minor triads.

(c) Key profiles: The musical prolongation (elaboration) of tonic sonorities and their pitch-class salience profiles, which stabilized profiles of tonal stability.

(4) Colonialism

The European colonization of other global regions, including South America, Africa, Australia, and India, which linked MmT to political power and contributed to its worldwide acceptance (Agawu 2009). Today, we need to raise awareness of this darker side of MmT and promote indigenous musics and musical diversity, while remaining aware of the dangers of revivalism (Livingston 1999) and remembering that authenticity is not preferable to hybridity (S. Weiss 2014).

The Modal-Tonal Transition

A psychohistoric stage model can shed light on how MmT emerged historically. That includes the controversial question of how the modal system of medieval chant and Renaissance polyphony was somehow displaced by MmT as "Dorian came closer to minor, Lydian and Mixolydian approached major" (Lowinsky 1961, 2). In a nutshell, we might claim that since only two of the many possible trichords in the chromatic scale were consonant enough to function as tonal centers (the major and minor triads), and the new tonality involved psychologically connecting an entire passage of music to a single tonic sonority, the number of modes was effectively reduced to two (Parncutt 2011).

The slow transition from modality to tonality in the fifteenth to seventeenth centuries, as it happened in both music itself and music theory, can be compared with the faster early twentieth-century transition from tonality to post-tonality in Western art music. Both transitions were partly organic, drawing on and developing existing compositional principles (as if musical structure was a kind of growing biological organism), and partly revolutionary, introducing radically new principles of tonal organization. The organic aspect meant maintaining and developing existing scalar and voice-leading structures, whereas the revolutionary principle meant either hearing the music relative to a chord (the tonic triad) rather than a tone (in the first case) or no longer hearing any such relationships (in the second). Schoenberg (1975) explained: "It was the first step toward a style which has since been called the style of 'atonality' . . . Although the word revolution had not, at this time (about 1907), exclusively the ominous political flavor which is attributed to it today, I always insisted the new music was merely a logical development of musical resources." (Forte 1978, 137).

The revolutionary post-tonal principle took various forms, starting from the idea that one could base music on scales that are more ambiguous than familiar major and minor scales or not based on perfect 5th/4th intervals (scales that were neither diatonic nor pentatonic in the usual way; e.g., the whole-tone or diminished scale). The most revolutionary idea was to avoid perceptible pitch anchors of any kind systematically and to strive for atonality. That brought new structural principles (e.g., those of serialism) to the fore, not all of which were perceptually relevant for listeners.

Returning to the Renaissance transition, and given that both modes and keys are abstract theoretical constructs, it may make more sense to talk about what happened in the actual music: "To understand tonal relationships in Renaissance music is to understand the ground from which our own sense of tonal relationships originally sprang. The primary tonal elements of Renaissance music are pitch-classes and triads, to all intents and purposes acoustically the same as those of eighteenth- and nineteenth-century music, and there is much in the detail of tonal relationships in Renaissance polyphony that is comfortably familiar. The sonic surface is sometimes faintly exotic, often charmingly vague and undirected to our ears, but hardly alien" (Powers 1981, 428).

Renaissance composers increasingly wrote, and Renaissance musicians increasingly performed, what we today call major and minor triads in root position. Listeners became increasingly familiar with the sound of those sonorities, and it became increasingly possible for composers and improvisers to prolong those triads in a Schenkerian sense, allowing MmT to emerge.

That being the case, medieval church modes were hardly "supplanted" by major and minor keys. Composers did not "steadily advance toward the ultimate goal of tonality," as formulated by Robert Wienpahl (1971, 408). Instead, church modes and (not yet theoretically identified) major and minor keys existed in parallel for centuries. During that time, they could be regarded as different possible interpretations of the same musical surface: "seventeenth-century pitch structure is neither fully modal nor fully tonal" (Rivera 1989, 25).

Against this background, we can explain why MmT has only two modes compared to the twelve of Heinrich Glarean (1547/1965). The principles that determined the number of church modes were quite different from those determining the number of later tonalities. Glarean expanded the medieval eight-mode system (with finals on D, E, F, and G, and authentic and plagal versions of each) to include twelve modes (adding C and A to the list of possible finals). There were six possible finals because, although the diatonic scale comprises seven tones, only six were suitable, the 5th interval above one of them (B) being diminished (creating a forbidden tritone).

In medieval chant, modes on C and A may have been avoided due to the proximity of the dissonant tone B to the final. A melody often ends with stepwise progression to the final from either above or below. In any case, chants did not always include all seven diatonic tones. They were sometimes confined to a hexachord and could be transposed to another pitch within the diatonic scale without changing the intervals (e.g., from C to F). In such cases, four finals were sufficient.

MmT has only two tonalities because there are only two possible tonic sonorities. The major and minor triads are the most consonant trichords because they uniquely combine two properties: high harmonicity (including a perfect 5th or 4th) and low roughness (lacking 2nds). A passage in a major or minor key is a prolongation of the tonic triad, whose pitch-salience profile correlates with the prevalence profile of pitch classes in the passage. The pitch-salience profile also correlates with the scale-step stability profile (the key profiles of Krumhansl and Kessler 1982).

These claims are consistent with Schenker's intuition that a passage of music in a major or minor key is perceived as a prolongation of its tonic triad. But they are not aesthetic arguments for or against MmT. Whether MmT is a good or a bad thing is a different question. The intention here is merely to explain how MmT works and why. For better or for worse, most Western music is still, after centuries of adventurous tonal exploration, mainly based on major and minor chords and scales.

Attentional Focus in the Arts

To shed light on MmT's psychological origins, it helps to divide it into two aspects. On the one hand, there is the recognition, categorization, and cognitive processing of large numbers of complex pitch patterns. On the other hand, there is the perception of such patterns relative to reference pitches (tonal centers). That suggests a link to other arts.

Regarding the first point, there could be an intrinsic human desire to recognize and understand patterns. That could drive both the visual and the auditory arts. Musical emotion could be a reward for successfully recognizing auditory patterns—comparable with the emotional effect of abstract visual art. According to Huron (2016), "stereophonic reproduction provides the listener with additional cures that aid in auditory scene analysis. If the brain rewards itself for forming coherent auditory images and if stereophonic localization cues contribute to the listener's ability to form vivid and coherent auditory images, then we can understand why listeners might experience stereophonic sound as sounding more pleasant than monophonic sound" (22).

Laurel Trainor (2015) proposed similarly that "musical pitch and timing structure adapted to preexisting features of auditory processing that evolved for auditory scene analysis." Valorie Salimpoor and colleagues (2015) offered details of how such a boot-strapping process might work: "At a fundamental level, listening to music involves tracking a series of sound events over time. Because humans are experts in pattern recognition, temporal predictions are constantly generated, creating a sense of anticipation. We summarize how complex cognitive abilities and cortical processes integrate with fundamental subcortical reward and motivation systems in the brain to give rise to musical pleasure" (86).

But there is something strange about the idea of auditory scene analysis as the origin of music or an aspect of it such as MmT. Auditory scene analysis has a clear function: it helps us identify sound sources and interact with them, which promotes survival and reproduction. Why would humans universally play games with auditory scene analysis that do not have any clear evolutionary benefit? We surely do not need to practice the skills of auditory scene analysis, since everyday interaction with the acoustic world already involves constant practice of that kind.

The perception of musical pitch patterns relative to reference pitches is hard to reduce to a general psychological or artistic principle. Researchers in music cognition have argued that tonality reduces the cognitive load on the listener (e.g., Bharucha 1987; Deutsch and Feroe 1981; Krumhansl 2000; Tillmann 2005). If so, one would expect to find analogous processes in other arts. But we do not—or if we do, the connection is tenuous.

In the visual arts, spatial relationships may be perceived relative to one or more focal points that spontaneously attract attention or manipulate the gaze. The effect can be created in diverse ways, including diagonal or converging lines or contrasts of dark versus light. Are such focal points analogous to tonal centers in music? Do they reduce the cognitive load on the observer, increasing appreciation of the painting? The results of Charles Miller (2011) went in the direction of confirming that hypothesis, whereas those of John Mullennix and colleagues (2018) did not.

Both in real life and in the visual arts, visual attention is normally focused on objects (Cupchik et al. 2009). That suggests that other aspects of the visual field are perceived relative to those objects, by analogy to musical tonality. In everyday audition, attention is similarly on the source of sound, which in music might be a singer or instrument, perceived relative to other objects or sound sources. In psychology, there is some evidence that perception is generally relative to reference points (Rosch 1975) or landmarks (Presson and Montello 1988). There is also the cognitive idea that perceiving a pattern relative to reference pitches reduces the cognitive load, explaining preferences for tonal over atonal passages of music (cf. Tillmann et al. 2008).

The term "framing" is frequently applied in analyses of literature. In fact, any kind of art may be understood to be confined to, or perceived within, a frame—a concept similar to schema in psychology (Wolf and Bernhart 2006). In that very general sense, musical tonality is a kind of frame or schema, but the comparison does not allow us to draw more specific conclusions. No commonly found structure in painting or literature is as specific and as popular as MmT.

Environmental affordances may be relevant. In an ecological approach, perception in all senses focuses on affordances, enabling the organism to interact successfully with the environment of which it is a part. This idea might help us understand relationships between different art forms. Common subjects in paintings are also common environmental affordances: other people (including the appearance of their bodies), landscapes (implying human dependence on the natural environment), and still life (flowers, fruit, utensils). Harmonic tonality is based on the harmonic series that is constantly heard in human speech and other environmental sounds; its important role in MmT and emotional connotations seem to be anchored in these affordances. Humans need social interaction, supported and enabled by language, to survive and reproduce.

To what extent can a stage play, novel, or film be compared to a piece of music? There are interesting connections between sonata form (with exposition, development, recapitulation) and a classic three-act narrative structure (setup, confrontation, resolution). In the first act, the characters are introduced, analogous to the introduction of musical themes in sonata form. The setting is also presented, analogous to the presentation of a musical style. In the second act, there is confrontation between the protagonists and their situation changes, analogous to the thematic development in the development section of sonata form. When musical themes appear in different tonalities and in modified forms, that is reminiscent of characters appearing in different situations. But the durations are different: a typical sonata form movement lasts for ten to thirty minutes, whereas a play lasts for two hours. In terms of content, it would be far-fetched to try to link the three acts of a classical narrative play to the three movements of a classical sonata or symphony.

Connections between visual and auditory arts are suggested by common terminologies. In both cases, there is composition, organization, form, ordering, line, shape, space, implied movement, and formal structure. Terms that focus more on visual arts include design, color, viewpoint, depth of field, and symmetry. The term "texture" refers to perceived surface quality, but in music it also has a more specific meaning: the number of simultaneous voices. Possible links to musical tonality include the theme of an art exhibition and the narrative thread or hero's journey (Batty 2010). Like tonality, themes and threads make it easier to understand and appreciate the content. But these preliminary observations are not specific enough to shed interesting light on the structure of musical tonality.

20 Progression and Modulation

Successive chords in MmT often have one or two tones in common and hence medium to high pitch commonality. Progressions of two chords may be preferred if the first chord implies the second, the implied *pitches in the first chord being* realized *in the second. That leads to asymmetries: rising 4th intervals between successive roots happen more often than falling 4ths, and falling 3rds happen more often than rising 3rds. But there are also phenomena that pitch commonality cannot readily explain. In major tonalities, V and IV are more prevalent than iii and vi; in minor tonalities, V and iv are more prevalent than III and VI. That suggests the existence of a cognitive hierarchy in which the tonic I is more important than the dominant V, the dominant more important than the subdominant IV (or ii), and the subdominant more important than other diatonic triads or functions. In such a hierarchy, the origin of the tonic is clear: it is the prolonged sonority or the sonority to which all other sonorities relate. The dominant implies the tonic because subsidiary pitches in the dominant triad are realized in the tonic triad. The tones of the subdominant (or predominant) are chosen such that all diatonic scale degrees can harmonized by tonic, dominant, or subdominant chords. The perceived size of a modulation from one tonality to another depends on scalesteps that are not common to the two tonalities and their salience. In this way, patterns of implied pitches and pitch salience in chords can explain asymmetries in both chord progressions and modulations.*

Introduction

Chords are perceived to progress well when they stand in perceptible harmonic relationships with each other (pitch commonality) and the voices lead appropriately (pitch distance). Good voice leading in common-practice tonality is largely about reducing the size of melodic intervals: the smaller the intervals traversed within each voice, the smaller the overall perceived distance between the chords. Larger intervals between successive tones in a voice are possible if followed by a step in the opposite direction. Contrary motion between parts is favored over similar (e.g., Forte 1962).

A progression from one chord to another may also depend on the similarity of the pitch-class sets represented by the chord. The idea is relatively abstract and was originally intended for understanding analyses of the atonal repertoire. Robert Morris (1979, 449) offered the following example. How similar are a diminished-7th chord (in semitones: [0369]) and the all-interval tetrachord ([0146])? The interval vector of the dim7 chord is <004002> (i.e., it contains four minor 3rds and two tritones), whereas the interval vector of the all-interval tetrachord is <111111> (as its name implies, the all-interval tetrachord includes all six intervals). Clearly, the two interval vectors are very different, which justifies regarding the chords as very different.

But when the diminished-7th chord and the all-interval tetrachord are realized in music, possibly in very different voicings, the differences that we perceive may depend on quite different factors. If we perceive the all-interval tetrachord as much more dissonant, the reason might not be that it includes all six chromatic intervals up to a tritone, but rather that it in particular includes intervals of one and two semitones (producing roughness). Moreover, the chord's pitch pattern is very different from the harmonic series. The chord is not smooth, harmonic, or familiar and is therefore highly dissonant. By comparison, the diminished 7th seems relatively consonant, with less roughness and more harmonicity.

Now consider the perception of chord progressions in tonal contexts. Perceptible harmonic relationships between successive chords can be analyzed either top-down, starting with theoretical assumptions about harmonic progressions in major and minor keys, or bottom-up, assuming nothing about context and harmonic function and instead invoking simple psychoacoustic principles. The top-down approach relies on musical knowledge and learning and in that sense may be called a nurture approach. The bottom-up approach is closer to what psychologists call nature, although it may also involve learning if Terhardt's learning-based theory of pitch perception is applied. As so often, the truth lies between nature and nurture. Our perception of harmonic relationships depends on a mixture of direct perception of tones within chords and their relationships, and our experience of similar sounds in the tonal music to which we have been exposed.

For example, chord progressions often define tonalities—especially in conventional cadences such as ii-V-I. From a psychological viewpoint, this process can be understood by combining the top-down cognitive-structural approach of Krumhansl (1990) and Lerdahl and Jackendoff (1983) with the more bottom-up psychoacoustic approach of Huron and Parncutt (1993) and Parncutt and Bregman (2000).

Harmony and Melody

The melody or main voice of a four-part arrangement is normally in the soprano, although countless exceptions could be listed: medieval counterpoint in which the tenor carries the tune and discants are higher in pitch; Western pop songs sung by men with higher instrumental accompaniment; pop, rock, and folk numbers in which a second singer harmonizes the melody by adding 3rd and 6th intervals above it. Exceptions aside: Why is the melody usually in the soprano?

We may first ask why we perceive one voice in a multi-voice texture as melody at all. Why not hear all voices as equally important? The theory of auditory scene analysis offers a possible answer. An important function of auditory scene analysis is to separate the auditory foreground from the background so that attention can be focused on the foreground. This mechanism is also used to separate melodies from accompaniments in music.

In a four-part chord progression, it is often possible for a musician to hear out the voices and transcribe them, even if the voices themselves are not equally audible. In general, the outer voices are more audible than the inner voices because they are masked from one side rather than both. The soprano and bass voices are masked from only one side, whereas the alto and tenor are masked from both sides. That is not true for all partials, but it is true at least for the fundamentals, which tend to be the most audible partials. The degree to which one tone masks another depends on the interval between them; if it is large enough, there is no masking. That can explain why our attention tends to focus on the soprano and bass lines, and why rules of counterpoint tend to apply more strictly to soprano and bass voices.

But why do we hear the melody in the soprano and not the bass? The soprano does not mask the bass more than the bass masks the soprano; on the contrary, at higher sound levels, a lower pure tone masks a higher pure tone more than a higher tone masks a lower, a phenomenon known as upward spread of masking (Oxenham and Plack 1998). The effect was explicitly included in the algorithm of Terhardt et al. (1982b).

But there is another explanation for the tendency of the soprano voice to carry the melody. In MmT, the bass voice determines the harmonic function of the chord. Therefore, it cannot at the same time carry an independent melody, unless it is appropriate for the inversions of the chords to be determined by the melody. Therefore, the melody can be presented most audibly and independently in the upper voice.

That being the case, there may be no "innate, peripheral origin for higher-voice superiority" as Trainor et al. (2014, 60) claimed. Instead, there is a tradition (at least in Western music of the past few centuries) of performing the highest voice more loudly, regardless of whether it is the root, because that makes it easier to follow the melody.

Amplifying the upper voice also reduces the roughness of the sonority. Roughness is produced by pairs of partials (pure tones) within the sonority that are close to each other in frequency. The roughness of such a pair, and hence their contribution to total roughness, depends on beating in the amplitude envelope, which in turn depends strongly on their relative amplitude. A large difference in amplitude means much less beating in the temporal envelope and much less roughness (Terhardt 1968).

It is also possible for a soprano voice to have a lower sound level than other voices and still be heard because it is effectively masked from only one side. High solo violins, woodwinds, or voices may be heard relatively easily over loud, dense orchestral or choral textures.

Chord Prevalence

Data on the frequency of occurrence of chord functions relative to major and minor keys were provided by Helen Budge (1943) in compositions of eighteenth- and nineteenth-century composers. including Couperin, Rameau, Scarlatti, J. S. Bach, Händel, Gluck, C. P. E. Bach, Haydn, Mozart, Cherubini, Beethoven, Weber, Rossini, Schubert, Mendelssohn, Chopin, Schumann, Liszt, Wagner, Verdi, Brahms, Moussorgsky, and Tschaikowsky. One interesting finding was that III chords happen remarkably seldom, although they have two tones in common with both I and V, in both major and minor keys. In table 1 of Budge (1943, 10), the I chord in root position (which includes i in a minor key) made up 22 percent of all counted chords, whereas III in root position (which includes iii in a major key) made up only 2 percent. Chord V made up 9 percent, although it has only one tone in common with I, whereas III has two tones in common with I.

Clearly, pitch commonality with the tonic triad cannot account for these findings. Instead, there is a remarkable focus on I, V, and IV—consistent with the theories of Rameau and Riemann that these are the most important chords or functions in MmT.

A possible explanation for the relative rareness of iii and vi in a major key, and III and VI in minor, involves psychological complexity. Given the enormous number of compositional possibilities that the major-minor system offers, listeners may prefer music that focuses on a small number of primary chords to reduce the overall complexity. To establish the prevailing scale and tonic most clearly and simply, the main chords should include all seven tones of the diatonic scale and no others.

If we assume the scale is diatonic and the chords are triads, the minimum number of triads that satisfies this criterion is three. Chords I, V, and IV include nine tones altogether, which is greater than the required seven, whereas two triads can include no more than six tones, which is not enough. Chords I, V, and IV represent two tones of the diatonic scale twice each, and the other five once each. The repeated tones are $\hat{1}$, which is part of both I and IV, and $\hat{5}$, which is part of both I and V—the two tones that are most stable in a major or minor tonality. That seems like an optimal solution.

Asymmetries in Chord Progressions

A psychoacoustic theory of chord progression is initially independent of the order of chords. If chord A goes with chord B, the relationship is heard regardless of which chord comes first. But some chord progressions tend to move in one direction more than another, and the very word "progression" implies directionality. Such progressions are asymmetrical if the relationship between the chords is changed when the order is changed.

Often, roots move by rising 4ths. Sometimes, there is a chain of them. The jazz standard *All the Things You Are* by Jerome Kern Oscar Hammerstein II, which begins vi-ii-V-I-IV . . . , is full of such chains. It's hard to find a jazz standard that does not feature ii-V-I progressions that tonicize different keys. Progressions of that kind are also familiar from common-practice tonal music.

Schoenberg explained that some progressions are *stronger* than others (Goehr 1973). Progressions whose roots fall through 5ths (e.g., V-I) and 3rds (e.g., I-vi) are strong by comparison to progressions in the opposite direction (e.g., I-V, I-iii). In a strong progression, the second chord is perceived as a goal of motion and is perceived to replace the first chord. In a superstrong progression, the root rises or falls by a step (e.g., I-ii), and the two chords have no tones in common. A weak progression is a prolongation of the first chord.

If the second event in a sequence elaborates the first and is heard retrospectively relative to the first, we may speak of a tensing motion (Lerdahl 2020, 12). Consider the weak progression I-iii (in C major: C-Em). It can be heard as a prolongation of I; the iii chord is created by adding the leading tone as a passing tone, so the chord seems like a passing chord, and is slightly dissonant relative to the original chord. Whatever terminology is used, the second chord is heard as subsidiary.

Schoenberg's idea of strong, weak, and superstrong progressions is backed up by empirical data. Roland Eberlein (1994) assembled a representative sample of mainstream tonal Western music that was composed between 1700 and 1850. The music was in four or more parts and represented the late Baroque, classic, and early romantic periods. The sample included works by J. S. Bach (7 chorales; *Kleine harmonische Labyrinth*), Handel (Trio Sonata Op. 5 No. 5), Mozart (*Missa Brevis*), Beethoven (Mass in C Major), and Mendelssohn (Motets Op. 78). Eberlein reduced these works to chord progressions by hand, ignoring dissonant passing notes and focusing on harmonic functions. The identified chords were limited to major, minor, and diminished triads, and major-minor, major, minor, diminished, and half-diminished-7th chords. He then counted how often given chord types progressed to given other chord types, the root moving through a given rising or falling interval.

Table 20.1 is an extract from Eberlein's results, focusing on root-position major and minor triads. The statistics confirm the predominance of rising 4ths between roots of diatonic triads by comparison to falling 4ths in his sample. But progressions in the opposite direction are also common. In a Schenkerian approach, a IV-I progression (e.g., F-C) can be regarded as a prolongation of I, especially if it is preceded by V (V-IV-I). The progression ♭VII-IV-I in rock (e.g., B♭-F-C) can similarly be regarded as a prolongation of I.

Konstantinos Giannos and colleagues (2022) presented original and retrograde chord progressions and asked listeners to rate their consonance. The

Table 20.1
Prevalence of selected progressions of two triads in a representative sample of Western classical music of the eighteenth and nineteenth centuries according to Eberlein (1994, 422–423).

Progression	Rising 4th between roots	Falling 4th
Major-major	64	19
Major-minor	60	1
Minor-minor	21	5
Minor-major	5	20
Total	**150**	**45**

original progressions were either cadential (e.g., IV7-II7/♯$\hat{4}$-V7-I) or modulating. In both cases, they ended with a V-I progression. Retrograde progressions comprised the same chords in reverse order. The original progressions were rated more consonant, and the authors concluded that directionality plays a significant role in C/D perception.

De Clercq and Temperley (2011) counted two-chord progressions in a corpus of harmonic transcriptions of rock. They found that whereas V progressed more often to I than to IV, consistent with classical music (to I, 788 times; to IV, 392 times), IV progressed more often to I than to V (to I, 1162; to V, 514), falling 4ths outnumbering rising in that case. That departure from classical music is surprising, given that classical harmony and rock harmony are remarkably similar in other ways: "harmonies in rock are overwhelmingly in root position and . . . major triads are more common than minor triads (the latter difference comparable to that found in common-practice)" (De Clercq and Temperley 2011, 67). Rock may deliberately avoid V-I progressions because they sound too classical—just as rock singers avoid classical styles of voice production.

David Sears and David Forrest (2021) suggested that although differences of this kind may point to stylistic characteristics of rock, rock as a whole may nevertheless conform to conventional expectations. Their "RS-200/Billboard (popular) data" database comprised the Rolling Stone Corpus RS-200 of De Clercq and Temperley (2011) and the McGill Billboard data set of John Burgoyne and colleagues (2011)—the latter using automatic chord recognition from audio. When Sears and Forrest analyzed the database for the "top ten conventional 3-gram types" (the most common three-chord progressions), the progression IV-V-I happened 474 times, followed by V-IV-I (264), I-IV-V (209), V-I-IV (202), vi-IV-V (172), and ii-V-I (109; table 2, 148). When the same progressions were ranked in order of "*pMI*$_c$, a statistical association measure that privieges patterns whose members co-occur more frequently than the background probability would suggest" (147), the order from most to least frequent was IV-V-I, vi-IV-V, ii-V-I, V-I-IV, ♭VI-♭VII-i, V-IV-I, ii^7-V^7-I, I-IV-V, V^7/V-V^7-I, IV^{M7}-iii^7-ii^7 (where V^7/V = II^7). Taken as a whole, these data suggest that whereas rock music sometimes noticeably contradicts classical norms, consistent with its status as protest music (Damodaran 2016), "the tonal norms basic to the pop music from which rock emerged are the same norms common to the system of common-practice tonality" (Everett 2004). For Sears and Forrest (2021), that includes persistent preferences for rising 4ths between successive roots: "For the popular corpus, seven of the top ten types featured tonal progressions, reinforcing claims about the influence of the major-minor diatonic system in popular music . . . By comparison, progressions characterized by flat-side triads, the 'harmonic code [of] rock' . . . are notably scarce" (147).

Harmonic Implication-Realization

The theory of virtual pitch, combined with Narmour's (1990) theory of implication and realization, can account for asymmetries in triadic progressions. Narmour argued that certain melodic movements imply subsequent movements. For example, a rising leap implies a following step in the other direction. Chord progressions may work in a similar way if the pitches evoked by a chord include virtual pitches at non-chord pitch classes, CEG implying F and A, ACE implying D and F.

Consider the pitches implied by the first chord in a two-chord progression. The chord CEG implies F because the root (C) and the 5th (G) of the chord correspond to the 3rd and 9th harmonics of F, respectively. The chord implies A because its 3rd (E) corresponds to A's 3rd harmonic, and its 5th (G) to its 7th harmonic. The chord implies D if C is perceived as the 7th harmonic of D, and E as the 9th. Extending Narmour's concept, we might predict that a listener will feel a kind of satisfaction or closure if one of more these implied pitches are realized in the next chord—for example, if the chord following C major (CEG) is F major (CFA), A minor (ACE), or D minor (DFA). That could explain why falling-3rd and rising-4th progressions (strong) and rising-2nd progressions (superstrong) tend to be preferred over progressions in the opposite direction.

Here, the word "imply" is being used in two different ways: the chord CEG *simultaneously* implies virtual pitches at missing fundamentals F, A, and D; and the virtual pitches F, A, and D *successively* imply that the next chord will contain them. These pitches are expectations that, if realized in the next chord, cause the progression to move forward.

The feeling of a chord progression moving forward ("progressing") can be an effect of either nature or nurture. Independent of context, if we expect a C-major triad to progress to F major, A minor, or D minor (e.g., I-IV, I-iv, I-ii), there are two possible reasons: implication and realization (nature) or musical familiarity (nurture). Polyphonic music may have been influenced by the described implication-realization effect for centuries, with the consequence that rising-4th and falling-3rd progressions were preferred, even if the implication-realization effect was not directly perceived. As people became familiar with these asymmetries, they started to expect them. Chord progressions in the opposite direction (I-V, I-iii, ii-I) are not perceived as moving backwards. Rather, they have a more static quality, and the second chord is more likely to be perceived as a prolongation of the first.

The implication-realization effect of expecting a pitch that is implied by a chord (e.g., the pitch F implied by the chord CEG) to be realized in the next may be especially relevant for singers or performers of melodic instruments. When performing one pitch after another (e.g., the tones E to F, or G to F) within a chord progression (e.g., C major to F major), a singer or instrumentalist may use implied pitches (missing fundamentals) as a guide for intonation. Good intonation relies on imagining (audiating) what will happen next before executing it (Jordan 1987) while simultaneously producing the current sound (E. Johnson and Klonoski 2003). In the progression C-F-C (I-IV-I in C major) harmonized by four voices (SATB) as CGCE-FACF-CGCE, the sopranos imagine the F in the second chord while singing the E in the first. That process is facilitated in this case by the F that is implied as a missing fundamental within the first chord. The sopranos then imagine the third chord while singing the second, which is more straightforward, the third chord being the same as the first (so it is still available in echoic memory). Such effects are less relevant for keyboard or guitar players, which could explain why such asymmetries happen less often in rock.

The Dominant

One of the most exciting things about classical MmT is dominant prolongation, the dominant usually being V or V7. A well-known example is the closing measures of Prelude No. 1 in C from J. S. Bach's Well-Tempered Clavier (vol. 1). The final dominant harmony lasts seven measures built on G2 in the bass. During that time, expectation of tonic resolution increases, enhancing the feeling of resolution or satisfaction when the tonic finally arrives in the fourth-last measure. Another example by J. S. Bach can be heard in Suite No. 3 in C Major for Solo Cello, BWV 1009, Prelude (taking advantage of the open G string).

In the classical style of Haydn, Mozart, and Beethoven, the main theme of a piece returns in the tonic key toward the end of a piece or movement, and dominant harmony prepares the listener for that return. William Caplin (1998) explained: "The final harmony of the development is normally the dominant of the home key. This harmony best prepares for the recapitulation, which almost always begins with the home-key tonic. The dominant usually appears as the final harmony of a half-cadential progression and represents the goal of the last themelike unit of the development. The harmony is typically prolonged by a substantial standing on the dominant in order to reinforce expectations for its resolution to tonic" (141).

A well-known example of this kind of dominant is the end of the third movement (Scherzo: Allegro) of Beethoven's Symphony No. 5. The dominant (a G-major triad in the key of C minor) is prolonged, followed by prolongation of the tonic minor, in preparation for the triumphal return of the tonic C major at the start of the fourth movement (Allegro). A more striking case is the prolonged dominant in the second movement of Brahms' Requiem, "Denn alles Fleisch ist wie Gras." The dominant harmony creates a special kind of expectation that is hardly likely to be an arbitrary feature of a randomly chosen musical style. In a totally different style, Maurice Ravel arrived dramatically at a dominant harmony (C♯ relative to an implied tonic on F♯) in measure 65 of the first movement (*Ondine*) *of Gaspard de la Nuit* for piano solo. Decades later, Roy Orbison and Bill Dees used a prolonged dominant to dramatic effect near the end of their 1964 hit "Oh, Pretty Woman."

Following a dominant preparation, the tonic return—whether it happens as expected or is somehow delayed or changed—is a characteristically emotional moment. Leonard Meyer (1956) explained the emotion in terms of implication and realization. The dominant creates strong expectations, whereas the subdominant is more stable.

Despite harmonic developments in jazz and rock in the twentieth century, MmT in those genres is still dominated by tonic, dominant, and subdominant functions, even if they are fuzzily defined (e.g., IV and ii being almost interchangeable), and even if, contra Riemann, not all harmonies in a progression can be clearly classified that way (tonal styles since 1900 having evolved in diverse directions). Remarkably, there is still consensus about the relative importance of Riemann's three main harmonic functions (I being the most important and IV the least); their relative stability (I being the most stable and V the least); the order in which they are "normally" heard (IV-V-I); and the ambiguity of IV (better called "predominant" and including ii, IV, II, iv, Neapolitan 6th, augmented 6ths, and others). Krumhansl's (1990) concept of tonal hierarchies, like that of Lerdahl (1988), relies similarly on this hierarchical structure, and her concept is based on empirical data from experiments in which participants had presumably heard more popular music than classical.

Origin of the Dominant

Melody is based on speech, and speakers breathe. In that sense, phrasing and closure are musical universals (e.g., Lynch et al. 1995; Lerdahl and Jackendoff 1983). Each phrase comes to an end before the next phrase starts (or at the same time, in elision). At a deeper hierarchical level, groups of phrases come to an end, and eventually a whole piece comes to an end. According to Igor Stravinsky (1947), "All music is nothing more than a succession of impulses that converge toward a definite point of repose. That is as true of Gregorian chant as it is of a Bach fugue, as true of Brahms' music as it is of Debussy's. The general law of attraction is satisfied in only a limited way by the traditional diatonic system, for that system possesses no absolute value" (36; as cited in W. Thomson 1958).

Inspired by Caplin (1998), David Sears and colleagues (2018) statistically analyzed a corpus of Western classical musical scores and concluded that "terminal note and chord events from perfect authentic cadences are more predictable than (1) non-cadential events featuring tonic harmony in root position and supporting any scale degree in the soprano, and (2) non-cadential events featuring any other harmony and any other scale degree in the soprano . . . the reported findings support the role of expectancy in models of cadential closure, with the most complete or closed cadences also serving as the most expected or probable" (48).

The cadential-closure effect investigated by Sears et al. was based on musical experience. How can we account for the origin of the effect in the history of music perception? Real-time music perception cannot be separated from historic changes in musical syntax, insofar as the latter indirectly affects the former. That is an example of the more general principle that nature and nurture cannot be separated.

The dominant is usually defined in music theory as scale step $\hat{5}$ or chord V, but there is also a more general definition. In the tradition of Riemann, a dominant chord is one that has dominant function: it is heard to be unstable, and the expected subsequent tonic chord resolves the instability. On that more general basis, a dominant chord may be defined as a chord that (1) is clearly different from the tonic chord and (2) implies the tonic chord in advance. That being the case, we may ask how and why V implies I in MmT. What exactly is implied by a dominant, and what is the origin of the implication? Any chord can generate expectations based on how it is normally used. What is so special about the dominant?

Riemann (1872) presented the idea that a I-IV-$I^{6/4}$-V-I progression can be interpreted in the tradition of Hegel and Hauptmann as thesis-antithesis-thesis-synthesis-thesis, or perhaps more simply as thesis-antithesis-thesis. Focusing on the three examples of tonic harmony (I) in that progression, he argued that "tonic thesis becomes a tonic antithesis and finally a tonic synthesis" (Mooney 2000, 85). His abstract ideas of "thetic fifth" and "thetic third" can be invoked to explain why the dominant is built on $\hat{5}$ and not $\hat{4}$, but the idea of linking the cadence to the dialectic seems arbitrary and far-fetched.

Rameau (1722) proposed that a perfect cadence leads from scale-degree $\hat{5}$—the 3rd harmonic of $\hat{1}$—to its source (the tonic, root, fundamental, or 1st harmonic), from which it is generated. Imagery of that kind is interesting but hardly scientific.

In the classical V7-I cadence, scale-degree $\hat{7}$ (the leading tone) rises to $\hat{1}$ (the tonic) at the same time as $\hat{4}$ (the dissonant 7th above the dominant) resolves to $\hat{3}$ (the 3rd of the tonic), in contrary motion. The two resolutions can also happen one after the other, creating the $\hat{1}-\hat{7}-\hat{4}-\hat{3}$ schema that happened frequently in the music of the 1770s (Gjerdingen 1988). Sounding both resolutions at the same time reinforces the feeling of closure. The argument can be expressed more generally. There are two semitones in the C-major scale:

B-C and E-F. Because of the strong relationship that we perceive between successive tones a semitone apart (due to their proximity—a general principle in auditory scene analysis), the most final-sounding progression that ends with a C-major triad is one that includes both these semitone transitions. The obvious choice is G7-C. D♭7-C is also possible but less likely because it is not confined to a regular diatonic scale.

But that explanation of the bias toward falling 5ths between roots relies on the addition of a 7th to V. There is also a rising–falling asymmetry between chord roots separated by perfect 4th intervals in *triadic* diatonic music or in more complex music when only triads are counted (table 20.1), suggesting that the asymmetry is historically older than 7th chords. One often hears triadic rising-4th progressions in the music of seventeenth-century composers such as John Dowland, Girolamo Frescobaldi, Orlando Gibbons, Samuel Scheidt, or Heinrich Schütz.

The bias toward rising 4ths between successive roots can be explained if the V-I progression that often ends a polyphonic phrase is a harmonization of $\hat{7}-\hat{1}$ or $\hat{2}-\hat{1}$ in the melody: "As a cadence, both [the triad on $\hat{5}$ and the 7th chord built on $\hat{5}$] are most often resolved to the tonic triad because of the presence in both of the leading tone and the supertonic" (Randel 1986, "Dominant," 236). But $\hat{7}-\hat{1}$ can also be harmonized iii-I, and $\hat{2}-\hat{1}$ can be harmonized ii-I. Besides, this idea cannot easily explain prolonged dominant (V) preparations in nineteenth-century music.

Agmon (1994) proposed that a progression of two chords is preferred or "makes a stronger effect" (245) if the root of the second chord is not sounded in the first chord. In that sense, the root of the second chord sounds new, and the principle may be called the *root-newness* principle. In the chord progression from G major to C major, or G major to A minor, the root of the second chord is new in that it is not included in the chord of G major. That sounds intuitively correct, but the idea is not anchored within a more general theory.

Another possible explanation for the origin of V-I cadences involves medieval two-part voice leading, in which the resolution of the major 6th to the octave in contrary motion (e.g., GE-FF) was prominent. Dahlhaus (1968) argued that this resolution became V-I when other voices were added (e.g., GE-FF → GCE-FCF → CGCE-FACF; see also Stern 1981). But centuries elapsed between the 6–8 resolution and the V-I cadence, and during that time, the fourteenth-century double leading-tone cadence, which can be explained in a similar way (GE-FF → GBE-FCF), was forgotten. Why might the major 6th to octave resolution have been maintained for so many centuries? After all, "superficial similarities between medieval and later phenomena (things that sound like triads and V-I cadences) have placed us in danger of ignoring major differences in the grammar and syntax which govern their use" (M. Bent 1998, 30–31). It is likely that "medieval composers could not conceive sonorities of three, four, five parts as extensions of the underlying dyadic grammar" (M. Bent 1998, 33). It may therefore be far-fetched to propose a causal link between a feature of two-part medieval counterpoint (the 6–8 resolution) and a feature of classical harmony (the V-I progression), given the enormous changes in musical syntax that happened in the intervening centuries.

How, then, can we explain the enduring importance of V as dominant and, more generally, the hierarchy of I, V, and IV? The question can be broken down into subquestions. An initial issue is why are there three main chords at the top of the tonal hierarchy and not some other number. The answer to that question could be rather simple. If one wanted to harmonize a diatonic melody with just two chords, one's choices would be rather limited. If the scale was CDEFGAB and one chord was CEG, the other would be BDFA, a half-diminished 7th. If one chord was DFA, the other would be CEGB, a major-7th chord. If one was EGB, the other would be FACD, an added-6th chord. If one was FAC, the other would be GBDE, an added-6th chord. If one was GBD, the other would be FACE, a major-7th chord. If one was ACE, the other would be GBDF, a Mm7 chord. Those are all possible ways to harmonize diatonic melodies with just two chords, but they do not happen very often in real music because the 7th chords are relatively dissonant. The three-chord solution represented by chords I, IV, and V is preferable because all three chords are, or can be, more consonant. Four chords are unnecessary.

The next question is: Why do the three chords comprise the specific notes that they do? In C major, the tonic is CEG, the dominant is GBD, and the subdominant is usually FAC or DFA. Why those notes and not others?

Agmon (1995) proposed a possible solution. In a first step, he explained: "There are three principles on the basis of which the primary triads I, IV, and V may be selected as prototypes. The first principle is self-evident: the tonic triad, by virtue of its special, referential status, is a prototype. (Referential and prototypical are synonymous for all practical purposes.)" (210). So far, so good. He continued: "The second principle may be termed the principle of symmetry. It states that . . . if one selects any triad as prototype, one must also select its mirror image" (210).

But symmetry of that kind is not a fundamental music-theoretical or psychological principle. Listeners do not perceive symmetry relations between chords. Major and minor triads are perceived as related because they have the same interval vector—not because they are mirror images of each other. Another example: chords [047T] (Mm7) and [036T] (ø7) have the same interval vector <012111>, both being members of pitch-class set {0359}, and may therefore be perceived as similar—but they are not perceived as symmetrical.

Agmon (1995) continued: "The third and last principle states that two maximally similar triads cannot, simultaneously, be selected as harmonic prototypes (in other words, no two prototypes may occupy adjacent positions on the circle of thirds)" (201). In that case, I agree. Two triads with two tones in common can only harmonize four scale tones. If we want to harmonize all seven scale tones with a small number of chords, we should select chords that have only one tone in common. That criterion can explain the importance of 4th/5th relations between chord roots.

One way to explain the origin of the dominant is to focus on dominant function: the dominant implies the tonic. In mainstream tonal music, we strongly expect the tonic to follow the dominant. Of course, the expectation is culturally learned, and there are many exceptions. But to understand the exceptions, we need to understand the rule. Riemann, Rameau, and other music theorists to the contrary, the question can be answered without recourse to number ratios, numerology, or any kind of sophisticated mathematics. We need to remember the following points:

- From the seventeenth to the nineteenth centuries, there was a general tendency for the roots of triads to rise by 4ths or fall by 3rds. This tendency did not depend on the addition of 6ths or 7ths to the chords, or the resolution from the 7th of the dominant to the 3rd of the tonic (see table 20.1).
- The pitches that we hear in music do not necessarily correspond to the notes in the score. Pitches also differ in perceptual salience. A chord typically evokes several pitches with different saliences.
- Statistical asymmetries in chord progressions suggest that a chord implies a following chord if relatively weak pitches in the first chord are realized as relatively strong pitches in the second—a trivial example of an implication-realization relationship.

The implication-realization approach to understanding musical pitch-time structure is promising because it generalizes. It predicts not only melodic effects, as in Narmour's theory, but also harmonic effects, when weak virtual pitches are regarded as implications. It predicts not only the tendency toward rising 4ths between successive roots in tonal chord progressions, but also the tendency toward falling 3rds.

If a G-major triad (GBD) is the dominant and is heard to imply two virtual pitches (at E and C), these implications are realized if the chord if followed by a C-major triad (CEG)—keeping one tone (G) constant and moving the other two. That makes the C chord a candidate for the tonic. If a G-major triad is heard to imply three virtual pitches (at E, C, and A), it can function as the dominant of A minor, corresponding to Schoenberg's superstrong progression (with no common tone). In the key of A minor (or Aeolian mode), G major is the subtonic or ♭VII (similar to the minor dominant v; here, E minor); as such, it can act as a modal dominant (cf. Biamonte 2010). If the tonic is C, the progression from G major to A minor (V-vi) is the familiar interrupted or deceptive cadence. If the G triad is understood to imply only one virtual pitch at E, the implication is realized if it progresses to E minor, with two tones in common. In the key of C major, that is V-iii, whereas in G major, it is I-vi.

These ideas can explain why progressions V-I and I-IV are common. They can also explain why the superstrong progression IV-V is common—but only in major keys. In the key of C major, IV is FAC, which implies D, B♭, and G as missing fundamentals. Two of these three implied pitches are part of V (GBD). In the key of C minor, iv is FA♭C, which implies D♭ and B♭; none of the three tones in V (GBD) are anticipated in this way.

The same implication-realization idea can explain why, in a controlled psychoacoustic experiment (Parncutt 1989), successive tones in the central musical pitch register sound more similar if they rise than if they fall through an octave. Similarly, Piet Vos and Jim Troost (1989, figure 3) found that roughly 70 percent of octave intervals in Western and non-Western melodies are rising. The second most salient pitch in a harmonic complex tone is an octave higher than the fundamental; it may act as an implication that is realized in a rising octave leap. If the tones are pure, the asymmetry may be reversed (see figure 13.2). Many examples of this octave asymmetry could be listed. The first two tones of the melody *Over the Rainbow* (by Harold Arlen and Yip Harburg in the 1939 film *Wizard of Oz*) rise through an octave ("Somewhere"). So do the first two tones of "High on a hill" at the start of *Lonely Goatherd* from the 1959 Rodgers and Hammerstein musical *Sound of Music*. The start of *You Needed Me* (1978) by Randy Goodrum is another example ("I cried a tear"). Exceptions include the falling melodic octave at the start of the third movement of J. S.

Bach's Italian Concerto BWV 971 in F major. There are also repeated falling octaves in the chorus of *Mein Herr* from the 1972 film production of the 1966 musical *Cabaret* by John Kander and Fred Ebb. Some of the examples in this paragraph are from the start of a phrase, where one in any case expects a rising interval. A possible alternative explanation for the bias toward rising intervals in that case is the arch-like shape of musical phrases (Huron 1996).

Since musical melodies are almost never performed with pure tones, we can hardly compare octave leaps between pure versus complex tones in music. If pure-tone melodies were common, we might, according to implication-realization theory, expect falling octaves to be more common than rising, the second most salient pitch of a pure tone being an octave lower than the fundamental (provided there is noise in the background; Houtgast 1976). Again, the arch-like shape of melodic phrases might be relevant.

The implication-realization idea generalizes in another way, functioning in different ways in different musical styles. The delayed resolution of a prolonged dominant can be compared to a beat drop in electronic dance music (EDM): "Buildups come after breakdown sections and they lead toward cores and beat drops. The buildups and drops in EDM have been described by Snoman as 'emotional waves,' in a manner reminiscent of Ernst Kurth's 'force waves.' Buildup sections are transitional, in between moments of rest (breakdowns) and moments of the highest energy in a track (cores)" (J. W. Smith 2019, 32–33). At a beat drop, a previously implied pitch in the bass is realized by a strong bass beat. In many cases, the implied pitch is simply the root of the current chord in a lower octave register: instead of a V-I progression, a beat drop may feature a falling octave in the bass. The emotional effect is comparable with that of a classical prolonged dominant leading to a recapitulation.

Origin of the Subdominant

If the above theory of the origin of the dominant is correct, and assuming a hierarchical framework in which the tonic is more stable than the dominant, we can now ask: What tones make up the subdominant?

A third primary harmony is needed because the tonic and dominant triads do not include all diatonic scale tones. Taken together, the tonic and dominant triads include five out seven tones (in C major, GBD and CEG include C, D, E, G, and B). To harmonize the remaining tones (F and A, or scale steps $\hat{4}$ or $\hat{6}$), we need a third triad that includes those two tones.

The tones F and A together belong to two triads: F major and D minor. Similarly, in the key of C minor, the tones F and A♭ belong to two triads: F minor and D diminished. This simple idea explains the origin of subdominant harmony in both major and minor keys. If non-diatonic tones are allowed, in the minor key, we have the option of a D♭-major triad: a Neapolitan harmony with subdominant function.

The main points in this account of the origin of tonic, dominant, and subdominant functions can be formulated in another way, focusing on the main constraints:

- *Optimal complexity*: Western listeners like consonant sounds, but they also like a certain amount of complexity. They prefer standard diatonic scales because seven tones per octave allow for innumerable melodic combinations, and those combinations are maximally consonant, the scale being a continuous segment on the circle of 5ths. Western listeners like major and minor triads for similar reasons: the three tones allow for a lot of variety, but the intervals between them maximize consonance.
- *The tonic scale degree as root*: Assuming a preference for a small number of primary harmonies and a clear tonal center, the tonal center should be the root of one of the primary harmonies.
- *Interrelation of primary harmonies*: To maximize consonance, the primary harmonies should sound related to each other.
- *Harmonization of all scale tones*: The primary harmonies should be different enough to allow for all scale degrees to be harmonized. Pairs of primary harmonies should therefore have no more than one tone in common. The chords C major and E minor together cover only four out of seven tones of the C-major scale, whereas C major and F major cover five; so, C and F are preferable. That could explain why 4th intervals between the roots of primary harmonies are preferred over 3rds or 2nds. That is a more tangible explanation than one based on the circle of 5ths as a cognitive structure.
- *Creating closure*: The dominant should create an expectation that the tonic will follow.

Modulation: An Ecological Approach

A piece of music in MmT starts in a given key (or tonality). The key then changes (modulates), often many times. In the classical tradition, the piece returns to the original key. Between the two keys, there is a modulating chord progression. Sometimes, there is a pivot chord that is diatonic in both the previous and the following tonalities.

Heinrich Schenker proposed that great classical works never modulate because at the deepest background level they are perceived relative to one and the same key. Modulations only happen at higher hierarchical levels, closer to the surface. Passing modulations in the middleground or foreground are not real modulations. Despite this influential idea, the idea of modulation is still central to the theory of tonal music, and it begs the question: What does it mean to say that a passing modulation has happened?

To answer this, we need to know how long it takes listeners to forget a previous tonality when the music modulates to a new one. Research on pitch memory suggests musicians can retain pitches for about two minutes (Rakowski and Rogowski 2007). Nicholas Cook (1987) suggested that listeners might remember a previous tonality for one minute. In a systematic study of tension ratings while listening to different modulating passages, Morwaread Farbood (2016) reduced that estimate to twenty seconds. These findings challenge the psychological reality of classical forms and suggest how the concept of passing modulation in music theory can be defined or operationalized. A passing modulation is one in which the previous key has not yet been completely forgotten, such that returning to the earlier key reduces harmonic tension rather than increasing it.

In standard music theory, the key of C major is related to the key of G major because the two scales have several pitch classes in common—six, in fact. Only one pitch class differs (F in C major ≠ F♯ in G major). C major and G major are also held to be related because they are adjacent on the circle of 5ths, but that idea is problematic, there being no direct empirical evidence for the circle's psychological reality. Instead of assuming that such cognitive structures exist, it may be preferable to focus on direct perception such that "perceiving, acting and knowing [can] be studied as emergent properties of an organism–environment system" (Richardson et al. 2008, 159; see also Mos 1988).

Relationships between musical keys (tonalities, before and after modulations) depend on the pitch commonality of their scales. Take C major and G minor. In that case, the circle of 5ths idea does not make a simple prediction because one key is major and the other minor. The perceived distance is greater than one step on the circle of 5ths because the change from major to minor increases the distance. It is presumably also greater than the distance between C major and B♭ major, B♭ being the relative major of G minor. The theory of pitch commonality sidesteps this issue and simply observes that G minor has two flats in its key signature, suggesting that it is perceived as two steps away from C major, which has neither sharps nor flats. If the sharpened leading tone is important, or if the scale is understood to be harmonic rather than natural, G minor is three steps away from C major. The estimate can be improved by weighting scale steps according to their stability (Huron and Parncutt 1993) and taking the correlation coefficient between key profiles (Krumhansl 1990; Krumhansl and Kessler 1982).

The high correlation between Krumhansl's major and minor key profiles and the pitch-class salience profiles of the corresponding major and minor tonic triads suggests a new definition and explanation for the phenomenon of modulation. Schenker's idea of tonicization can be regarded as ecological (focusing on the perceptible structure and affordances of the stimulus), whereas the music-theoretical idea of modulation is cognitive (focusing on mental representations). In an ecological approach:

- When we perceive music, we perceive the stimulus directly. The patterns that we perceive arise when perceptible aspects of the stimulus are compared with prior experience of similar stimuli. Prior experience includes typical chord progressions and how they relate to tonic triads (invariants of MmT).
- When we perceive a passage to be in a key, we perceive it as a prolongation or an elaboration of that key's tonic triad, such that everything we hear somehow relates or points to that triad, represented by its pitch class salience profile. When we perceive a modulation, we perceive a change in the referential tonic triad. After the modulation, we perceive the music relative to the new triad.
- Perception is radically situated (cf. R. G. Barker 1968). We can only predict how people will behave in a given situation if we understand the situation. Music is similar. Its pitch structures imply spatial movement in three dimensions (Eitan and Granot 2006; Kurth 1917). We perceive the spatial movement relative to a reference sound (the tonic) rather than an abstract structure.

A directly perceived musical stimulus is compared with memory for similar stimuli. If the only music people hear in their daily lives is MmT, being primed to expect MmT is the same as being primed to expect music. A possible priming stimulus is simple repetition of short pitch-time patterns, such a phrase that is repeated exactly (Deutsch et al. 2008). In typical real-life settings, that prime is combined with other primes such as perceptible objects (loudspeakers, musical instruments) or situations (concerts, nightclubs).

If perception is active, perception and action are inseparable. In the case of MmT, when tonal structure becomes ambiguous (such as during a modulation), it is the triadic reference that becomes ambiguous. To resolve the ambiguity, we actively search for a new triadic reference while at the same time engaging with the music's inherent emotionality. To understand that process, the idea of cognitive structures may be too abstract. The meaning of music may instead reside in the specific musical structures that exist in real time and in memory, and their interrelationships.

In that sense, tonality may not be represented by a cognitive structure. Instead, a piece of music may be perceived to be in a key when the familiar pitch-time patterns characteristic of that key (its invariances) are recognized. Salient pitch classes in the tonic triad are reinforced by repetition or the rising leading-tone effect.

Changing Tone Profiles

Krumhansl (1990) modeled musical modulation by tracking tone profiles as they change from one chord to the next. The perceived closeness of two tonalities corresponded to the correlation coefficient between the corresponding profiles.

In an experiment, listeners heard a chord progression followed by one of twelve chromatic octave-complex tones (to avoid effects of pitch register). They were then asked how well the probe tone followed the progression. The profile of twelve mean ratings was the tone profile following that progression, at that moment in the music. If the most recently heard chords had clearly defined a key (e.g., predominant, dominant, tonic), the tone profile was a key profile (as illustrated in figure 1.1).

To investigate how the tone profile changes during a chord progression, the experiment was repeated after every chord in the progression. In each trial, listeners heard all the chords up to that point followed by a probe tone. The correlation between the real-time profile and the profile of the original key was then calculated to check how far the music had modulated. One can also correlate the real-time profile against the target key at the end of the progression to see how the music is approaching that key. On the assumption that there are twenty-four major and minor keys, one can quasi-continuously correlate the tone profile with all possible keys to discover which is the closest at a given moment.

Terhardt's pitch theory allows this process to be modeled (Huron and Parncutt 1993). The tone profile at any given point in the progression is predicted by adding the calculated pitch-salience profiles of the immediately preceding chords. To account for recency (more recent chords have a bigger effect on the profile), each profile is weighted exponentially for memory decay. The first chord in the progression may also have an advantage (primacy effect).

In a cognitive approach, the listener's cognitive representation is a tone profile from one moment to the next, which includes the salience or stability of each chromatic scale step. If this profile is similar to a key profile, the tonality is clear. If it is relatively flat (without clear peaks), the tonality is unclear. That can account for the experience of modulation as changing tonalities—an important aspect of MmT.

Asymmetries in Modulations

In a classical sonata in a major key, the first modulation is to V, the dominant. Often, there are two thematic subjects, of which the second is presented in the key of V. That suggests that key V is perceived to be closer to key I than key IV. In minor keys, the first modulation is often to III, the mediant or relative major, suggesting that III is perceived to be closer to i than VI.

These asymmetries can be understood by considering the perceptual salience of the individual non-diatonic (chromatic) tones that provoke or enable the modulation. A classical modulation from a major key to its dominant (V/I) involves scale degree $\sharp\hat{4}$, whereas a modulation to the subdominant (IV/I) involves $\flat\hat{7}$. Scale degree $\sharp\hat{4}$ tends to be less salient (less noticeable to the listener) than $\flat\hat{7}$, on average, because $\sharp\hat{4}$ is usually the major 3rd of chord II, whereas $\flat\hat{7}$ is the root of ♭VII, or the minor 3rd of v; the 3rd of a minor triad is more salient than the 5th (see figure 15.1). Recall also that in Krumhansl's key profiles, $\hat{3}$ is rated higher, and hence more stable, in a minor than a major key (see figure 1.1). Therefore, modulating to IV sounds more distant than modulating to V.

A Mm7 chord can be built on any scale degree (as a secondary dominant or applied dominant), and the 3rd above the root of a secondary dominant is often notated as a sharp relative to prevailing scale. Secondary dominant chords are commonplace and do not sound very surprising (or chromatic). They do not necessarily to cause the music to modulate to another key, and perhaps they do not even trigger passing modulations. The chromatically altered scale step lies a major 3rd above the root, and is not very salient.

The tendency of music in a minor key to modulate to III rather than VI can be explained without reference to virtual pitch. The rising-3rd modulation to III unsharpens the leading tone (♯7 becomes ♮7). The falling-3rd modulation to VI does the same and, in addition, shifts $\hat{2}$ to $\flat\hat{2}$, introducing a salient new tone that can be the root of the new IV, the 3rd of the new ii, or the 7th of the new V. Therefore, the modulation to VI sounds more distant.

Remarkably, these asymmetries are opposite to asymmetries found for chord progressions (see chapter 19). In chord progressions, rising perfect 4ths (e.g., V-I) and falling diatonic 3rds (e.g., I-vi) happen more often than falling perfect 4ths and rising diatonic 3rds, respectively. In modulations, falling perfect 4ths (e.g., I-V) and rising diatonic 3rds (e.g., i-III) happen more often than rising perfect 4ths or falling diatonic 3rds, respectively.

21 Structural Cognition

A detailed consideration of tonality's cognitive structures is beyond the scope of a book on psychoacoustic foundations. But cognitive and psychoacoustic aspects interact, and the results of empirical studies can often be accounted for by a mixture of both. Cognitive representations of tonal passages can be visualized as tree structures or networks, inspired by syntactic structures in the linguistic theory of Noam Chomsky. Musical pitch structures exist on both surface and deep levels; they specify which pitches and sonorities act as references for other pitches and sonorities. Like Schenkerian analyses, Fred Lerdahl's tree structures are intrinsically ambiguous: for a given passage, different possible solutions arise that music listeners might experience or theorists might consider.

Psychological Reality

A psychoacoustic approach to understanding the perception of musical chord progressions starts by focusing on the stimulus. It initially ignores the musical experience of the listener and the musical structures that the listener might intuitively understand based on prior musical experience or knowledge. It considers all pitches that may be evoked in a chord, their relative salience, and relationships that can be perceived between successive chords (pitch commonality and pitch distance). Such an approach may be regarded as bottom-up by comparison to perception of chords relative to tonality, which is more top-down.

But musical chords heard in psychoacoustic experiments are not heard in isolation. A chord in one experimental trial is heard relative to the previous trial, and even without that comparison, a chord may be heard relative to an imagined tonal context. Listeners in such experiments usually need a musical background to participate successfully, and they use that background to make sense of the experimental task.

In a chord progression, some chords are perceived to be more important than others: some chords are more likely to be heard as stable references, relative to which the others are perceived. Familiarization with chord progressions involves acquiring the tendency to hear some chords as stable references, and when many people classify chords as more or less important according to similar rules, the tendency may be described as a skill and related to cognitive parsing of language (natural language processing).

Structural cognition is fundamental to music. Most music seems to have it, like all languages do. In MmT, whenever a chord in a progression is perceived as superordinate or subordinate to another chord, and each chord is prolonged (becoming a hierarchical level of its own), we can speak of cognitive structures of musical pitch.

The psychological reality of musical structures can hardly be denied. Intuitively, music would be meaningless without them, just as language would be meaningless without cognitive structures for grammar and meaning. Many music theorists have assumed that cognitive structures of musical pitch exist, and some have demonstrated their existence empirically (e.g., Lerdahl and Krumhansl 2007). Structurally based predictions of note-by-note changes in musical tension (Bigand et al. 1996; Lerdahl and Krumhansl 2007) provided additional empirical evidence for the psychological reality of such structures. Petr Janata and colleagues (2002) found evidence for their neural foundations, providing further support. The generative nature of cognitive structures in both language and music is also intuitively clear: a relatively simple structure can generate a large or possibly infinite number of linguistic or musical realizations.

Musical expectations that reflect cognitive structures or musical syntax have been thoroughly investigated both behaviorally and physiologically (e.g., Cenkerová and Parncutt 2015; Levitin and Tirovolas 2009; Loui et al. 2009; Margulis 2005; Patel et al. 2008; Pearce et al. 2010; Rohrmeier and Koelsch 2012; Salimpoor et al. 2015). Harmonic priming experiments have combined behavioral, neurophysiological, and computational methods to study the perception of melodies (Marmel et al. 2010), chord pairs (Bharucha and Stoeckig 1986), and chord sequences (Bigand et al. 1999; Tillmann et al. 2006). In diverse ways, these studies have shown that pitch-time patterns that happen repeatedly are spontaneously learned and subsequently expected.

Some modernist composers have tried to remove structure from music, or disguise or obscure it, such that listeners no longer perceive it (Dibben 1994). In such cases, listeners may instead be aware of what the sounds specify: "physical source, physical space and proximity, genre, musical function, performance skill, emotional attributes and social context" (Dibben 2001, 161). That idea corresponds to a performative theory of meaning in which "meaning is constructed through performative acts" (184) and casts doubt on the status or universality of cognitive structures in music.

Even outside of modernist traditions, it has not always been possible to demonstrate the existence of long-term structural hierarchies in music empirically. Listeners may attend more to the musical surface than the underlying structure (Deliège et al. 1996; Tillmann and Bigand 2004), liking a piece of music just as much or even more when the order of its sections is changed (Karno and Konečni 1992; Tillmann and Bigand 1996). That may be true, even when music theorists regard the order of sections in the music as crucial, and musicians and critics alike would be shocked if the order was changed in performance.

The auditory system is remarkably good at processing musical or linguistic patterns in shorter time windows of less than thirty seconds and poor at processing such patterns over longer timespans or at integrating information from shorter windows into longer ones. Barbara Tillmann and Emmanuel Bigand (2004) commented that "for short time-spans, simulations with connectionist models suggest that the influence of local structures on music processing can emerge from the simple accumulation of activation patterns over time (weighted by temporal delay), without necessarily extracting a hierarchical organization" (219). They went on to ask: "What is the real benefit for the musical brain in processing large-scale structure? Why should the brain consume attentional capacity for global processing if local musical structures are rich enough for emotional responses to music? To some extent, this question extends beyond the musical domain. Indeed, when listening to a speaker, it sometimes occurs that weakness in or the complete absence of large-scale structure of a discourse can entirely be compensated for by fascinating eloquence in short time windows. . . . More generally, time may tend to be processed moment by moment when such moments lead to an extremely rich aesthetic, intellectual, spiritual, or emotional experience" (219).

If findings of that kind seem to contradict the intuitions of musicians and music theorists, a possible explanation is that psychological research methods are somehow insensitive to longer-term effects. However, evidence for that interpretation is also lacking.

The concept of cognitive structure is problematic in an approach based on ecological psychology and 4E cognition. In an approach that focuses on direct perception, the existence or ontological status of cognitive structures can be doubted (Costall 1984). The function of musical elements relative to cognitive structures is generally ambiguous, and the analyses proposed by music theorists are typically selected from several possible alternatives, all of which could be perceived by listeners.

Linguistic Analogies

Noam Chomsky (1965) proposed the existence of a universal grammar for language—a set of structural rules for language that are common to all humans and unique to them. Chomsky's argument relied on the poverty of the stimulus: the idea that the linguistic stimuli to which infants are exposed are too impoverished for them to develop intuitive grammatical understanding without access to a preexisting, genetically determined model. But evidence from statistical language learning—not to mention the personal experience of any infant's parent or carer—contradicts the poverty-of-the-stimulus argument. Infants repeatedly hear similar simple patterns of syllables and words and try to reproduce them countless times—more often, it seems, than would be necessary to pick up the basics of grammar such as the difference between nouns, verbs, and prepositions.

Language is less ambiguous than music—otherwise, we would have difficulty communicating. Nonetheless, linguistic utterances are often ambiguous because they are grammatically incomplete (people usually do not communicate in grammatically complete sentences). Even complete sentences are often ambiguous—on different levels simultaneously. Chomsky's example "Colorless green ideas sleep furiously" is grammatically clear but semantically ambiguous. If I said, "The Pope sent me a bull," it would be unclear whether he sent me an animal or a text. But the meaning of a sentence can be ambiguous, even if there is no ambiguity in either the meaning of the words or their grammatical function ("Some man loves every woman").

Humans are predisposed by a mixture of nature (genetically determined aspects of behavior) and nurture (preverbal experience) to construct quasi-grammatical representations of culture-specific sonic relations (both linguistic and musical) during social interactions. The function of those interactions may be primarily social, the specific linguistic and musical outcomes being mere byproducts. The distinction between nouns, verbs, and prepositions exists in the outside world, in the ways that people (including infants and their carers) interact with physical objects and with each other. Therefore, grammatical knowledge is not necessarily innate, and it may be completely learned.

Similarly, the feeling of approaching the end of a musical phrase, when the pitch contour curves down or the tonic triad is sounded, could be entirely learned. What might be innate is a tendency to regard one musical or linguistic sound object as more important than another and then to perceive the less important relative to the more important. That could be the foundation for building cognitive hierarchies, and it is related to Schenker's concept of prolongation.

The Role of Learning

In ethnomusicological and psychological approaches to music cognition, members of a cultural group learn the structures of their music by exposure and interaction. That involves many repetitions, music being intrinsically repetitive.

Given the diversity of musical structures, we can hardly speak of a universal grammar of music. Whereas it is surely correct that the tonal-musical grammar proposed by Lerdahl and Jackendoff (1983) mixes universal aspects (e.g., those related to or based on auditory scene analysis) with culture-specific aspects, their claim at the start of their *Rule Index* that "Rules marked with an asterisk are idiom-specific. We believe the rest are universal" (345) is problematic if it implies that most of the rules are universal.

In a study of psychological grouping in dance entitled "Steps toward a universal grammar of dance," Isabelle Charnavel (2019) proposed "six principles of change that determine group boundaries in dance (change of body part, orientation, level, direction, speed, quality)" (abstract). One could argue that these principles, which were inspired by Gestalt psychology and also refer to Lerdahl and Jackendoff, enable universals of dance to be learned from universals of the human body and the physical and social environments with which it interacts. Charnavel's finding that "grouping is a general cognitive capacity applying across domains and modalities, and shows how specific grouping principles are stated in relation to modality-specific and domain-specific dimensions" (abstract) can be interpreted that way. These are the general ways in which musical structures are learned, and they are reminiscent of the universal grouping principles of auditory scene analysis. But they do not necessarily imply the prior existence of a universal grammar of dance that is analogous to Chomsky's universal grammar of language.

Listeners to passages of music in MmT experience the same music differently, depending on their expertise and listening experience. That is additional evidence that the structures are learned. Different listeners may get different fuzzy or implicit impressions of what the current tonality is and how a given tone or chord relates to it.

Pitch Space and Key Profile

It is difficult to measure perceived relationships between tones, chords, and tonalities directly. To understand how those relationships work psychologically, it can help to create a model for them and test that model by comparing empirical data (such as musical tension ratings) with predictions based on the model. If there is agreement between the predictions and the measurements, confidence in the model grows. If an entirely different model makes similar predictions, our confidence in the model is reduced.

A pitch space (cf. Deutsch and Feroe 1981) is an example of such a model. The basic space of a major key according to Lerdahl (1988) is shown in table 21.1. I have added the term "hierarchical depth" to the figure to show why some scale degrees are more stable than others and to what extent. The inverse of hierarchical depth is embedding distance, which Lerdahl contrasted with step distance.

A pitch-space path is a path between two locations in pitch space, such as between two chords in a progression, each of which is perceived relative to the prevailing tonality, but also relative to each other. The distance depends on the level: for example, a perfect-4th interval "has a step value of 1 in triadic space, 2 in diatonic space, and 4 in chromatic space" (Lerdahl 1988, 322). In an empirical psychological approach, we may ask if or to what extent these different spaces exist, how clearly they are defined, and to what extent they depend on musical experience.

The idea of pitch spaces can help us understand the harmonic relationships that are perceived between chords in progressions. For example, a V

Table 21.1
The basic space of Lerdahl (1988) for the key of C major, showing hierarchical depth (number of levels) and its inverse, embedding distance.

Level *a*	C											
Level *b* (5th space)	C							G				
Level *c* (triad space)	C				E			G				
Level *d* (diatonic space)	C		D		E	F		G		A		B
Level *e* (chromatic space)	C	D♭	D	E♭	E	F	F♯	G	A♭	A	B♭	B
Hierarchical depth	5	1	2	1	3	2	1	4	1	2	1	2
Embedding distance	0	4	3	4	2	3	4	1	4	3	4	3

chord relative to an opening tonality (V/I) may be perceived as the tonic of the dominant key (I/V). It may be perceived as a pivot chord with a different function relative to preceding and following passages. According to some music theories, that is a standard way to execute a modulation or key change. The same chord might then progress to ii/V, which can also be heard as iv/I, depending on whether one is listening in longer or shorter timespans.

The psychological reality of such cognitive structures was demonstrated in Krumhansl's first key-profile experiment (Krumhansl and Shepard 1979). A simple ascending scale generated a hierarchically structured pattern of expectations called key profiles, one for major keys and one for minor. That discovery motivated attempts to explain the psychological origin of the profiles. The method of Krumhansl and Kessler (1982) suggested the origin could be found in cadential chord progressions—an idea that was later confirmed by psychoacoustic modeling (see chapter 9). That being the case, the data of Krumhansl and Shepard (1979), in which listeners heard scales rather than chord progressions, can be explained if the key profiles are learned from chord progressions and then applied to unaccompanied melodies—especially when the melodies imply tonal chord progressions.

Two contrasting approaches to modeling Krumhansl's key profiles produce high correlation coefficients of $p \sim 0.95$ across twelve pitch classes in both major and minor keys. First, the key profiles can be compared with the hierarchical depth of each chromatic-scale degree in Lerdahl's (1988) basic space (table 21.1; minor tonality can be represented in the same way but with the 3rd and 6th scale degrees shifted down by a semitone). Second, the key profiles can be compared with the pitch-salience profile of the tonic triad alone (figure 1.1; Parncutt 2011). Of these two models, Lerdahl's is a more convincing model of tonal process (*how* music works) because it includes detailed assumptions about how tonality works, whereas Parncutt's is a more convincing model of tonal origins (*why* music is like it is) because it does not include prior assumptions about tonality and in that way avoids circular reasoning.

Inherent Ambiguity

Whereas it is clear that music is perceived to have structure, and in that sense cognitive structures of music may be said to exist, musical structures are also generally ambiguous. If the pitch(es) of individual musical sounds are also ambiguous, we can say that ambiguity is inherent in both bottom-up and top-down approaches to tonal musical structure.

Kofi Agawu (1994) mentioned a familiar example of musical ambiguity: the first four notes of Beethoven's Symphony No. 5. Here, both the rhythmic structure (duple or triple?) and the pitch structure (where is the tonic?) are deliberately ambiguous. In this case, the ambiguity seems to contribute to the musical drama.

More generally, one and the same melody can be divided into phrases and subphrases in different ways. Some listeners may hear a phrase starting on the beat, others before it (anacrusis). The same piece of music may have a tempo (tactus) of either 60 or 120 events per minute, depending on how it is heard.

In MmT, we typically hear a metrical structure at the same time as we hear a tonal structure. One and the same tone can be rhythmically stable and tonally unstable (e.g., an accented appoggiatura). In other words, two simultaneous structures can contradict. If we ask which of two sound events is more likely to be perceived as a psychological reference, the answer might be event A in one case and event B in the other. The problem is not only that musical elements generally have multiple functions (just as people have multiple identities), but also that there is no objective way of determining those functions.

In functional harmony, tones and chords often have ambiguous meanings. A supertonic chord can be perceived as a subdominant and vice versa—as Rameau explained. Riemann's idea of harmonic function is similarly ambiguous. Chords that normally belong to a minor key can be introduced into a major key (e.g., the minor subdominant, or a Neapolitan or augmented sixth chord), making the tonality ambiguous (harmonic mixture). Music theorists disagree on the status of chromatic chords (chords containing non-scale tones), some preferring shorter-term modulations (lasting for a few seconds), others longer-term modulations or none at all. The tone F in a context of C major or C minor can be heard as the melodic upper neighbor of E or E♭, the 7th of G (V chord), the 3rd of D (ii chord), or the 5th of B♭ (♭VII chord). Some theorists of the past have regarded a cadential 6/4 chord as a consonant second-inversion tonic, others as a dissonant dominant with a double suspension. In fact, the chord mixes these two functions, as Heinrich Koch explained (Mickelsen 1977, 21).

Structural analyses and reductions can assign musical elements to different hierarchical levels in different ways. In the progression I-IV-I, the IV chord can be regarded either as a subdominant (potentially leading to V) or a neighbor chord (created by prolonging I). In fact, it is both. Similarly, just about every chord in a chord progression can have more than one such function. Musical analyses that insist on just one interpretation when several are possible are standing on thin ice.

Often, it is not clear which structure is heard most often by the average person. Structures perceived by nonmusicians (the usual focus of music psychology) are not necessarily the same as structures heard by expert listeners, or structures that should be heard according to given music-theoretical principles, or structures that were (apparently) intended by composers.

The subjectivity of the analyst plays an important role—as it should, otherwise the analysis would be invalid, in the same sense that a psychological test is invalid if it does not measure what it is purported to measure. But even if one can argue convincingly on the basis of a rule system and/or musical intuition that one analytic interpretation is more likely or more valid than another, the outcome is nevertheless generally uncertain. There is no absolute criterion for choosing among different possible cognitive structures for a piece of music and deciding which interpretation is correct.

The pioneering approach of Lerdahl and Jackendoff (1983) addressed this issue systematically with preference rules, but ultimately could not overcome the basic difficulty. Ambiguity was explicit in their approach, in which:

> in many places in a piece of music the preference rules come into conflict with one another. A grouping boundary may be indicated in one place by a pitch discontinuity, and in an immediately adjacent position by a change in note duration, and in order to avoid the absurdity of splitting a piece of music into single note groups (a result that Lerdahl and Jackendoff's own preference rules warn against) a choice has to be made as to which of the conflicting rules should be ignored and which

> observed . . . properly speaking ambiguity is the result of two or more conflicting structural interpretations each of which is quite definite in itself. . . . Rather than taking the ambiguous result of conflicting structural tendencies as its object, Lerdahl and Jackendoff's theory should address the level at which each conflicting interpretation is singular and definite. (Clarke 1986, 9–10)

Robert Fink (1999) investigated the distinction between the musical surface and its deeper hierarchical levels. He admitted that it would be impossible to prove that distinctions between the levels exist. Given the enormous body of work in both music theory and music cognition that has assumed the surface–depth distinction, it is pragmatic to assume that the distinction is real, such that both surface and depth play important roles in musical experience. By the same token we should be careful not to place too much emphasis on structure. "Hierarchic music theories ask us to renounce the pleasures of the surface for the defensive security of the depths. We may not all want to make the exchange. Perhaps, after all, beauty is only skin-deep" (Fink 1999, 137).

Personally, I am constantly hearing harmonic functions in tonal music. I hear chords relative to tonalities and often notice that a given chord can be perceived relative to two different tonal centers. But my perception may be biased by the approach to music theory that I first learned (Lovelock 1974), my experience as a pianist (imagining structure relative to the keyboard, playing a mixture of classical and jazz/pop), the kinds of music that I like or often hear, or my interest in writing choral arrangements. If other people were hearing different harmonic functions in the same passage, either consciously or not, I would be unaware of the difference. In such cases, it would be unclear which structure was correct—less still, which corresponds to the composer's intention (if there wa a composer, and if she or he had any such intention).

The uncertainty is not limited to which analysis of a given musical passage is the most music-theoretically appropriate or which is the most likely to be heard. Listeners are not necessarily perceiving chords relative to tonalities at all. They may be perceiving chords more directly as independent sound objects that relate to one another in a more direct way. In a study of sensitivity to key change in chorale sequences, Thompson and Cuddy (1989) suggested that "harmony and melody operate somewhat independently in the implication of key structure"; the representation of musical pitch may be only "partially hierarchical" (151). That idea is consistent with the more general claim that the perception of pitch relations in harmonic progressions depends on a mixture of two quasi-independent components: nature (quasi-universal sensory aspects) and nurture (learned hierarchical structures).

VI The Big Picture

22 Theoretical Evaluation

This book has explored links between universal perceptual principles and culture-specific pitch structures, drawing on relevant empirical findings and provoking interesting questions about the structure, perception, and cognition of Western music. More general issues arose about how competing theories or explanations might be evaluated relative to each other, especially when predictions are quantitatively equally successful. To evaluate an interdisciplinarity theory, it helps to consider its structure—how the parts relate to the whole. In my attempts to explain MmT, I have invoked different but related dualities and taxonomies intended as conceptual frameworks to orient the reader and to clarify underlying assumptions and implications. They include psychoacoustics versus cognition, humanities versus sciences, and physical versus subjective reality. Understanding such relationships and their interconnections helps us to evaluate and improve the theory critically. The theory can also be evaluated against more general principles of theory construction.

Psychoacoustics versus Cognition

A psychoacoustic or sensory account of musical structure focuses on relationships between sound as physically measured and sound as experienced in the style of Terhardt. A cognitive explanation focuses on music as information, and how the mind/brain processes that information, in the style of Krumhansl. That is one of the central distinctions made in this book. Questions such as why most tonal Western music is based on major and minor scales or why some chord progressions or modulations are more common than others can be addressed in either a psychoacoustic or a cognitive approach. In both cases, plausible answers can be found that can be backed up by quantitative modeling. Such models are necessarily correlative, such that causation cannot be inferred.

Which approach is more appropriate? Are the foundations of MmT more psychoacoustic or more cognitive in nature? Are the structures of MmT dependent more on the psychoacoustics of the complex tone perception or on the cognitive neuroscience of musical pitch? A possible answer is this: in real-time music listening, cognitive processes predominate (e.g., Bigand et al. 2003), but psychoacoustic modeling may better account for the origins of musical pitch structures.

The quantitative findings of Krumhansl (1990) can often be explained in either of these two ways: a psychoacoustic approach based on harmonic pitch pattern recognition, learned from outside of music (speech and environmental sounds), or a cognitive approach based on cognitive structures, learned from exposure to tonal music. Both approaches can generate accurate predictions (e.g., Parncutt 1994). Modeling the results of different priming experiments, Tom Collins and colleagues (2014) found that "multiple representational stages along a sensory–cognitive continuum combine to shape tonal expectations in music" (33). Given that both approaches depend on learning—Terhardt considering learning in speech and environmental sounds, Krumhansl in music—and that the same neural networks may be involved in both cases, the question is not so much one of nature versus nurture but what kinds of nurture. The published data of researchers such as Bharucha, Bigand, Cuddy, Milne, Russo, Temperley, Thompson, Tillmann, and Toiviainen might usefully be revisited from this perspective.

Positivism versus Relativism

The history of music theory is characterized by positivism—the belief that clear answers to questions about musical structure exist and that these answers, when found, will be true in all times and all places. Leading figures such as Pythagoras (represented by the Pythagoreans), Boethius, Zarlino, Rameau, Riemann, and Schenker tended toward positivism. Arnold Schoenberg was more balanced, and with the rise of musical modernism and ethnomusicology, relativism became a more powerful force. But positivism has always played an important role in music theory—as in music psychology.

The late twentieth century still featured strong tendencies toward positivism in both humanities and sciences. Positivist music historians insisted on the universal and permanent value of the notated canon. Positivist musical set theorists implied that musical structure is ultimately mathematical in nature. Positivist Schenkerians insisted that good music and good analyses are based on the *Ursatz*. There is also positivism in psychological or cognitive approaches. Apart from my own positivist contributions, we can find David Huron claiming that the foundation of voice-leading traditions lies in universal principles of auditory scene analysis, Fred Lerdahl comparing musical structure with a Chomsky-like tree diagram, Ernst Terhardt proposing that the foundations of harmony and tonality lie in universal principles of pitch perception, and Carol Krumhansl implying that the key profiles are a mathematical foundation for a theory of tonality.

The modernism of early twentieth-century music, along with advances in ethnomusicology, undermined music-theoretical and music-historical positivism. For the past century or so, many music theorists have understood the work of historical music theorists only relative to their cultural contexts—adopting a relativist approach to the positivism of the past and highlighting the cultural arbitrariness of music-theoretical "laws." Those scholars include Norman Cazden, Nicholas Cook, Carl Dahlhaus, Jean-Jacques Nattiez, and Philip Tagg. Leonard Meyer managed to balance positivist and relativist tendencies, whereas Eugene Narmour tended again in a positivist direction. In the past few decades, music-theoretical positivism seemed stronger in North America than in Europe, despite the discipline's epistemological diversity.

Looking at the big picture, we have learned a lot from both relativist and positivist approaches. The most appropriate way forward may be to move back and forth from one side to the other, striving for an appropriate balance.

Inherent Interdisciplinarity

An overwhelming amount of literature—old and new, and in contrasting disciplines—addresses questions about musical pitch structure and MmT. Whereas many academic disciplines are involved, the main two are music theory (related to humanities) and music psychology (related to sciences). Both disciplines have ancient roots, but in practice, music psychology dates to the nineteenth century (especially in central Europe), and cognitive music psychology (especially in the USA) dates to the 1960s. There is a great amount of additional literature on musical pitch structure and MmT in the disciplines of acoustics, mathematics, neuroscience, and philosophy (including aesthetics).

I have tried in this book to survey the relevant literature in both main areas. But I cannot do justice to all relevant disciplines, given the enormous and constantly expanding literature in each of the established disciplines that are relevant for this book: historical musicology, music theory, music aesthetics, music psychology, music cognition, music information retrieval, and music philosophy. Today, no one can claim to be an expert in more than one such discipline, but everyone has the opportunity to read widely and incorporate as much material as is reasonably possible when considering interdisciplinary questions.

The vast amount of literature reflects the social importance of MmT and, more generally, musical structure. An important driver is music's mysterious emotional power, which motivates the desire to know where music comes from and hence what it actually is.

The approach to cognitive music theory that I am advocating is based on a philosophy of reality that acknowledges not only the independent existence of subjectivity, which is no news to humanities scholars, but also its quantitative measurability, which is. It also involves a scientific approach based on experimentation and modeling in the tradition of logical positivism, which is no news to scientists; but when that involves inherently subjective parameters such as pitch, the approach is indeed new. Perhaps the expression "quantitative phenomenology" is right for this approach. If so, it represents a significant departure from the usual worlds that humanities scholars and scientists inhabit.

The following argument can be difficult to swallow for scholars and researchers on both sides of the humanities–sciences divide. If we want to understand musical pitch structure systematically, we have to measure pitch. We have to do that subjectively because pitch is inherently subjective. It is part of musical experience, which is also inherently subjective. In the process, we have to consider issues considered important for both humanities scholars and scientists. On the scientific side, we need to consider experimental method carefully, including design, stimulus control, and serial-position effects. On the humanities side, we need to understand the history of music theory and the ways in which music perception changed historically as a function of the music that was being made and heard at the time. We also need to consider the role of intuition and introspection in music analysis.

A truly cross-disciplinary approach is a difficult balancing act, given the gulf that separates and continues to separate the humanities (here: music theory) from the scientists (music psychology). And there are further divisions on each side, such as the mathematical versus non-mathematical music theorists, and the cognitive versus non-cognitive music psychologists. Given that each of these groups approaches questions of musical structure differently, and the main approach of this book does not clearly belong to any of them, difficulties of communication are unsurprising.

Consider the psychoacousticians and neuroscientists. Colleagues in both areas know a lot about the mind-body problem but may be reluctant to raise the issue, although it is clearly central. Many researchers are physicalists who believe that, in the end, everything can be reduced to physics—a position that I sympathize with, although I have argued against it. The physical world is indeed closed and complete, and everything that happens in a psychoacoustical or neuroscientific experiment can in some sense be regarded as physical.

Some scientists think that such philosophical or metaphysical questions belong to the humanities and are therefore beyond their scope. The trouble is that the subjective judgments of listeners in listening experiments are *inherently* subjective. Their judgments exist only in their subjective world of consciousness—however one conceives of that world. The data that they provide are always mediated by their subjective consciousness: if they fall asleep during the experiment, there will be no data, and if they lose concentration, the data will be affected.

The humanities–sciences contrast in theories of musical structure also involves the difference between descriptive and prescriptive/suggestive approaches (see chapter 2). Music psychology may try to account for how the average person (however defined) normally hears music, whereas music theory is often about how music could or should be heard by experts. A descriptive approach, which aims to describe without evaluation, is often taken for granted in the sciences, whereas a prescriptive approach is normal in the humanities. The difference can lead to misunderstandings and conflict.

Being primarily a scientist, I have adopted a descriptive approach for this book. But I am not opposed to prescriptive approaches that suggest how musical structure should be heard—what structures and relationships we should hear or should be able to hear—provided such claims do not imply the superiority of one kind of music over another. Within the descriptive category, some approaches take music theory as their starting point and test aspects of it empirically (Krumhansl), whereas others (Terhardt) try to avoid circularity by starting outside of music (e.g., in the area of speech, hearing, and psychoacoustics). All such approaches have advantages and disadvantages.

Kinds of Reality

To deal with the inherent subjectivity of quantitative data in psychophysics, we need to take a stand on mind-body dualism. To what extent can physical and subjective worlds be treated as separate? How do they depend on each other? These are not scientific questions, but the science cannot be done unless we tentatively answer them first (see chapter 4).

In an attempt to clarify arguments about the nature and origin of musical structures, this book has repeatedly invoked Karl Popper's three worlds. World 1 is the physical world of sound, including frequencies in hertz. World 2 is the experiential world of sensations and emotions, including perceived pitch, or how high tones seem to us, both absolutely and relative to each other. World 3 is the abstract world of thoughts and information, including musical scores—whether printed on paper, stored in computers, or memorized in the mind of a musician.

There is an interesting connection between the psychoacoustic–cognitive dichotomy and Popper's worlds. Psychoacoustic approaches focus on the relationship between Worlds 1 and 2, as measured and quantified in psychoacoustic experiments and models. In that case, World 1 is represented by temporal and spectral representations of musical sounds (waveforms, spectra), whereas World 2 comprises patterns of perceived pitch, loudness, timbre, rhythm, and so on. By contrast, cognitive approaches investigate information processing in the mind/brain—that is, World 3. Both cognitive models of MmT and the pitch algorithm of Terhardt et al. (1982b) belong to World 3, but there is an interesting difference: the input to the cognitive models is the musical score (considering only fundamental frequencies), whereas the input to the pitch algorithm is the entire physical stimulus. In Krumhansl's (1990) discussions of musical scales and chord progressions, music as heard is tacitly assumed to correspond to music notation rather than to a complete physical description or running amplitude spectrum.

There is also an interesting relationship between the humanities–sciences dichotomy and Popper's worlds. Scientists tend to regard Popper's World 1 (physical) as fundamental, whereas humanities scholars are more interested in Popper's Worlds 2 (subjectivity) and 3 (knowledge and culture). Of course, that is a broad generalization, and there are many exceptions.

Psychoacoustics is one such exception. In this book, psychoacoustics is presented as a study of relationships between Worlds 1 and 2, which at first glance seems like an interaction between humanities and sciences. In fact, modern psychoacousticians regard themselves as scientists, but there is also an interesting nineteenth-century tradition of musically trained, phenomenologically oriented psychoacousticians (Hui 2012). Moreover, the discipline of psychoacoustics can be conceived of as belonging entirely to World 1 (cf. Cariani 2001; Patterson et al. 2002).

The domains within which claims made in this book are intended to be valid vary widely. In the introduction, I considered four levels: the physical level, the biological level, the Western level, and MmT. Again, this list overlaps with other dichotomies and taxonomies. Physics and biology are studied by scientists, whereas Western culture and MmT are studied by humanities scholars. The academic discipline known as music psychology or the cognitive sciences of music is an exception to the rule, in which Western culture and MmT are studied by scientists. There is also a link to different kinds or worlds of reality. The physical and biological levels are confined to Popper's World 1, whereas the Western level and MmT also involve Worlds 2 and 3.

Kinds of Learning

Structural aspects of music perception can be influenced by learning in three different ways:

(1) perceptual universals (e.g., familiarity of the auditory system with the harmonic series through exposure to harmonic complex tones in speech);
(2) structure of composed music (e.g., prevalence of major and minor triads in root position in mainstream tonal music; frequency of co-occurrence or transition probability of successive sounds); and
(3) music-theoretical thought (e.g., Rameau's approach to explaining chord roots).

I have focused on the first two, assuming the third effect to be smaller—at least in a first pass at the problem and assuming that we are talking about the music perception of the average person rather than that of musical experts. Music-theoretical insights tend to happen historically later than the developments in musical structure that inspired them. Chord roots are a good example: the idea emerged in the seventeenth century, but statistical analyses of polyphonic scores from different periods suggest that chord roots existed psychologically centuries earlier (Parncutt, Reisinger et al. 2019). In that case, we can assume that the effect of music-theoretical thought on musical development and perception was relatively small. Exceptions to this rule became commonplace in twentieth-century modernism, which is beyond my scope.

All three ways of learning involve another dichotomy: the nature–nurture distinction. Perceptual universals can depend on universals of either biology or perceptible environments. The structure of composed music can depend on either previous music with which composers and performers are familiar or universals of some other kind. Music-theoretical thought can explain musical structures in terms of either nature or nurture. Everything about musical structure may ultimately be learned, at different levels, perceptual universals being closer to nature and music-theoretical thought closer to nurture.

What Makes a Good Theory?

The history of science is full of crazy ideas and embarrassing mistakes—entertainingly documented by Bill Bryson (2004). We should be skeptical about all ideas, musical or otherwise, advanced by all thinkers (scientists and philosophers) of both the past and the present—even those who later became justifiably famous.

That is not to say that the history of music theory is uninteresting. On the contrary, it is a fascinating chapter in the history of ideas. Musicians have always been passionate about music and searched creatively for explanations of musical phenomena, and ideas from the past continue to inspire new research.

That raises interesting issues of evaluation. How can the theory presented in this book be evaluated? Is it yet another crazy idea or embarrassing mistake? Consider the following principles:

(1) *A good theory is parsimonious (Ockham's razor)*. A simple idea is easier to criticize than a complex one. Therefore, a simple idea is more likely to be true than a complex one. The target of criticism is more clearly defined if the idea is simpler. That sounds fine in principle, but what exactly is meant by simplicity or parsimony? Simplicity can be either mathematical or conceptual; I am assuming that both aspects are important, both in general and in regard to music theory. If the phenomena to be explained

are intrinsically complex, simple theories may be inappropriate. A better theory might be one whose complexity is optimal: not too much and not too little.

(2) *A good theory is transparent and testable.* All parts of the theory or model should be clearly operationalized. That enables the detailed elements of the theory to be independently and systematically tested. Spectral and temporal theories of pitch perception achieved this in different ways. In a temporal model, different physiological stages in the periphery and auditory pathways are separated. In Terhardt's model, spectral and virtual pitch are separated. In both cases, testing is facilitated by modularization.

(3) *A good theory makes surprising predictions and accounts for diverse empirical findings.* The theories of Cariani (2001) and Terhardt et al. (1982b) did both of those things. A testable model of pitch perception should take any kind of sound as input, always producing a plausible output. Physiologically and temporally based models of that kind include Marc Leman (2012), Ray Meddis and Michael Hewitt (1991a, 1991b), Ray Meddis and Lowel O'Mard (1997), and Roy Patterson et al. (1995). The predictions should also be consistent with empirical findings in psychoacoustics that are relevant for the perception of everyday sound, including speech and music. The predictions of spectral and temporal models are often similar due to the mathematical equivalence of periodicity in the time domain and harmonicity in the frequency domain. All of them can in principle predict that certain pitches will be heard in certain sonorities, and some will be more salient than others. Thus, the truth may lie between. For musical purposes, Terhardt's model has proven to be both useful and testable, but that is not clear evidence that it is more correct.

(4) *A good theory is internally consistent.* The idea of assigning central concepts to Popper's worlds is intended to satisfy this criterion. A scientific theory should be based on empirical data obtained using scientific methods and principles (e.g., controlled manipulation of independent variables in psychological experiments). Objective and subjective measures should be strictly separated because they are obtained from fundamentally different kinds of method or operationalization.

(5) *A good theory is falsifiable.* It should be possible to conduct a crucial experiment that would disprove the theory. If such experiments are carried out and fail to disprove the theory, we become more confident of the theory's

validity. Unfortunately, music theories are seldom falsifiable in this sense due to the inherent complexity and diversity of musical structures. The same can apply to the interpretation of psychoacoustic experiments. If successively presented tones at perfect 5th intervals are perceived to be more similar than tritones (Stoll and Parncutt 1987), the reason is not immediately clear. Is the effect due to pitch commonality, as the authors claimed, or is it due to musical experience or (equivalently) a tonal cognitive structure? Similarly unfalsifiable theories include the theory of cognitive structures of music advanced by Lerdahl and Jackendoff (1983)—comparable with syntactic hierarches in language with their different deep and surface structures (Chomsky 1957). Structural models in cognition nevertheless do two interesting things: they correspond to intuitions, and they predict the results of relevant experiments.

23 Universals and Diversity

This book has attempted to explain aspects of a politically dominant musical culture in terms of psychological quasi universals. Ethnomusicologists and scholars with a liberal intercultural orientation will rightly object. Might such a study indirectly promote racism or colonialism, or be seen to do so? Should the project be abandoned, or is it enough to be sensitive to the issues? Every musical style features a mixture of psychological universals and culture-specific elements, and MmT is no exception. Psychological research methods can be applied to any music, Western or non-Western, preferably in collaboration with local experts (cognitive ethnomusicology). Similarly, ethnomusicological methods can be applied to investigations of Western music. Contrary to nineteenth-century colonialist illusions of cultural superiority, MmT is one of many possible musical structural principles that are consistent with selected universal perceptual principles. For centuries, Europeans have been forcing MmT onto non-European cultures by violent military, political, and economic means. Today, partly as a result of this global cultural conquest, more people than ever before are listening to, enjoying, and identifying with MmT. But what about diversity, equity, and inclusion? To grapple with these issues, we need a multidisciplinary understanding that includes history and anthropology, psychology and sociology, politics and economics.

Musical Universals?

Robert Lundin (1947) surmised that "judging a musical interval consonant or dissonant . . . is determined by many conditions which operate in an individual's life history" (48). But he forgot to mention that C/D is itself a culture-specific concept. In Western MmT, consonance depends positively on harmonicity and negatively on roughness. But of those three phenomena—consonance, harmonicity, and roughness—none is a *musical* universal: none contributes to musical structure in all cultural contexts, although all may universally contribute to timbre and texture.

Schoenberg (1975) was surely wrong when he claimed that "all musical phenomena can be referred to the overtone series, so that all things appear to be the application of the more simple and more complex relationships of that series" (271; as cited in Väisälä 2002, 207). Outside of Western music and Westernized musical cultures, musical sounds are not necessarily preferred if they produce less psychoacoustic roughness or their spectra are more similar to a harmonic series. On the contrary, acoustic roughness is often aesthetically valued, for example in Balkan *Schwebungs-Diaphonie* (produced by singing small harmonic intervals; Brandl 2008; Vyčinienė 2012) and Asante ivory trumpet music (Kaminski 2009). Post-tonal systems of tonal organization in Western music may seem (or strive) to ignore roughness and harmonicity completely (or deliberately), but that does not mean listeners have stopped perceiving roughness or harmonicity. Instead, listeners are invited to attend to other perceptible aspects of musical structure.

If there are no musical universals, can we define music? If not, can musicology exist? The answer is yes if music is a category of human behavior that exists in all human cultures. Some forms of music lie closer to the middle of the category "music" than others, just as a chair is a more typical piece of furniture than a radio (Rosch and Mervis 1998). Similarly, in categorical pitch perception, some intervals are closer to the middle of the category "minor 3rd" than others: any interval that is within about a quarter tone of three semitones relative to 12-EDO can be recognized as a minor 3rd by Western musicians. A Westerner may regard Adhan, the Islamic call to prayer, as a clear example of song or melody, and hence of music, whereas for a Muslim, it is a recitation, music being forbidden in this context.

Human behaviors can be recognized as music on the basis of a list of sub-behaviors regarded as family resemblances, not all of which are necessary. Eleanor Rosch and Carolyn Mervis (1998) explained:

> The principle [of family resemblances] was first suggested in philosophy; Wittgenstein (1953) argued that the referents of a word need not have common elements in order for the word to be understood and used in the normal functioning of language. He suggested that, rather, a family resemblance might be what linked the various referents of a word. A family resemblance relationship consists of a set of items of the form AB, BC, CD, DE. That is, each item has at least one, and probably

several, elements in common with one or more other items, but no, or few, elements are common to all items. (18)

One kind of music may feature A and B, another B and C, and another C and D. We recognize them all as music because of family resemblances. Sensitivity to family resemblances is a psychological universal: we naturally form concepts in our minds, and the process does not necessarily rely on clearly defined features in the sense of universals. For example, even if "chair" is regarded as a prototype of the category "furniture," we use the word "furniture" to describe certain things found in houses that lack chairs because people traditionally sit on the floor or on cushions.

Recent studies have argued for the existence of musical universals or features of music that are shared by at least most musical styles (S. Brown and Jordania 2013; Mehr et al. 2018; Savage et al. 2015). Samuel Mehr and colleagues (2019) note that "music appears in every society observed; that variation in song events is well characterized by three dimensions (formality, arousal, religiosity); that musical behavior varies more within societies than across them on these dimensions; and that music is regularly associated with behavioral contexts such as infant care, healing, dance, and love; . . . that identifiable acoustic features of songs (accent, tempo, pitch range, etc.) predict their primary behavioral context (love, healing, etc.); that musical forms vary along two dimensions (melodic and rhythmic complexity); that melodic and rhythmic bigrams fall into power-law distributions; and that tonality is widespread, perhaps universal."

Perhaps the term "family resemblance" is preferable to the term "universal" in this case—but that is not the whole story. Humans also create concepts theoretically, and concepts tend to be embedded in broader theories. For a concept to be coherent, it needs to fit a person's background knowledge or more general ideas or theories about the world. The concept "music" is not only a list of features (some of which happen in music A but not B, and so on), but also something people all over the world do (musicking) that is understood relative to other things that people do. For some people, but not others, music is something that feels good but can also be sad, or it is something that expresses personal or social identity.

Misunderstandings about so-called universals can lead to prejudice and arrogance. Depending on our cultural standpoint or perspective, we might perceive one form of music to be more prototypical than another. Music that we know well might seem to us to have more universal musical features (taken from a list of quasi universals) than other forms of music, and therefore to lie closer to the middle of the category "music." Perhaps for that reason, many people today tacitly assume that MmT is more prototypically musical than other forms of music. From a purely scientific or statistical standpoint, that would be a difficult claim to support.

Said another way, musical quasi universals (often, those identified as universal by Western scholars) that are part of Western music do not make Western music superior. Many relevant psychological and biological universals exist that are not characteristic of MmT. Most of the behavioral phenomena studied by psychologists and biologists can be regarded as universals. All humans react to perceptual stimuli, have happy and sad feelings, judge and evaluate information, and so on.

Overtone singing is an interesting example. Overtone singers adjust the shape of their vocal tract to reinforce one or more harmonics, so that listeners hear the harmonics as separate tones. There are overtone singing traditions in many countries and regions, including Mongolia, Tuva, Tibet, Uzbekistan, Pakistan, and Kurdistan, but the technique only found its way into mainstream Western music relatively recently (Wiliford 1999). In the eighteenth century, Rameau was surprised to discover that he could hear out harmonics from a single harmonic complex tone, not knowing that was an everyday thing in other cultural traditions. Overtone singing is biologically and psychologically universal in the sense that almost anyone could potentially learn to perform it by changing the shape of their vocal tract while singing a tone. In any case, almost anyone can perceive separate overtones in overtone singing because almost anyone has a basilar membrane that separates high from low frequencies.

MmT is not inferior because it ignores overtone singing. Similarly, the variable tuning of Indonesian gamelan orchestras (which would be considered catastrophic if applied to Western orchestras) does not make Indonesian music inferior. All three styles—overtone singing, gamelan, and MmT—depend on the same physiological universal: the ability to perceive the partials of a complex tone separately. The example shows that a single physiological universal can enable or facilitate the development of very different musical styles, depending on cultural context. Overtone singing depends on an additional universal: the motor ability to manipulate the resonances of the vocal tract. Every speaking person has that ability, but it is used in a special way in overtone singing. Clearly, biological universals exist, and they do not necessarily determine or influence musical structure.

Given the cultural sensitivity of this topic, any claims of universality should be carefully defined and investigated. Consider the idea that all humans are sensitive to the consonance of major and minor triads. Without evidence, opinions on this issue vary. To find out more, Imre Lahdelma and colleagues (2021) presented Western musical sounds, including major and minor triads, to members of Khowar and Kalash tribes (native to Northwest Pakistan). The authors reported that "the major triad was described as sounding 'strange,' as 'European music,' and as something that is 'not our music'" (81). They concluded that major triads are not universally perceived as more consonant than minor—which might have surprised Riemann and other nineteenth-century harmonic dualists.

But that study was problematic. Western listeners were not asked to rate authentic musical examples from Khowar/Kalash culture (assuming that authentic examples exist or can be identified). The design was not interculturally balanced, making it seem colonialist, despite the best of intentions. Other intercultural psychological studies in the area of music cognition have made a similar mistake, presenting Western music to non-Western listeners but not vice versa (e.g., Fritz et al. 2009).

Intercultural studies have nevertheless produced interesting results. Western psychological ideas can be applied while remaining sensitive to the autonomy and diversity of the music under study. Whereas perception of music from one's own culture involves a mixture of sensory and cognitive processes, sensory processes may dominate when the culture is foreign, paradoxically putting perceptual universals in the foreground (e.g., Balkwill and Thompson 1999; Castellano et al. 1984; C. J. Stevens 2012).

Perceptual Universals

Human sensory systems depend on both genes and the environment. They are influenced by cultural transmission as they develop and mature in interaction with social and physical environments. To the extent that genetics and social/physical environments have universal or quasi-universal aspects, we can speak of perceptual universals.

Harmonicity and roughness are perceptual universals that play important roles in MmT. Sensitivity to roughness is an inevitable consequence of the physiology of spectral analysis: when partials of a complex sound are close in frequency, they interact within auditory filters, producing amplitude modulation. Sensitivity to harmonicity is equally universal because it is necessary for acoustic interaction. Without it, the organism could not recognize and monitor many sound sources (Bregman 1990). In particular, it would be difficult to follow the fundamental frequency contour of speech, which carries a lot of the meaning. Sensitivity to harmonicity is universal because it is necessary for regular speech and communication. Since the auditory system is constantly exposed to harmonic pitch patterns in speech, it is no surprise that those patterns are universally recognized.

But it is a different matter to ask whether roughness and harmonicity are perceived as good or bad in music, or as musical or unmusical. It does not follow that these or any other perceptual universals (should) give rise to musical universals. Helmholtz (1877) was right that "we see in the history of music how the same properties of the human ear serve as the foundation for very different musical systems" (as cited in Steege 2012, ch. 4, cambridge.org).

Sensitivity to harmonicity becomes relevant for music when a number of conditions are met. More than one tone must be sounded simultaneously (deliberately and regularly). Both or all simultaneous tones must be held long enough that the pitches and intervals between them can be clearly heard. Selected pitch intervals must occur repeatedly. Musicians must somehow associate those musical intervals with the intervals among the audible partials of a harmonic complex tone. There should be a reason for doing that.

Psychological universals that give rise to musical universals may be anchored in biological and physiological universals. In melody, the tendency for small intervals (steps) to happen more often, and large (leaps) less often, seems universal (P. G. Vos and Troost 1989). The reason may involve universal physical and physiological constraints of singing. Possibly, the finding does not apply to all music but only to sung melodies due to the technical difficulty of traversing large melodic leaps while at the same time singing in tune (cf. Huron's [2001] interpretation of Fitts' law). For similar reasons, the pitch range of the world's melodies is typically roughly one octave.

Cognitive Universals

A century ago, an educated liberal European might innocently have asked whether MmT (and, in the same breath, grammar in English or European languages) corresponds to the universal workings of the human mind. Today, given the obvious diversity of the world's musics and languages, such a question seems arrogant, colonialist, and possibly racist.

Surprisingly, one of the twentieth century's greatest campaigners for global justice across cultural and racial divides, Noam Chomsky, proposed the idea of a universal grammar that is the same for all languages, despite their diversity (Chomsky 1965, 1986). The idea remains disputed because so much can be explained by environmental interaction and statistical learning. Chomsky to the contrary, interactions between infants and carers are linguistically rich (not impoverished). Moreover, there can hardly be a musical parallel. Whereas Lerdahl and Jackendoff (1983) listed many potential universals in music perception that are reminiscent of Bregman's (1990) auditory scene analysis (e.g., the grouping together of different elements that are close together in pitch and time, and the hierarchical structuring of such groups), probably most ethnomusicologists reject the idea a universal grammar of music. However, they may agree that some nongrammatical aspects of musical structure, such as repetition, recapitulation, or call-response structures, arise independently in different cultures (I. Monson 1999).

On the one hand, the idea of universal grammar can be seen as culturally arrogant, especially if the details of grammar were originally inspired by English or European languages. On the other hand, it can also be used as an argument for cultural equality: "Taking a cue from the generativist linguistic theory of Universal Grammar as developed by Noam Chomsky, we argue that, at a very basic, abstract level or realization, all human beings are endowed with an innate capacity for culture acquisition, evolvement, development and/or abandonment. The foundational argument is that humans are fundamentally the same and that variations in culture cannot constitute sufficient grounds for any form of discrimination whatsoever" (Matambirofa 2018, 30). One might similarly argue that musical universals, or at least family resemblances, are evidence that all humans belong to the same family and should treat each other with respect and empathy.

There are relevant universals in early biological development. All perceptual systems are functioning in the third-trimester human fetus (including vision), and perceptual learning is an intrinsic part of that perceptual function (James 2010). The third-trimester fetus can discriminate a few lower partials in the mother's voice (Hepper and Shahidullah 1994), which could represent a kind of universal foundation for the major triad—a culture-specific aspect in Western music (harmonics 3-4-5 or 4-5-6). If that is true, Western music is arbitrarily taking advantage of one of many possible relevant perceptual universals. Universal aspects of prenatally perceptible pitch patterns, such as, for example, the sound of maternal heartbeats and footsteps, perhaps mixed with the sound of fetal heartbeats, could influence the development of music in any culture (Parncutt and Chuckrow 2019). One or more of these perceptual universals may be important for Western tonality, others for Chinese opera or Australian indigenous song.

Summarizing, music draws upon different relevant perceptual and cognitive universals. But there is much to draw upon, and different musical cultures select among the psychological possibilities in different ways and according to different criteria. There is no causal relationship between psychological and musical universals.

Tonal Affinity

Helmholtz (1877) proposed that the structure of tonal Western music depends on the partials that are audible in musical sounds and their simultaneous (roughness) and successive (affinity) interactions. But it was also clear to Helmholtz that the details of musical structure depend on arbitrary aspects of cultural history. Steege (2010) commented:

> Hermann von Helmholtz's theory of melodic "affinity," which he considered his primary contribution to music-theoretical thought, is inherently historicist in its conception. It seeks not only to explain local tonal relationships but also to provide a rationale for transhistorical change in musical structures. Yet this historical dimension, in many ways typical of its era, is complicated by a tenuous balance between the determinism demanded by rigorous scientific explanation and Helmholtz's evident desire to preserve a complex role for the "choice" of fundamental aesthetic principles. Such principles would include the very parameters of modern tonality, so the status of tonal "affinity," a seemingly given and predictable attribute of musical tones, coexists uneasily in Helmholtz's writing with a humanist appreciation for the unpredictability of cultural choice and change. It also throws the physiological basis of his music theory into conflict with his ostensibly progressive treatment of non-European musics. (abstract)

Perhaps there is no such conflict. Every musical style in the world is determined by a mixture of nature and nurture. The distinction is similar to the sensory–cognitive distinction, but not the same, sensory processes being partially learned and cognitive partially innate. Nature includes perceptual and cognitive universals as well as local geography and climate. Nurture includes all aspects of culture: language, religion, customs, ways of thinking, social structures. In that regard, MmT is no different from any other music.

Helmholtz's vision of music history involved both free choice and progress, in addition to acoustics and perception. In that sense, his idea was consistent with a modern approach that allows for both universals and culture-specific elements, in both perception and musical structure, and a fertile interaction between humanities and sciences: "Although it was vital to Helmholtz's interests both as a scholar and as a member of the ascendant bourgeoisie to allow considerable leeway for individuals to pursue aesthetic tendencies according to the free exercise of their reason, his attempt to construct a universal theory of melody greatly complicates the role of rational choice . . . Helmholtz's historiography in particular was constitutively hybrid—neither fully idealist or *weltgeschichtlich* in Hegelian style nor fully materialist-sensualist, but also not entirely free from the potentially deterministic assumptions of either mode of thought" (Steege 2012, 219).

Western Bias

Schenker and Riemann expressed their German nationalism in their love for German music. Some Europeans and North Americans (among others) may still tacitly dream that MmT is inherently superior to the musical systems of other cultures—a musical form of white supremacy.

They may be forgetting, or were perhaps never reminded, that the main reason for the worldwide popularity of MmT is probably not the music itself but rather Western political dominance (Agawu 2009) and "guns, germs, and steel" (Diamond 1997). Presumably, musical styles other than MmT could have been made equally popular if promoted in a similarly large geographic area (half of the world) over a similarly long period (half a millennium) by similarly violent political or economic means. The Spanish conquest of South America comes to mind, along with the British occupation of North America, India, Australia, and parts of Africa and the Middle East, French colonies in Indochina, Africa, and the Pacific, and so on. These power struggles were religiously accompanied by Christianity and acoustically accompanied by MmT.

Any music may have universal appeal, which is why we like to refer to music as a universal language. That is a problematic comparison, since music is not a language with lexical meanings. It might be truer to say that almost anyone in almost any culture can learn to like the music of almost any other culture, given the right kinds of exposure in the right kinds of situation. MmT and Western C/D are no exception. In addition, perhaps the multilevel cognitive structures that are typical of MmT universally facilitate perception, reducing cognitive load (Bharucha 1987; Tillmann et al. 2008). There is also a feel-good factor. The tonic may be a cognitive metaphor for home: in English at least, the main key of a piece is called the home key, and the lyrics of countless pop songs refer to love of one's home country, hometown, family, partner, or child. Returning to the home key may be associated with good feelings of returning home, consistent with the musical emotions of nostalgia (Barrett et al., 2010) and social belonging (Schäfer and Eerola 2020).

But MmT is not inherently superior in any of these ways. Like every other musical tradition, it is merely different. This book's focus on Western music does not imply that there is something special about MmT. If Westerners have that impression, it could be an effect of familiarity (cf. Peretz et al. 1998): the more we hear a certain kind of music, the more we like it.

The recent history of pseudo-Darwinist academic racism and the liberal backlash that it produced (e.g., Jeynes 2011) clarified that no human culture is inherently superior to any other. We are all *Homo sapiens*, and it is not even clear whether we are superior to the Neanderthals, with their larger brains: "Although debates about Neanderthal behavior exist, these are mostly about whether behavioral differences are the result of demographic and/or historical factors, or whether there are cognitive and/or intrinsic behavioral differences between these groups" (Neubauer and Hublin 2012: 577).

This book's focus on harmony and harmonic tonality is misleading. From a universal human perspective, music is mainly based on melody and rhythm, or song and dance, two structural aspects that may be considered equally important (or at least not different in importance). On the cross-cultural status of melody by comparison to harmony, Eugene Helm (2013) commented that:

> Melody remains the fount of all music, world without end, even though, for a thousand years, tunes of every kind in the west have been increasingly transported into the interior of that colossal edifice called harmony. . . . In some parts of the world there are still cultures and styles in which it can be inappropriate, if not barbarous or sacrilegious or dangerous, to create any kind of harmony . . . Much traditional non-Western melody is not harmonized at all. India and south Asia, for example, are part of the vast musical territories whose ancient and labyrinthine traditions of melody—even in parts of Bollywood—are only peripherally related to the Western idea of harmony. . . . many other non-harmonic musics are also founded on venerable systems as endlessly complex as any other aspects of their cultures . . . a lifetime of study and performance is required to perceive

the musical significance of their philosophies, their literatures, their religions. With intricacies and subtleties like these, who needs harmony? (1–3)

Rameau's Root

Rameau's concept of chord roots was based on a physical universal: the harmonic series as it exists within complex tones in the physical world (the sounding body, *corps sonore*). In a modern psychological approach, chord roots are based on a psychological universal: the familiarity of the auditory system with the harmonic series in auditory running spectra.

There is a psychohistorical sequence here. First comes a psychological universal, then a musical structure, and finally a music-theoretical idea. Clearly, if relevant psychological universals exist (the perception of harmonic pitch patterns in voiced speech sounds being an example), they are very old. In the history of Western music, new musical structures tended to emerge long before theorists noticed them and tried to explain them. For example, statistical analyses of polyphonic vocal repertoires (Parncutt, Reisinger et al. 2019) revealed evidence for chord roots in music of the fourteenth century (a kind of structure), although theorists did not begin to verbalize ideas of root and chordal inversion until the early seventeenth. It is anachronistic to use the term "root" when describing music of the sixteenth century and earlier, but one may reasonably speak of "perceived roots" or "predicted roots" on the basis of corresponding empirical data. We constantly use modern terminology when describing history. Musical modernism offers examples of composers applying quasi-arbitrary compositional principles that contradict or ignore psychological universals. In such cases, the tables are turned, and the music-theoretical idea precedes the structure.

Thomas Christensen (1993) observed that "what makes Rameau's theory such a fascinating and rewarding subject for investigation, I believe, is the rich dialectical interplay it manifests between musical and cultural forces, between the 'internal' problems of musical practice and pedagogy that he addressed, and the 'external' ideas and language indigenous to the French Enlightenment by which he solved them" (7). Said another way, chord roots existed in the music of the fifteenth to seventeenth centuries, long before Rameau explained them in terms of the harmonic series. His new mode of explanation was embedded within, and limited by, a culture-specific discourse across music and the other arts and sciences. Like other historic music theorists, Rameau was a creative thinker, aware of the tension between the way music seemed to work from their perspective and the way others around them were thinking about music. His theorizing was limited by general attitudes of his time toward theorizing, epistemology, metacognition, and teaching and learning, as well as perceptual universals (cf. Eberlein 1994).

Western History

In the West and elsewhere, musical structure changed historically as the social functions of music changed and composers and performers experimented with new music-structural ideas. Music-structural change was typically gradual: the innovations of one generation had to be accepted before those of the next became possible. Sounds that in one period were perceived as dissonant were perceived in later periods as consonant. The sound of major and minor triads needed to become familiar before 6ths or 7ths could be added.

How we perceive music depends on how music is, and how music is depends on rules and conventions for making it. Sometimes, rules are explicit (conscious); in that case, they were usually written and became part of the historical record. The history of such rules has not been the focus of the book. Instead, I have made the simplifying assumption that the musical structures of a given period, as described by music theorists or quantified by statistical analysis of a musical score database, are not primarily determined by music-theoretical rules, although of course rules always play a role. Rather, they primarily emerge from a combination of perceptual universals, local constraints (physical, social, political, historical), and traditions of performance and composition. Traditions include what performers and composers do intuitively without explicit description, hence without written records.

How important are compositional conventions relative to other, less tangible influences? From an empirical scientific viewpoint, the question is hard to answer. One might for example compare statistical analyses of pitch-time patterns in musical scores with predictions based on the explicit music theory of a given period, which is beyond the present scope. Instead, I have assumed that music theory usually follows musical practice with a considerable time lag. Theory tends to be a post-hoc explanation of musical structure and listeners' responses to it. Conversely, musical structure and listeners' responses are determined primarily by frequency of exposure to different pitch configurations in existing music.

Non-Western Examples

Whether the theory presented in this book has interesting implications for non-Western musics that have not adopted some form of MmT, I do not know. I am not an ethnomusicologist, but I do know enough about the music of the world and the politics and psychology of cultural discrimination to realize how problematic it is to apply Western ideas to non-Western music. That would be reminiscent of misled colonialist projects of the past, reawakening uncomfortable racist attitudes, events, and political processes in the history of musical thought and musical practice.

The history of Australian music is pertinent. After the British invasion in 1788, indigenous peoples were brutally suppressed. Gradually, their culture and identity were destroyed. Recent decades have seen innumerable small steps toward cultural renewal and the restoration of indigenous rights amid political attempts at reconciliation. In that context, we may consider modern didgeridoo performances in Northern parts of Australia, which are often accompanied by singing and clapsticks. In a first approximation, the didgeridoo is a half-closed pipe that produces mainly odd-numbered harmonics (like a clarinet, but lower, the drone frequency being typically 70 Hz). The instrument is also a truncated conical horn (N. Fletcher 1996). The partials are exactly harmonic and phase locked, as in any wind instrument, but when the instrument is overblown, the intervals between fundamental modes of vibration are compressed relative to a harmonic series. The relative amplitudes of the harmonics can be changed by the player. The spectrum also changes if the performer sings while performing, creating new timbres: "The didgeridoo resonance pulls the voice toward the didgeridoo harmonics similar to the stepped effect a sweeping band-pass filter produces. The resulting sound is more or less similar to the phonemes, 'ah,' 'ee' and 'oo'" (Ayers and Horner 2007, 22), reminiscent of a jaw harp or mouth harp. In the YouTube video "Aboriginal Songs and Dances from Bamyili, NT, 1980s" (accessed 7 September 2022), the melody focuses on the 6th and 7th harmonics. That being the case, we might claim that the structure of this music is influenced by harmonicity—a perceptual universal that also influenced Western MmT.

Meanwhile in Australia, and in many other parts of the world, Western MmT has become integrated into, and inseparable from, local culture. When Australian Aboriginal rock bands perform, they are usually performing MmT (cf. Dunbar-Hall 2017). MmT has become a means of maintaining and transmitting local cultures that originally had quite different musical traditions. The same applies in many countries; Hawaii is a case in point (Lau 2022). Many of today's Hawaiians regard the song *Hawaii Aloha* as their second anthem, although the musical structures are Western, the lyrics were written by an early missionary, the melody is by a nineteenth-century American composer, and the song extolls a European god. But today's Hawaii is also very culturally diverse. An example of that diversity is the Han Sheng Chinese Choir in Honolulu. It sings arrangements in which the melody sounds Chinese (with recognizably Chinese singing technique and voice production), but the accompaniment is more or less regular MmT. That combination is recognized as Chinese and has a long history in choral traditions within China.

Foreign researchers need not ask whether this music, or the music of indigenous Australian rock bands, is authentic. For the people who are actively involved in these traditions, the music, whatever its history, expresses who they are and what they stand for.

Conclusion

On the one hand, it is clear that perceptual and cognitive universals exist. They depend on universal aspects of the human body such as the frequency range of the vocal apparatus, the mechanics of walking, or the physiology of hearing. They also depend on universal aspects of human environments that affected evolution. As such, they unite all humans on the planet.

On the other hand, the musics of the world are strikingly diverse, as are different ways of thinking and talking about music in both musical and academic circles. The diversity can be observed from different cultural standpoints. Cultural diversity is paradoxically enabled by the flexibility of biological universals.

The ideas presented in this book have focused on Western musical traditions, seen from a Western standpoint—if only for practical reasons. My horizons are limited. The contradiction that is implied by attempting to explain culture-specific musical phenomena with psychological universals remains problematic and has not been entirely resolved.

Music in MmT can nevertheless be meaningful for people from very diverse cultural backgrounds (both Western and non-Western). For better or for worse, a combination of political history and technological developments (including the rise of the Internet) means that most people in the world might today be regarded as Westernized, at least in part. If nothing else, that observation justifies and motivates a careful analysis of MmT.

References

Abramson, M., and Goldinger, S. D. (1997). "What the reader's eye tells the mind's ear: Silent reading activates inner speech." *Perception & Psychophysics* 59(7): 1059–1068.

Adler, G. (1885). "Umfang, Methode und Ziel der Musikwissenschaft." *Vierteljahresschrift für Musikwissenschaft* 1: 5–20.

Aebersold, J. (2000). *The Jazz Handbook: 50 Years of Jazz Education*. New Albany, IL: James Aebersold Jazz.

Agawu, K. (1994). "Ambiguity in tonal music: A preliminary study." In *Theory, Analysis and Meaning in Music*, edited by A. Pople, 86–107. Cambridge: Cambridge University Press.

Agawu, K. (2009). "Tonality as a colonizing force in African music." Distinguished lecture, CIRMMT, Schulich School of Music, McGill University, Montreal, Canada.

Agmon, E. (1991). "Linear transformations between cyclically generated chords." *Musikometrika* 3: 15–40.

Agmon, E. (1994). "Principles of chord progression." In *Proceedings of the Third International Conference on Music Perception and Cognition*, edited by E. Deliège, 245–246. Liège, Belgium: Centre de Recherches et de Formation Musicales de Wallonie.

Agmon, E. (1995). "Functional harmony revisited: A prototype-theoretic approach." *Music Theory Spectrum* 17(2): 196–214.

Ahlbäck, S. (2018). "The hidden modes: A computer-assisted approach to tonality analysis of Swedish Folk Music." In *Proceedings of the Eighth Folk Music Analysis International Workshop, Thessaloniki, Greece*, edited by A. Holzapfel and A. Pikrakis, 74–76. Thessaloniki, Greece: Aristotle University of Thessaloniki.

Altenmüller, E., Schmidt, S., and Zimmermann, E., eds. (2013). *The Evolution of Emotional Communication: From Sounds in Nonhuman Mammals to Speech and Music in Man*. Oxford: Oxford University Press.

Ambrazevičius, R. (2005). "Psychoacoustic and cognitive aspects of musical scales and their manifestation in Lithuanian traditional singing." Doctoral dissertation, Lithuanian Academy of Music and Theatre, Vilnius, Lithuania.

Ambrazevičius, R. (2017). "Dissonance/roughness and tonality perception in Lithuanian traditional Schwebungsdiaphonie." *Journal of Interdisciplinary Music Studies* 8: 39–53.

Ambrazevičius, R., Budrys, R., and Višnevska, I. (2015). *Scales in Lithuanian Traditional Music: Acoustics, Cognition, and Contexts*. Kaunas, Lithuania: Kaunas University of Technology.

Athanasopoulos, G., Eerola, T., Lahdelma, I., and Kaliakatsos-Papakostas, M. (2021). "Harmonic organisation conveys both universal and culture-specific cues for emotional expression in music." *PLoS One* 16(1): e0244964.

Apel, W. (1958). *Gregorian Chant.* Bloomington: Indiana University Press.

Argstatter, H. (2016). "Perception of basic emotions in music: Culture-specific or multicultural?" *Psychology of Music* 44(4): 674–690.

Arom, S. (2004). *African Polyphony and Polyrhythm: Musical Structure and Methodology.* Cambridge: Cambridge University Press.

Arom, S., and Voisin, F. (1998). "Theory and technology in African music." In *Garland Encyclopaedia of World Music,* vol. 1 Africa, edited by R. Stone, 254–270. New York, NY: Garland (as cited in R. Ambrazevičius et al. 2015).

Ashby, W. R. (1956). *An Introduction to Cybernetics*. London: Chapman & Hall.

Assmann, P. F., and Summerfield, Q. (1990). "Modeling the perception of concurrent vowels: Vowels with different fundamental frequencies." *Journal of the Acoustical Society of America* 88(2): 680–697.

Aures, W. (1985). "Ein Berechnungsverfahren der Rauhigkeit." *Acustica* 58(5): 268–281.

Ayers, L., and Horner, A. (2007). "Didgeridoo synthesis using timbre morphing." In *Proceedings of the Australasian Computer Music Conference,* edited by A. Riddell and A. Thorogood, 18–24. Fitzroy, Australia: Australasian Computer Music Association.

Babbitt, M. (1965). "The use of computers in musicological research." *Perspectives of New Music* 3(2): 74–83.

Bai, X., Ma, X., and Tao, Y. (2016). "The response effects of Chinese and Western music on emotion." *Acta Psychologica Sinica* 48(7): 757–769.

Bain, J. (2003). "Theorizing the cadence in the music of Machaut." *Journal of Music Theory* 47(2): 325–362.

Bain, J. (2005). "Tonal structure and the melodic role of chromatic inflections in the music of Machaut." *Plainsong & Medieval Music* 14(1): 59–88.

Balkwill, L. L., and Thompson, W. F. (1999). "A cross-cultural investigation of the perception of emotion in music: Psychophysical and cultural cues." *Music Perception* 17(1): 43–64.

Balzano, G. J. (1980). "The group-theoretic description of 12-fold and microtonal pitch systems." *Computer Music Journal* 4(4): 66–84.

Banks, W. P., and Pockett, S. (2007). "Benjamin Libet's work on the neuroscience of free will." In *The Blackwell Companion to Consciousness*, edited by M. Velmans and S. Schneider, 657–670. Chichester, West Sussex, UK: Blackwell.

Barbera, C. A. (1977). "Arithmetic and geometric divisions of the tetrachord." *Journal of Music Theory* 21(2): 294–323.

Barbour, J. M. (2004). *Tuning and Temperament: A Historical Survey*. North Chelmsford, MA: Courier Corporation.

Barker, A. (1978). "Music and perception: A study in Aristoxenus." *Journal of Hellenic Studies* 98: 9–16.

Barker, R. G. (1968). *Ecological Psychology: Concepts and Methods for Studying the Environment of Human Behavior*. Stanford, CA: Stanford University Press.

Barlow, H. (1990). "Conditions for versatile learning, Helmholtz's unconscious inference, and the task of perception." *Vision Research* 30(11): 1561–1571.

Barrett, F. S., Grimm, K. J., Robins, R. W., Wildschut, T., Sedikides, C., and Janata, P. (2010). "Music-evoked nostalgia: affect, memory, and personality." *Emotion* 10.3 (2010): 390.

Batty, C. (2010). "The physical and emotional threads of the archetypal hero's journey: Proposing common terminology and re-examining the narrative model." *Journal of Screenwriting* 1(2): 291–308.

Bello, J. P., Daudet, L., and Sandler, M. B. (2006). "Automatic piano transcription using frequency and time-domain information." *IEEE Transactions on Audio, Speech, and Language Processing* 14(6): 2242–2251.

Bendor, D., and Wang, X. (2005). "The neuronal representation of pitch in primate auditory cortex." *Nature* 436: 1161–1165.

Bent, I. (2011). "The problem of harmonic dualism: A translation and commentary." In *Oxford Handbook of Neo-Riemannian Music Theories*, edited by E. Gollin and A. Rehding, 167–193. New York, NY: Oxford University Press.

Bent, M. (1998). "The grammar of early music: Preconditions for analysis." In *Tonal Structures in Early Music*, edited by C. C. Judd, 15–59. New York, NY: Routledge.

Berger, A. M. B. (2005). *Medieval Music and the Art of Memory*. Berkeley: University of California Press.

Berlyne, D. E. (1971). *Aesthetics and Psychobiology*. New York, NY: Appleton-Century-Crofts.

Bernstein, J. G., and Oxenham, A. J. (2003). "Pitch discrimination of diotic and dichotic tone complexes: Harmonic resolvability or harmonic number?" *Journal of the Acoustical Society of America* 113: 3323–3334.

Bernstein, L. (1939). "The absorption of race elements into American music." Senior thesis, Harvard University, Cambridge, MA.

Bertossa, F., Besa, M., Ferrari, R., and Ferri, F. (2008). "Point zero: A phenomenological inquiry into the seat of consciousness." *Perceptual and Motor Skills* 107(2): 323–335.

Beyer, C. (2020). "Edmund Husserl." In *Stanford Encyclopedia of Philosophy*, edited by E. N. Zalta. https://plato.stanford.edu/

Bharucha, J. J. (1987). "Music cognition and perceptual facilitation: A connectionist framework." *Music Perception* 5(1): 1–30.

Bharucha, J. J. (1996). "Melodic anchoring." *Music Perception* 13(3): 383–400.

Bharucha, J. J., and Stoeckig, K. (1986). "Reaction time and musical expectancy: Priming of chords." *Journal of Experimental Psychology: Human Perception and Performance* 12(4): 403.

Biamonte, N. (2010). "Triadic modal and pentatonic patterns in rock music." *Music Theory Spectrum* 32(2): 95–110.

Biberger, T., and Ewert, S. D. (2016). "Envelope and intensity based prediction of psychoacoustic masking and speech intelligibility." *Journal of the Acoustical Society of America* 140(2): 1023–1038.

Bigand, E., Delbé, C., Poulin-Charronnat, B., Leman, M., and Tillmann, B. (2014). "Empirical evidence for musical syntax processing? Computer simulations reveal the contribution of auditory short-term memory." *Frontiers in Systems Neuroscience* 8: 94.

Bigand, E., Madurell, F., Tillmann, B., and Pineau, M. (1999). "Effect of global structure and temporal organization on chord progression." *Journal of Experimental Psychology: Human Perception and Performance* 25: 184–197.

Bigand, E., and Parncutt, R. (1999). "Perceiving musical tension in long chord sequences." *Psychological Research* 62(4): 237–254.

Bigand, E., Parncutt, R., and Lerdahl, F. (1996). "Perception of musical tension in short chord sequences: Influence of harmonic function, sensory dissonance, horizontal motion, and musical training." *Perception & Psychophysics* 58: 125–141.

Bigand, E., Poulin, B., Tillmann, B., Madurell, F., and D'Adamo, D. A. (2003). "Sensory versus cognitive components in harmonic priming." *Journal of Experimental Psychology: Human Perception and Performance* 29(1): 159.

Bower, C. M. (2002). "The transmission of ancient music theory into the Middle Ages." In *Cambridge History of Western Music Theory*, edited by T. W. Christensen, 136–167. Cambridge: Cambridge University Press.

Bowling, D. L., and Purves, D. (2015). "A biological rationale for musical consonance." *Proceedings of the National Academy of Sciences* 112(36): 11155–11160.

Bowling, D. L., Purves, D., and Gill, K. Z. (2018). "Vocal similarity predicts the relative attraction of musical chords." *Proceedings of the National Academy of Sciences* 115(1): 216–221.

Brandl, R. (2008). "New considerations of diaphony in Southeast Europe." In *European Voices: Multipart Singing in the Balkans and the Mediterranean*, edited by A. Ahmedaja and G. Haid, 281–298. Vienna: Böhlau.

Braus, I. (1995). "Retracing one's steps: An overview of pitch circularity and Shepard tones in European music, 1550–1990." *Music Perception* 12(3): 323–351.

Bregman, A. S. (1990). *Auditory Scene Analysis*. Cambridge, MA: MIT Press.

Bregman, A. S. (1993). "Auditory scene analysis: Hearing in complex environments." In *Thinking in Sound*, edited by S. McAdams and E. Bigand, 10–36. Oxford: Clarendon Press.

Bregman, A. S. (2005). "Auditory scene analysis and the role of phenomenology in experimental psychology." *Canadian Psychology* 46(1): 32.

Britz, T. (2022). "Does Amber Heard really have the world's most beautiful face? An expert explains why the Golden Ratio test is bogus." *The Conversation*, July 24, 2022.

Brown, H., and Butler, D. (1981). "Diatonic trichords as minimal tonal cue-cells." *In Theory Only* 5(6): 38–55.

Brown, S. (2001). "Are music and language homologues?" *Annals of the New York Academy of Sciences* 930(1): 372–374.

Brown, S., and Jordania, J. (2013). "Universals in the world's musics." *Psychology of Music* 41(2): 229–248.

Browne, R. (1981). "Tonal implications of the diatonic set." *In Theory Only* 5(6–7): 3–21.

Bryden, J. R., and Hughes, D. G. (1969). *An Index of Gregorian Chant*. Cambridge, MA: Harvard University Press.

Bryson, B. (2004). *A Short History of Nearly Everything*. New York, NY: Broadway.

Budge, H. (1943). "A study of chord frequencies." Doctoral dissertation, Teachers College, Columbia University, New York, NY.

Buechner, J. (2010). "Are the Gödel incompleteness theorems limitative results for the neurosciences?" *Journal of Biological Physics* 36(1): 23–44.

Bukofzer, M. F. (1947). *Music in the Baroque Era: From Monteverdi to Bach*. New York, NY: Norton.

Burgoyne, J. A., Wild, J., and Fujinaga, I. (2011). "An expert ground truth set for audio chord recognition and music analysis." In *Proceedings of the 12th International Society for Music Information Retrieval Conference*, edited by A. Klapuri and C. Leider, 633–638. Miami, Florida, USA: University of Miami.

Burns, E. M., and Ward, W. D. (1978). "Categorical perception—phenomenon or epiphenomenon: Evidence from experiments in the perception of melodic musical intervals." *Journal of the Acoustical Society of America* 63(2): 456–468.

Butler, D. (1989). "Describing the perception of tonality in music: A critique of the tonal hierarchy theory and a proposal for a theory of intervallic rivalry." *Music Perception* 6(3): 219–241.

Butler, D. (1992). *Musician's Guide to Perception and Cognition*. New York, NY: Schirmer.

Butler, D., and Brown, H. (1984). "Tonal structure versus function: Studies of the recognition of harmonic motion." *Music Perception* 2(1): 6–24.

Callender, C., Quinn, I., and Tymoczko, D. (2008). "Generalized voice-leading spaces." *Science* 320(5874): 346–348.

Caplin, W. E. (1998). *Classical Form: A Theory of Formal Functions for the Instrumental Music of Haydn, Mozart, and Beethoven*. New York, NY: Oxford University Press.

Carey, N., and Clampitt, D. (1989). "Aspects of well-formed scales." *Music Theory Spectrum* 11(2): 187–206.

Cariani, P. (1999). "Temporal coding of periodicity pitch in the auditory system: An overview." *Neural Plasticity* 6(4): 147–172.

Cariani, P. (2001). "Temporal codes, timing nets, and music perception." *Journal of New Music Research* 30(2): 107–135.

Carruthers, P. (2011). *The Opacity of Mind: An Integrative Theory of Self-knowledge*. Oxford: Oxford University Press.

Carterette, E. C., and Kendall, R. A. (1994). "On the tuning and stretched octave of Javanese gamelans." *Leonardo Music Journal* 4: 59–68.

Castellano, M. A., Bharucha, J. J., and Krumhansl, C. L. (1984). "Tonal hierarchies in the music of North India." *Journal of Experimental Psychology: General* 113(3): 394.

Cazden, N. (1954). "Hindemith and nature." *Music Review* 15: 288–306.

Cenkerová, Z., and Parncutt, R. (2015). "Style-dependency of melodic expectation: Changing the rules in real time." *Music Perception* 33(1): 110–128.

Ceulemans, A. E. (1990). "Fétis et les rapports entre l'harmonie et la mélodie." *Revue Belge de Musicologie/Belgisch Tijdschrift voor Muziekwetenschap* 44: 123–131.

Chaigne, A., and Askenfelt, A. (1994). "Numerical simulations of piano strings. I. A physical model for a struck string using finite difference methods." *Journal of the Acoustical Society of America* 95(2): 1112–1118.

Chalmers, D. J. (1995). "Facing up to the problem of consciousness." *Journal of Consciousness Studies* 2: 200–219.

Charnavel, I. (2019). "Steps toward a universal grammar of dance: Local grouping structure in basic human movement perception." *Frontiers in Psychology* 10: 1364.

Cheveigné, A. (1998). "Cancellation model of pitch perception." *Journal of the Acoustical Society of America* 103: 1261–1271.

Cheveigné, A. (1999). "Pitch shifts of mistuned partials: A time-domain model." *Journal of the Acoustical Society of America* 106(2): 887–897.

Cheveigné, A., and Kawahara, H. (2002). "YIN, a fundamental frequency estimator for speech and music." *Journal of the Acoustical Society of America* 111: 1917–1930.

Chew, E. (2000). "Towards a mathematical model of tonality." Doctoral dissertation. Massachusetts Institute of Technology, Cambridge, MA.

Chew, G. (1983). "The spice of music: Towards a theory of the leading note." *Music Analysis* 2(1): 35–53.

Chew, G. (1989). "The perfections of modern music: Consecutive fifths and tonal coherence in Monteverdi." *Music Analysis* 8(3): 247–273.

Chodos, A. T. (2018). "The blues scale: Historical and epistemological considerations." *Jazz Perspectives* 11(2): 139–171.

Chmiel, A., and Schubert, E. (2019). "Unusualness as a predictor of music preference." *Musicae Scientiae* 23(4): 426–441.

Chomsky, N. (1957). *Syntactic Structures*. The Hague, Netherlands: Mouton & Co.

Chomsky, N. (1959). "A review of B. F. Skinner's *Verbal Behavior*." *Language* 35: 26–58.

Chomsky, N. (1965). *Aspects of the Theory of Syntax*. Cambridge, MA: MIT Press.

Chomsky, N. (1986). *Knowledge of Language: Its Nature, Origin and Use*. New York, NY: Praeger.

Christensen, T. (1987). "Eighteenth-century science and the *corps sonore*: The scientific background to Rameau's principle of harmony." *Journal of Music Theory* 31: 23–49.

Christensen, T. (1993). *Rameau and Musical Thought in the Enlightenment*. Cambridge: Cambridge University Press.

Christensen, T. (2019). *Stories of Tonality in the Age of François-Joseph Fétis*. Chicago: University of Chicago Press.

Chuang, C. K., and Wang, W. S. Y. (1978). "Psychophysical pitch biases related to vowel quality, intensity difference, and sequential order." *Journal of the Acoustical Society of America* 64(4): 1004–1014.

Chung, A. (2019). "What is musical meaning? Theorizing music as performative utterance." *Music Theory Online* 25(1).

Clark, A., and Chalmers, D. (1998). "The extended mind." *Analysis* 58(1): 7–19.

Clarke, E. F. (1986). "Theory, analysis and the psychology of music: A critical evaluation of Lerdahl, F. and Jackendoff, R., A Generative Theory of Tonal Music." *Psychology of Music* 14(1): 3–16.

Clarke, E. F. (1989a). "Mind the gap: Formal structures and psychological processes in music." *Contemporary Music Review* 3(1): 1–13.

Clarke, E. F. (1989b). "Issues in language and music." *Contemporary Music Review* 4(1): 9–22.

Clement, B. (2019). "Diatonic and chromatic tonicization in rock music." *Journal of Music Theory* 63(1): 1–34.

Clough, J., and Douthett, J. (1991). "Maximally even sets." *Journal of Music Theory* 35: 93–173.

Cohen, A. (1961). "Further investigation of the effects of intensity upon the pitch of pure tones." *Journal of the Acoustical Society of America* 33(10): 1363–1376.

Cohen, A. J. (1991). "Tonality and perception: Musical scales primed by excerpts from The Well-Tempered Clavier of J. S. Bach." *Psychological Research* 53(4): 305–314.

Cohen, M. A., Grossberg, S., and Wyse, L. L. (1995). "A spectral network model of pitch perception." *Journal of the Acoustical Society of America* 98(2): 862–879.

Cohn, R. (1997). "Neo-Riemannian operations, parsimonious trichords, and their 'tonnetz' representations." *Journal of Music Theory* 41(1): 1–66.

Collins, T., Tillmann, B., Barrett, F. S., Delbé, C., and Janata, P. (2014). "A combined model of sensory and cognitive representations underlying tonal expectations in music: From audio signals to behavior." *Psychological Review* 121(1): 33.

Conway, C. M., and Christiansen, M. H. (2001). "Sequential learning in non-human primates." *Trends in Cognitive Sciences* 5(12): 539–546.

Cook, N. (1987). "The perception of large-scale tonal closure." *Music Perception* 5(2): 197–205.

Cook, N. (1990). *Music, Imagination, and Culture*. Oxford: Oxford University Press.

Cook, N. (1998). *Music: A Very Short Introduction*. Oxford, UK: Oxford University Press.

Cook, N. (2001). "On qualifying relativism." *Musicae Scientiae* 5(suppl. 2): 167–189.

Cook, N. (2007). *The Schenker Project: Culture, Race, and Music Theory in fin-de-siècle Vienna*. New York, NY: Oxford University Press.

Cook, N. (2013). *Beyond the score: Music as performance*. Oxford, UK: Oxford University Press.

Cook, N., and Everest, M., eds. (1999). *Rethinking Music*. Oxford: Oxford University Press.

Cooke, D. (1959). *The Language of Music*. London: Oxford University Press.

Costall, A. P. (1984). "Are theories of perception necessary? A review of Gibson's The Ecological Approach to Visual Perception." *Journal of the Experimental Analysis of Behavior* 41(1): 109–115.

Crocker, R. L. (1962). "Discant, counterpoint, and harmony." *Journal of the American Musicological Society* 15(1): 1–21.

Crowder, R. G. (1984). "Perception of the major/minor distinction: I. Historical and theoretical foundations." *Psychomusicology* 4(1–2): 3.

Cupchik, G. C., Vartanian, O., Crawley, A., and Mikulis, D. J. (2009). "Viewing artworks: Contributions of cognitive control and perceptual facilitation to aesthetic experience." *Brain and Cognition* 70(1): 84–91.

Cutting, C. B. (2018). "Microtonal analysis of 'blue notes' and the blues scale." *Empirical Musicology Review* 13(1–2): 84.

Dahlhaus, C. (1968). *Untersuchungen über die Entstehung der harmonischen Tonalität. Bärenreiter*. Translated by Robert O. Gjerdingen as *Studies on the Origin of Harmonic Tonality*. Princeton, NJ: Princeton University Press, 1990.

Dai, H. (2010). "Harmonic pitch: Dependence on resolved partials, spectral edges, and combination tones." *Hearing Research* 270(1–2): 143–150.

Dalla Bella, S., Peretz, I., Rousseau, L., and Gosselin, N. (2001). "A developmental study of the affective value of tempo and mode in music." *Cognition* 80(3): B1–B10.

Damodaran, S. (2016). "Protest and music. Oxford research encyclopedia of politics." https://oxfordre.com/politics/display/10.1093/acrefore/9780190228637.001.0001/acrefore-9780190228637-e-81;jsessionid=14F192A5DBD06C95FAFC3014F78C72EC

Danielou, A. (1943). *Introduction to the Study of Musical Scales*. London: India Society.

Darwin, C. J., Ciocca, V., and Sandell, G. J. (1994). "Effects of frequency and amplitude modulation on the pitch of a complex tone with a mistuned harmonic." *Journal of the Acoustical Society of America* 95(5): 2631–2636.

Davis, J. A. (1995). "Philosophical positivism and American atonal music theory." *Journal of the History of Ideas* 56(3): 501–522.

Day-O'Connell, J. (2007). *Pentatonicism from the Eighteenth Century to Debussy*. Rochester, NY: University of Rochester Press.

De Boer, E. (1956). "Pitch of inharmonic signals." *Nature* 178: 535–536.

De Clercq, T., and Temperley, D. (2011). "A corpus analysis of rock harmony." *Popular Music* 30: 47–70.

De Klerk, D. (1979). "Equal temperament." *Acta Musicologica* 51: 140–150.

de la Motte, D. (1976). *Harmonielehre* [*Theory of Harmony*]. Kassel, Germany: dtv/Bärenreiter.

Dean, R. T., Milne, A. J., and Bailes, F. (2019). "Spectral pitch similarity is a predictor of perceived change in sound-as well as note-based music." *Music & Science* 2: 1–14.

Deliège, I., Mélen, M., Stammers, D., and Cross, I. (1996). "Musical schemata in real-time listening to a piece of music." *Music Perception* 14(2): 117–159.

Demany, L., and Armand, F. (1984). "The perceptual reality of tone chroma in early infancy." *Journal of the Acoustical Society of America* 76(1): 57–66.

Demany, L., Monteiro, G., Semal, C., Shamma, S., and Carlyon, R. P. (2021). "The perception of octave pitch affinity and harmonic fusion have a common origin." *Hearing Research* 404: 108213.

Dennett, D. (1991). *Consciousness Explained*. Boston, MA: Little Brown.

Dennett, D. (2003). "Who's on first? Heterophenomenology explained." *Journal of Consciousness Studies* 10(9–10): 19–30.

d'Errico, F. (2003). "The invisible frontier. A multiple species model for the origin of behavioral modernity." *Evolutionary Anthropology* 12(4): 188–202.

Deutsch, D., and Feroe, J. (1981). "The internal representation of pitch sequences in tonal music." *Psychological Review* 88(6): 503.

Deutsch, D., Lapidis, R., and Henthorn, T. (2008). "The speech-to-song illusion." *Journal of the Acoustical Society of America* 124(4_supplement): 2471.

Devaney, J., Mandel, M. I., Ellis, D. P. W., and Fujinaga, I. (2011). "Automatically extracting performance data from recordings of trained singers." *Psychomusicology* 21(1–2): 108–136.

DeWitt, L. A., and Crowder, R. G. (1987). "Tonal fusion of consonant musical intervals: The oomph in Stumpf." *Perception & Psychophysics* 41(1): 73–84.

Diamond, J. (1997). *Guns, Germs, and Steel: The Fates of Human Societies*. New York, NY: Norton.

Dibben, N. (1994). "The cognitive reality of hierarchic structure in tonal and atonal music." *Music Perception* 12(1): 1–25.

Dibben, N. (2001). "What do we hear, when we hear music? Music perception and musical material." *Musicae Scientiae* 5(2): 161–194.

Dissanayake, E. (2001). "Antecedents of the temporal arts in early mother-infant interaction." In *The Origins of Music*, edited by N. L. Wallin, B. Merker, and S. Brown, 389–410. Cambridge, MA: MIT Press.

Dobbins, P. A., and Cuddy, L. L. (1982). "Octave discrimination: An experimental confirmation of the 'stretched' subjective octave." *Journal of the Acoustical Society of America* 72(2): 411–415.

Dodds, M. R. (2003). "Plainchant at Florence's cathedral in the Late Seicento: Matteo Coferati and shifting concepts of tonal space." *Journal of Musicology* 20(4): 526–555.

Dowling, W. J. (1978). "Scale and contour: Two components of a theory of memory for melodies." *Psychological Review* 85(4): 341.

Duffin, R. W. (2007). *How Equal Temperament Ruined Harmony (and Why You Should Care)*. New York, NY: Norton.

Duke, R. A. (1985). "Wind instrumentalists' intonational performance of selected musical intervals." *Journal of Research in Music Education* 33(2): 101–111.

Dunbar-Hall, P. (2017). "'We have survived': Popular music as a representation of Australian Aboriginal cultural loss and reclamation." In *The Resisting Muse: Popular Music and Social Protest*, edited by I. Peddie, 119–131. Abingdon, Oxon, UK: Routledge.

Dunn, P. G., de Ruyter, B., and Bouwhuis, D. G. (2012). "Toward a better understanding of the relation between music preference, listening behavior, and personality." *Psychology of Music* 40(4): 411–428.

Durkheim, E. (1912). *Les formes élémentaires de la vie religieuse*. Paris: Alcan.

Eberlein, R. (1994). *Die Entstehung der tonalen Klangsyntax* [*The Origin of Tonal-Harmonic Syntax*]. New York, NY: Peter Lang.

Edwards, P. (1967). *The Encyclopedia of Philosophy*, vol. 8. New York, NY: Macmillan.

Egan, J. P., and Hake, H. W. (1950). "On the masking pattern of a simple auditory stimulus." *Journal of the Acoustical Society of America* 22(5): 622–630. xxyy

Eggen, J. H. (1986). "The strike note of bells." Unpublished master's thesis, Eindhoven University of Technology, Eindhoven, Netherlands. research.tue.nl

Einarsson, A., and Ziemke, T. (2017). "Exploring the multi-layered affordances of composing and performing interactive music with responsive technologies." *Frontiers in Psychology* 8: 1701.

Eitan, Z., and Granot, R. Y. (2006). "How music moves: Musical parameters and listeners images of motion." *Music Perception* 23(3): 221–248.

Ellis, A. (1885). "On the musical scales of various nations." *Journal of the Society of Arts* 33: 485–527 (as cited in A. Hui 2012).

Erfanian Saeedi, N., Blamey, P. J., Burkitt, A. N., and Grayden, D. B. (2016). "Learning pitch with STDP: A computational model of place and temporal pitch perception using spiking neural networks." *PLoS Computational Biology* 12(4): e1004860.

Esteve-Gibert, N., and Prieto, P. (2018). "Early development of prosody–meaning interface." In *The Development of Prosody in First Language Acquisition*, edited by P. Prieto and N. Esteve-Gibert, 228–246. Amsterdam: John Benjamins.

Etcoff, N. L., and Magee, J. J. (1992). "Categorical perception of facial expressions." *Cognition* 44(3): 227–240.

Euler, L. (1739). *Tentamen Novae Theoriae Musicae Ex Certississmis Harmoniae Principiis Dilucide Expositae* [*A New Scientific Music Theory That Is Clearly Based on the Most Reliable Principles of Harmony*]. Saint Petersburg Academy. digitale-sammlungen.de.

Evans, E. F. (1981). "The dynamic range problem: place and time coding at the level of cochlear nerve and nucleus." In *Neuronal Mechanisms of Hearing*, edited by J. Syka and L. Aitkin, 69–85. Boston, MA: Springer.

Everett, W. (2004). "Making sense of rock's tonal systems." *Music Theory Online* 10(4).

Ewell, P. (2020). "Music theory's white racial frame." *Music Theory Spectrum* 43(2): 324–329.

Farbood, M. M. (2016). "Memory of a tonal center after modulation." *Music Perception* 34(1): 71–93.

Ferrandino, M. (2022). "Multi-centric complexes in pop-rock music." *Intégral* 35: 27–43.

Fétis, F.-J. (1844). *Traité complet de la théorie et de la pratique de l'harmonie*. Paris: Maurice Schlesinger.

Fink, R. (1999). "Going flat: Post-hierarchical music theory and the musical surface." In *Rethinking Music*, edited by N. Cook and M. Everest, 102–137. Oxford: Oxford University Press.

Fitch, W. T. (2006). "The biology and evolution of music: A comparative perspective." *Cognition* 100(1): 173–215.

Fitch, W. T., and Reby, D. (2001). "The descended larynx is not uniquely human." *Proceedings of the Royal Society of London, Series B: Biological Sciences* 268(1477): 1669–1675.

Fletcher, H. (1940). "Auditory patterns." *Reviews of Modern Physics* 12(1): 47–65.

Fletcher, N. (1996). "The didjeridu (didgeridoo)." *Acoustics Australia* 24: 11–16.

Forrest, D., Neff, S., and Reef, N. (2021). "American Music Theory, 1955–2017." Oxford Bibliographies. https://www.oxfordbibliographies.com/display/document/obo-9780199757824/obo-9780199757824-0073.xml

Forte, A. (1962). *Tonal Harmony in Concept and Practice*. New York, NY: Holt, Rinehart and Winston.

Forte, A. (1973). *The Structure of Atonal Music*. New Haven, CT: Yale University Press.

Forte, A. (1978). "Schoenberg's creative evolution: The path to atonality." *Musical Quarterly* 64(2): 133–176.

Forte, A., and Gilbert, S. E. (1982). *Introduction to Schenkerian Analysis*. New York, NY: Norton.

Foutakis, P. (2014). "Did the Greeks build according to the golden ratio?" *Cambridge Archaeological Journal* 24(1): 71–86.

Fransson, F., Sundberg, J., and Tjerlund, P. (1974). "The scale in played music." *Swedish Journal of Musicology* 56: 49–54.

Friberg, A., Bresin, R., and Sundberg, J. (2006). "Overview of the KTH rule system for musical performance." *Advances in Cognitive Psychology* 2(2): 145.

Frick, R. W. (1985). "Communicating emotion: The role of prosodic features." *Psychological Bulletin* 97(3): 412–429.

Fritz, T., Jentschke, S., Gosselin, N., Sammler, D., Peretz, I., Turner, R., Friederici, A. D., et al. (2009). "Universal recognition of three basic emotions in music." *Current Biology* 19(7): 573–576.

Fuller, R. (1991). "A study of microtonal equal temperaments." *Journal of Music Theory* 35(1–2): 211–237.

Fuller, S. (1986). "On sonority in fourteenth-century polyphony: Some preliminary reflections." *Journal of Music Theory* 30(1): 35–70.

Fürniss, S. (2000). "Cadres de quartes, fluctuations d'intervalles et mutations pentatoniques dans la polyphonie vocale aka [Frameworks of fourths, interval fluctuations and pentatonic mutations in Aka vocal polyphony]." *Musicaie Scientiae* 4(1): 65–72.

Fuster, J. M. (1997). "Network memory." *Trends in Neurosciences* 20(10): 451–459.

Fyk, J. (1997). "Intonational protention in the performance of melodic octaves on the violin." In *Music, Gestalt, and Computing*, edited by M. Leman, 421–430. Berlin, Germany: Springer.

Gabrielsson, A., and Juslin, P. N. (1996). "Emotional expression in music performance: Between the performer's intention and the listener's experience." *Psychology of Music* 24(1): 68–91.

Galembo, A., Askenfelt, A., Cuddy, L. L., and Russo, F. A. (2001). "Effects of relative phases on pitch and timbre in the piano bass range." *Journal of the Acoustical Society of America* 110(3): 1649–1666.

Gandour, J., Wong, D., and Hutchins, G. (1998). "Pitch processing in the human brain is influenced by language experience." *Neuroreport* 9(9): 2115–2119.

Gauldin, R. (1983). "The cycle-7 complex: Relations of diatonic set theory to the evolution of ancient tonal systems." *Music Theory Spectrum* 5: 39–55.

Giannos, K., Athanasopoulos, G., and Cambouropoulos, E. (2022). "Musical consonance and dissonance: The relevance of the horizontal dimension." Presented at Psychology and Music—Interdisciplinary Encounters, Belgrade, Serbia, October 26–29, 2022.

Gibson, E., and Rader, N. (1979). "Attention." In *Attention and Cognitive Development*, edited by G. A. Hale and M. Lewis, 1–21. Springer. link.springer.com.

Gibson, J. J. (1966). *The Senses Considered as Perceptual Systems*. Boston, MA: Houghton Mifflin.

Gibson, J. J. (1979). *The Ecological Approach to Visual Perception*. New York, NY: Psychology Press.

Giordano, N. (2015). "Explaining the Railsback stretch in terms of the inharmonicity of piano tones and sensory dissonance." *Journal of the Acoustical Society of America* 138(4): 2359–2366.

Gjerdingen, R. O. (1988). *A Classic Turn of Phrase: Music and the Psychology of Convention*. Philadelphia: University of Pennsylvania Press.

Glarean, H. (1547/1965). *Dodecachordon*. Translated by C. A. Miller. Rome: American Institute of Musicology.

Glasberg, B. R., and Moore, B. C. (1990). "Derivation of auditory filter shapes from notched-noise data." *Hearing Research* 47(1–2): 103–138.

Glotzbach, P. A., and Heft, H. (1982). "Ecological and phenomenological contributions to the psychology of perception." *Noûs* 16(1): 108–121.

Goehr, A. (1973). "The theoretical writings of Arnold Schoenberg." In *Proceedings of the Royal Musical Association* 100: 85–96.

Goldenberg, Y. (2008). *Prolongation of Seventh Chords in Tonal Music*. Lewiston, ME: Edwin Mellen Press.

Goldman, R. F. (1965). *Harmony in Western Music*. New York, NY: Norton.

Goldstein, J. L. (1967). "Combination tones and auditory nonlinearity." *Journal of the Acoustical Society of America* 41(6): 1603–1603.

Golembiewski, J. (2013). "Lost in space: The place of the architectural milieu in the aetiology and treatment of schizophrenia." *Facilities* 31(9–10): 427–448.

Gollin, E. (2004). "From tonoi to modi: A transformational perspective." *Music Theory Spectrum* 26(1): 119–129.

Grabner, H. (1944). *Handbuch der funktionellen Harmonielehre*. Berlin: Max Hesses Verlag.

Greenhough, M., Williams, R. P., and Wright, S. J. (1973). "Some experiments on the second effect of pitch shift in three-tone complexes." *Journal of Sound and Vibration* 26(2): 277–282.

Greenwood, D. D. (1961). "Auditory masking and the critical band." *Journal of the Acoustical Society of America* 33(4): 484–502.

Greenwood, D. D. (1990). "A cochlear frequency-position function for several species—29 years later." *Journal of the Acoustical Society of America* 87(6): 2592–2605.

Grey, J. M. (1977). "Multidimensional perceptual scaling of musical timbres." *Journal of the Acoustical Society of America* 61(5): 1270–1277.

Griesinger, D. (2015). "Acoustic quality, proximity, and localization in concert halls: The role of harmonic phase alignment." *Psychomusicology* 25(3): 339–344.

Guettler, K. (2002). "On playing 'harmonics' (flageolet tones)." *Catgut Acoustical Society Journal* 4(5): 5–7.

Guldbrandsen, E. E., and Johnson, J., eds. (2015). *Transformations of Musical Modernism*. Cambridge, UK: Cambridge University Press.

Hall, D. A., and Plack, C. J. (2009). "Pitch processing sites in the human auditory brain." *Cerebral Cortex* 19(3): 576–585.

Hall, J. L. (1974). "Two-tone distortion products in a nonlinear model of the basilar membrane." *Journal of the Acoustical Society of America* 56(6): 1818–1828.

Hall, M. D., and Pastore, R. E. (1992). "Musical duplex perception: Perception of figurally good chords with subliminal distinguishing tones." *Journal of Experimental Psychology: Human Perception and Performance* 18(3): 752–762.

Hanson, H. (1960). *Harmonic Materials of Modern Music: Resources of the Tempered Scale*. New York, NY: Appleton-Century-Crofts.

Harasim, D., Moss, F. C., Ramirez, M., and Rohrmeier, M. (2021). "Exploring the foundations of tonality: Statistical cognitive modeling of modes in the history of Western classical music." *Humanities and Social Sciences Communications* 8(1): 1–11.

Harrison, P., and Pearce, M. T. (2020). "Simultaneous consonance in music perception and composition." *Psychological Review* 127(2): 216.

Hartmann, W. M. (1993). "On the origin of the enlarged melodic octave." *Journal of the Acoustical Society of America* 93(6): 3400–3409.

Hartmann, W. M., and Doty, S. L. (1996). "On the pitches of the components of a complex tone." *Journal of the Acoustical Society of America* 99(1): 567–578.

Hartmann, W. M., and Goupell, M. J. (2006). "Enhancing and unmasking the harmonics of a complex tone." *Journal of the Acoustical Society of America* 120(4): 2142–2157.

Hartt, J. C. (2010). "Rehearing Machaut's motets: Taking the next step in understanding sonority." *Journal of Music Theory* 54(2): 179–234.

Hatfield, G. (2002). "Psychology, philosophy, and cognitive science: Reflections on the history and philosophy of experimental psychology." *Mind & Language* 17(3): 207–232.

Hauptmann, M. (1868). *Die Natur der Harmonik und der Metrik.* Leipzig: Breitkopf und Härtel.

Haverkamp, M. (2019). "How does what we hear sound? The qualia problem in acoustics." In *Proceedings of the 23rd International Congress on Acoustics, Aachen, Germany*, edited by M. Vorländer and J. Fels, 1418–1425. publications.rwth-aachen.de.

Heil, P., and Peterson, A. J. (2017). "Spike timing in auditory-nerve fibers during spontaneous activity and phase locking." *Synapse* 71(1): 5–36.

Heinbach, W. (1988). "Aurally adequate signal representation: The part-tone-time-pattern." *Acustica* 67(2): 113–121.

Heisenberg, M. (2009). "Is free will an illusion?" *Nature* 459(7244): 164–165.

Helm, E. E. (2013). *Melody, Harmony, Tonality: A Book for Connoisseurs and Amateurs.* Lanham, MD: Scarecrow Press.

Helmholtz, H. von (1856). "Ueber Combinationstöne." *Annalen der Physik und Chemie* 99: 497–540 (as cited in B. Steege 2012).

Helmholtz, H. von. (1858). "Über die physiologischen Ursachen der musikalischen Harmonie." *Vorträge und Reden* 1: 79–115 (as cited in Steege 2012, 22).

Helmholtz, H. von (1877). *Die Lehre von den Tonempfindungen* [*Theory of Tone Sensation*]. 4th ed. Braunschweig, Germany: Vieweg.

Henderson, R. V. (1969). "Solmization syllables in musical theory, 1100 to 1600." Dissertation, Columbia University, New York, NY.

Henning, G. B., and Grosberg, S. L. (1968). "Effect of harmonic components on frequency discrimination." *Journal of the Acoustical Society of America* 44(5): 1386–1389.

Hepper, P. G., and Shahidullah, B. S. (1994). "Development of fetal hearing." *Archives of Disease in Childhood—Fetal and Neonatal Edition* 71(2): F81–F87.

Herbert, R. (2016). *Everyday Music Listening: Absorption, Dissociation and Trancing.* London, UK: Routledge.

Hermes, D. J. (1988). "Measurement of pitch by subharmonic summation." *Journal of the Acoustical Society of America* 83(1): 257–264.

Hill, K., Barton, M., and Hurtado, A. M. (2009). "The emergence of human uniqueness: Characters underlying behavioral modernity." *Evolutionary Anthropology* 18(5): 187–200.

Hiller Jr, L. A., and Isaacson, L. M. (1957). "Musical composition with a high speed digital computer." Paper presented at the 9th annual meeting of the Audio Engineering Society, New York, October 8–12, 1957.

Hindemith, P. (1937). *Unterweisung im Tonsatz. I. Theoretischer Teil* [*Instruction in Composition I: Theoretical Part*]. Mainz, Germany: Schott.

Holtmeier, L. (2011). "The reception of Hugo Riemann's music theory." In *Oxford Handbook of Neo-Riemannian Music Theories*, edited by E. Gollin and A. Rehding, 3–54. New York, NY: Oxford University Press.

Honing, H. (2006). "Computational modeling of music cognition: A case study on model selection." *Music Perception* 23(5): 365–376.

Honing, H., ed. (2018). *The Origins of Musicality: Interdisciplinary Perspectives on the Capacity to Perceive, Appreciate, and Make Music.* Cambridge, MA: MIT Press.

Hood, M. (1960). "The challenge of 'bi-musicality.'" *Ethnomusicology* 4(2): 55–59.

Houtgast, T. (1976). "Subharmonic pitches of a pure tone at low S/N ratio." *Journal of the Acoustical Society of America* 60(2): 405–409.

Houtsma, A. J., Rossing, T. D., and Wagemakers, W. M. (1987). *Auditory Demonstrations* (CD with explanatory booklet). New York, NY: Acoustical Society of America.

Hubbard, T. L. (2022). "The Pythagorean comma and preference for a stretched octave." *Psychology of Music* 50(2): 670–683.

Huffman, C. (2018). "Pythagoras." In *Stanford Encyclopedia of Philosophy*, edited by E. N. Zalta. https://plato.stanford.edu/

Hui, A. (2012). *The Psychophysical Ear: Musical Experiments, Experimental Sounds, 1840–1910*. Cambridge, MA: MIT Press.

Huovinen, E. (2006). "Colloquy: Two arguments for the mental reality of diatonicism." *Music Theory Spectrum* 28(1): 141–153.

Huron, D. (1991). "Tonal consonance versus tonal fusion in polyphonic sonorities." *Music Perception* 9(2): 135–154.

Huron, D. (1996). "The melodic arch in Western folksongs." *Computing in Musicology* 10: 3–23.

Huron, D. (2001). "Tone and voice: A derivation of the rules of voice-leading from perceptual principles." *Music Perception* 19(1): 1–64.

Huron, D. (2006). *Sweet Anticipation: Music and the Psychology of Expectation.* Cambridge, MA: MIT Press.

Huron, D. (2016). *Voice Leading: The Science Behind a Musical Art.* Cambridge, MA: MIT Press.

Huron, D., and Parncutt, R. (1993). "An improved model of tonality perception incorporating pitch salience and echoic memory." *Psychomusicology* 12: 152–169.

Huron, D., and Veltman, J. (2006). "A cognitive approach to medieval mode: Evidence for an historical antecedent to the major/minor system." *Empirical Musicology Review* 1(1): 33–55.

Hurwitz, R. I. (1970). "An investigation in the perception of root in harmonic intervals." Doctoral dissertation, Indiana University, Bloomington, IN.

Husserl, E. (1900). *Logische Untersuchungen: Prolegomena zur reinen Logik* [*Logical Investigations: Pure Logic*]. Hamburg: Felix Meiner Verlag.

Hutchins, S., and Palmer, C. (2008). "Repetition priming in music." *Journal of Experimental Psychology: Human Perception and Performance* 34: 693–707.

Hutchinson, W., and Knopoff, L. (1978). "The acoustic component of Western consonance." *Journal of New Music Research* 7(1): 1–29.

Hyer, B. (2011). "What *is* a function?" In *Oxford Handbook of Neo-Riemannian Music Theories*, edited by E. Gollin and A. Rehding, 92–139. New York, NY: Oxford University Press.

Jackendoff, R., and Lerdahl, F. (2006). "The capacity for music: What is it, and what's special about it?" *Cognition* 100(1): 33–72.

Jacoby, N., Undurraga, E. A., McPherson, M. J., Valdés, J., Ossandón, T., and McDermott, J. H. (2019). "Universal and non-universal features of musical pitch perception revealed by singing." *Current Biology* 29(19): 3229–3243.e12.

James, D. K. (2010). "Fetal learning: A critical review." *Infant and Child Development* 19(1): 45–54.

Janata, P., Birk, J. L., Van Horn, J. D., Leman, M., Tillmann, B., and Bharucha, J. J. (2002). "The cortical topography of tonal structures underlying Western music." *Science* 298(5601): 2167–2170.

Jeynes, W. H. (2011). "Race, racism, and Darwinism." *Education and Urban Society* 43(5): 535–559.

Johnson, E., and Klonoski, E. (2003). "Connecting the inner ear and the voice." *Choral Journal* 44(3), 35–40.

Johnson, P. (2004). "'Expressive intonation' in string performance: Problems of analysis and interpretation." In *The Music Practitioner*, edited by J. W. Davidson, 79–90. Oxford: Ashgate.

Johnson-Laird, P. N., Kang, O. E., and Leong, Y. C. (2012). "On musical dissonance." *Music Perception* 30: 19–35.

Jonas, O. (1982). *Introduction to the Theory of Heinrich Schenker*. New York, NY: Longman.

Jordan, J. M. (1987). "The pedagogy of choral intonation: Efficient pedagogy to approach an old problem." *Choral Journal* 27(9): 9–16.

Jürgensen, F. (2011). "Cadential accidentals in the *Buxheim Organ Book* and its concordances: A mid-fifteenth-century context for 'musica ficta' practice." *Acta Musicologica* 83(1): 39–68.

Juslin, P. N., and Laukka, P. (2003). "Communication of emotions in vocal expression and music performance: Different channels, same code?" *Psychological Bulletin* 129(5): 770–814.

Juslin, P. N., and Sloboda, J. A., eds. (2010). *Handbook of Music and Emotion: Theory, Research, Applications*. Oxford: Oxford University Press.

Kaernbach, C., and Bering, C. (2001). "Exploring the temporal mechanism involved in the pitch of unresolved harmonics." *Journal of the Acoustical Society of America* 110(2): 1039–1048.

Kallman, H. J. (1982). "Octave equivalence as measured by similarity ratings." *Perception & Psychophysics* 32(1): 37–49.

Kamien, R. (1986). "Subtle enharmonic relationships in Mozart's music." *Journal of Music Theory* 30(2): 169–183.

Kaminski, J. S. (2009). "Fusion theory (Verschmelzungstheorie) and Asante ivory trumpet music: Comparative musicology and African music in the twenty-first century." *Muzyka* 54: 37–50.

Kant, I. (1781). *Critik der reinen Vernunft* [*Critique of Pure Reason*]. Riga: Hartknoch.

Kantorski, V. J. (1986). "String instrument intonation in upper and lower registers: The effects of accompaniment." *Journal of Research in Music Education* 34(3): 200–210.

Karno, M., and Konečni, V. J. (1992). "The effects of structural interventions in the first movement of Mozart's Symphony in G Minor K. 550 on aesthetic preference." *Music Perception* 10(1): 63–72.

Karrick, B. (1998). "An examination of the intonation tendencies of wind instrumentalists based on their performance of selected harmonic musical intervals." *Journal of Research in Music Education* 46(1): 112–127.

Kessler, E. J., Hansen, C., and Shepard, R. N. (1984). "Tonal schemata in the perception of music in Bali and in the West." *Music Perception* 2: 131–165.

Kingsley, P. (1995). *Ancient Philosophy, Mystery, and Magic: Empedocles and Pythagorean Tradition*. Oxford: Clarendon Press.

Kisilevsky, B. S., Hains, S. M. J., Brown, C. A., Lee, C. T., Cowperthwaite, B., Stutzman, S. S., Swansburg, M. L., et al. (2009). "Fetal sensitivity to properties of maternal speech and language." *Infant Behavior and Development* 32: 59–71.

Kisilevsky, B. S., Hains, S. M. J., Lee, K., Xie, X., Huang, H., Ye, H.-H., Zhang, K., et al. (2003). "Effects of experience on fetal voice recognition." *Psychological Science* 14: 220–224.

Klapuri, A. (2004). "Signal processing methods for automatic transcription of music." Dissertation, Tampere University, Tampere, Finland.

Kleppinger, S. V. (2011). "Reconsidering pitch centricity." *Theory and Practice* 36: 65–109.

Koch, H. C. (1782). *Versuch einer Anleitung zur Composition* (A guide to composition). Leipzig: Böhme (as cited in L. G. Ratner 1956).

Kohut, H. (1957). "Observations on the psychological functions of music." *Journal of the American Psychoanalytic Association* 5(3): 389–407.

Kolinski, M. (1967). "Recent trends in ethnomusicology." *Ethnomusicology* 11(1): 1–24.

Komar, A. J. (1971). *Theory of Suspensions: A Study of Metrical and Pitch Relations in Tonal Music*. Princeton, NJ: Princeton University Press (as cited in S. Larson 1997).

Kopiez, R. (2003). "Intonation of harmonic intervals: Adaptability of expert musicians to equal temperament and just intonation." *Music Perception* 20(4): 383–410.

Krahenbuehl, D., and Schmidt, C. (1962). "On the development of musical systems." *Journal of Music Theory* 6: 32–65.

Krenek, E. (1940). *Music Here and Now*. Translated by B. Fles. New York, NY: Norton (as cited in W. E. Thomson 1999).

Krishnaswamy, A. (2003). "On the twelve basic intervals in South Indian classical music." Paper presented at the 115th Convention of the Audio Engineering Society, New York, October 10–13, 2003 (as cited in R. Ambrazevičius et al. 2015).

Krumbholz, K., Patterson, R. D., Seither-Preisler, A., and Lammertmann, C. (2003). "Neuromagnetic evidence for a pitch processing center in Heschl's gyrus." *Cerebral Cortex* 13: 765–772.

Krumhansl, C. L. (1990). *Cognitive Foundations of Musical Pitch*. New York, NY: Oxford University Press.

Krumhansl, C. L. (2000). "Rhythm and pitch in music cognition." *Psychological Bulletin* 126(1): 159–179.

Krumhansl, C. L. (2004). "The cognition of tonality—As we know it today." *Journal of New Music Research* 33(3): 253–268.

Krumhansl, C. L., and Kessler, E. J. (1982). "Tracing the dynamic changes in perceived tonal organization in a spatial representation of musical keys." *Psychological Review* 89(4): 334.

Krumhansl, C. L., and Shepard, R. N. (1979). "Quantification of the hierarchy of tonal functions within a diatonic context." *Journal of Experimental Psychology: Human Perception and Performance* 5(4): 579.

Kubik, G. (2010). *Theory of African Music*. Chicago: University of Chicago Press.

Kubovy, M., and Jordan, R. (1979). "Tone-segregation by phase: On the phase sensitivity of the single ear." *Journal of the Acoustical Society of America* 66(1): 100–106.

Kuhn, T. (1968). *The Structure of Scientific Revolutions*. Chicago: University of Chicago Press.

Kurth, E. (1917). *Grundlagen des linearen Kontrapunkts: Einführung in Stil und Technik von Bach's melodischer Polyphonie*. Bern, Switzerland: Drechsel (as cited in G. Chew 1983).

Laden, B. (1993). "A parallel learning model of musical pitch perception." *Journal of the Acoustical Society of America* 93(4): 2402–2402.

Lahdelma, I., Athanasopoulos, G., and Eerola, T. (2021). "Sweetness is in the ear of the beholder: Chord preference across United Kingdom and Pakistani listeners." *Annals of the New York Academy of Sciences* 1502(1): 72–84.

Lahdelma, I., and Eerola, T. (2016). "Single chords convey distinct emotional qualities to both naïve and expert listeners." *Psychology of Music* 44(1): 37–54.

Langner, G. (1997). "Temporal processing of pitch in the auditory system." *Journal of New Music Research* 26(2): 116–132.

Lantz, M. E., Cui, A. X., and Cuddy, L. L. (2020). "The role of duration and frequency of occurrence in perceived pitch structure." *PLoS One* 15(9): e0239582.

Larson, A. (2006). "Textual density in the choral music of Eric Whitacre." *Choral Journal* 47(6): 22–33.

Larson, S. (1997). "The problem of prolongation in 'tonal' music: Terminology, perception, and expressive meaning." *Journal of Music Theory* 41(1): 101–136.

Larson, S., and VanHandel, L. (2005). "Measuring musical forces." *Music Perception* 23(2): 119–136.

Lau, F. (2022). "Music beyond its making." Keynote address at the 13th Conference on Interdisciplinary Musicology, Hong Kong, China, and Edinburgh, Scotland, June 9, 2022.

Lecanuet, J.-P. (1996). "Prenatal auditory experiences." In *Musical Beginnings: Origins and Development of Musical Competence*, edited by I. Deliège and J. Sloboda, 37–54. Oxford: Oxford University Press.

Lee, J., and Green, D. M. (1994). "Detection of a mistuned component in a harmonic complex." *Journal of the Acoustical Society of America* 96(2): 716–725.

Leman, M. (2000). "An auditory model of the role of short-term memory in probe-tone ratings." *Music Perception* 17(4): 481–509.

Leman, M. (2012). *Music and Schema Theory: Cognitive Foundations of Systematic Musicology*. Berlin, Germany: Springer.

Lerdahl, F. (1988). "Tonal pitch space." *Music Perception* 5(3): 315–349.

Lerdahl, F. (1992). "Cognitive constraints on compositional systems." *Contemporary Music Review* 6: 97–121.

Lerdahl, F. (1996). "Calculating tonal tension." *Music Perception* 13(3): 319–363.

Lerdahl, F. (2001). *Tonal Pitch Space*. New York, NY: Oxford University Press.

Lerdahl, F. (2020). *Composition and Cognition: Reflections on Contemporary Music and the Musical Mind*. Berkeley: University of California Press.

Lerdahl, F., and Jackendoff, R. (1983). *A Generative Theory of Tonal Music*. Cambridge, MA: MIT Press.

Lerdahl, F., and Krumhansl, C. L. (2007). "Modeling tonal tension." *Music Perception* 24(4): 329–366.

Lester, J. (1992). *Compositional Theory in the Eighteenth Century*. Cambridge, MA: Harvard University Press. (reviewed by D. Damschroder in Music Theory Spectrum, February 2, 1994).

Levitin, D. J., and Tirovolas, A. K. (2009). "Current advances in the cognitive neuroscience of music." *Annals of the New York Academy of Sciences* 1156(1): 211–231.

Lewin, D. (1968). "Inversional balance as an organizing force in Schoenberg's music and thought." *Perspectives of New Music* 6(2): 1–21.

Liberman, A. M., Harris, K. S., Hoffman, H. S., and Griffith, B. C. (1957). "The discrimination of speech sounds within and across phoneme boundaries." *Journal of Experimental Psychology* 54(5): 358.

Lieberman, P. (1973). "On the evolution of language: A unified view." *Cognition* 2(1): 59–94.

Liégois-Chauvel, C., Giraud, K., Badier, J.-M., Marquis, P. and Chauvel, P. (2001). "Intracerebral evoked potentials in pitch perception reveal a functional asymmetry of the human auditory cortex." *Annals of the New York Academy of Sciences* 930(1): 117–132.

Linnavalli, T., Ojala, J., Haveri, L., Putkinen, V., Kostilainen, K., Seppänen, S., and Tervaniemi, M. (2020). "Musical expertise facilitates dissonance detection on behavioral, not on early sensory level." *Music Perception* 38(1): 78–98.

Lippius, J. (1612/1977). *Synopsis musica novae.* Strasburg: Kieffer. Translated by B. V. Rivera in 1977. *Johannes Lippius: Synopsis of New Music.* Colorado Springs: Colorado College Music Press.

Livingston, T. E. (1999). "Music revivals: Towards a general theory." *Ethnomusicology* 43(1): 66–85.

Lloyd, J., Barz, G., and Brummel-Smith, K. (2008). *Oxford Handbook of Medical Ethnomusicology.* New York, NY: Oxford University Press.

Long, M. K. (2018). "Cadential syntax and tonal expectation in late sixteenth-century homophony." *Music Theory Spectrum* 40(1): 52–83.

Loosen, F. (1993). "Intonation of solo violin performance with reference to equally tempered, Pythagorean, and just intonations." *Journal of the Acoustical Society of America* 93(1): 525–539.

Loui, P., Wu, E. H., Wessel, D. L., and Knight, R. T. (2009). "A generalized mechanism for perception of pitch patterns." *Journal of Neuroscience* 29(2): 454–459.

Lovelock, W. (1974). *First Year Harmony.* Salhouse, UK: William Elkin Music Services.

Lowinsky, E. E. (1961). *Tonality and Atonality in Sixteenth-century Music.* Berkeley: University of California Press.

Lundin, R. W. (1947). "Toward a cultural theory of consonance." *Journal of Psychology* 23(1): 45–49.

Luo, H., and Poeppel, D. (2012). "Cortical oscillations in auditory perception and speech: evidence for two temporal windows in human auditory cortex." *Frontiers in Psychology* 3: 170.

Lütkenhöner, B., Seither-Preisler, A., Krumbholz, K., and Patterson, R. D. (2010). "Auditory cortex tracks the temporal regularity of sustained noisy sounds." *Hearing Research* 272(1–2): 85–94.

Lynch, M. P., Oller, D. K., Steffens, M. L., and Buder, E. H. (1995). "Phrasing in prelinguistic vocalizations." *Developmental Psychobiology: Journal of the International Society for Developmental Psychobiology* 28(1): 3–25.

Malmberg, C. F. (1918). "The perception of consonance and dissonance." *Psychological Monographs* 25(2): 93.

Manns, J. (1994). "On composing 'by the rules.'" *Journal of Aesthetics and Art Criticism* 52(1): 83–91.

Margulis, E. H. (2005). "A model of melodic expectation." *Music Perception* 22(4): 663–714.

Margulis, E. H. (2014). *On Repeat: How Music Plays the Mind.* New York, NY: Oxford University Press.

Marmel, F., Tillmann, B., and Delbé, C. (2010). "Priming in melody perception: tracking down the strength of cognitive expectations." *Journal of Experimental Psychology: Human Perception & Performance* 36(4): 1016–1028.

Martin, D. W., and Ward, W. D. (1961). "Subjective evaluation of musical scale temperament in pianos." *Journal of the Acoustical Society of America* 33(5): 582–585.

Martin, J. A. (2014). *Beauty and Holiness*. Princeton, NJ: Princeton University Press.

Mason, J. A. (1960). "Comparison of solo and ensemble performances with reference to Pythagorean, just, and equitempered intonations." *Journal of Research in Music Education* 8: 31–38.

Matambirofa, F. (2018). "Racism and discrimination in Zimbabwe: Cues from Noam Chomsky's universal grammar." In *Philosophy, Race and Multiculturalism in Southern Africa*, edited by F. Mangena and J. D. McClymont, 29–45. Washington, DC: Council for Research in Values and Philosophy.

Matsunaga, R., Hartono, P., Yokosawa, K., and Abe, J. I. (2020). "The development of sensitivity to tonality structure of music: Evidence from Japanese children raised in a simultaneous and unbalanced bi-musical environment." *Music Perception* 37(3): 225–239.

Mattheson, J. (1713). *Das neu-eröffnete Orchestre*. Hamburg: Schiller.

McAdams, S. (1984). "The auditory image: A metaphor for musical and psychological research on auditory organization." In *Cognitive Processes in the Perception of Art*, edited by W. R. Crozier and A. J. Chapman, 289–323. North-Holland: Elsevier.

McAdams, S., and Bregman, A. (1979). "Hearing musical streams." *Computer Music Journal* 3(4): 26–43.

McClain, E. G. (1979). "Chinese cyclic tunings in late antiquity." *Ethnomusicology* 23(2): 205–224.

McDermott, J. H., Keebler, M. V., Micheyl, C., and Oxenham, A. J. (2010). "Musical intervals and relative pitch: Frequency resolution, not interval resolution, is special." *Journal of the Acoustical Society of America* 128(4): 1943–1951.

McDermott, J. H., Lehr, A. J., and Oxenham, A. J. (2010). "Individual differences reveal the basis of consonance." *Current Biology* 20(11): 1035–1041.

McDermott, J. H., Schultz, A. F., Undurraga, E. A., and Godoy, R. A. (2016). "Indifference to dissonance in native Amazonians reveals cultural variation in music perception." *Nature* 535(7613): 547–550.

McGowan, J. (2011). "Psychoacoustic foundations of contextual harmonic stability in jazz piano voicings." *Journal of Jazz Studies* 7(2): 156–191.

McHose, A. I. (1947). *The Contrapuntal Harmonic Technique of the 18th Century*. Englewood Cliffs, NJ: Prentice-Hall.

McLachlan, N., Marco, D., Light, M., and Wilson, S. (2013). "Consonance and pitch." *Journal of Experimental Psychology: General* 142(4): 1142.

McLachlan, N. M., Marco, D. J., and Wilson, S. J. (2013). "The musical environment and auditory plasticity: Hearing the pitch of percussion." *Frontiers in Psychology* 4: 768.

McLachlan, N., and Wilson, S. (2010). "The central role of recognition in auditory perception: A neurobiological model." *Psychological Review* 117(1): 175–196.

McLucas, A. D. (2010). *The Musical Ear: Oral Tradition in the USA*. Oxford: Ashgate.

McRae, C. (2020). "Performative listening." In *Handbook of Listening*, edited by D. L. Worthington and G. D. Bodie, 399–407. Hoboken, NY: Wiley.

Meddis, R., and Hewitt, M. J. (1991a). "Virtual pitch and phase sensitivity of a computer model of the auditory periphery. I: Pitch identification." *Journal of the Acoustical Society of America* 89(6): 2866–2882.

Meddis, R., and Hewitt, M. J. (1991b). "Virtual pitch and phase sensitivity of a computer model of the auditory periphery. II: Phase sensitivity." *Journal of the Acoustical Society of America* 89(6): 2883–2894.

Meddis, R., and O'Mard, L. (1997). "A unitary model of pitch perception." *Journal of the Acoustical Society of America* 102: 1811–1820.

Mehr, S. A., Singh, M., Knox, D., Ketter, D. M., Pickens-Jones, D., Atwood, S., Lucas, C., et al. (2019). "Universality and diversity in human song." *Science* 366(6468).

Mehr, S. A., Singh, M., York, H., Glowacki, L., and Krasnow, M. M. (2018). "Form and function in human song." *Current Biology* 28(3): 356–368.

Menary, R. (2010). "Introduction to the special issue on 4E cognition." *Phenomenology and the Cognitive Sciences* 9(4): 459–463.

Meyer, L. B. (1956). *Emotion and Meaning in Music*. Chicago: Chicago University Press.

Meyer, L. B. (1957). "Meaning in music and information theory." *Journal of Aesthetics and Art Criticism* 15(4): 412–424.

Meyer, L. B. (1973). *Explaining Music*. Berkeley: University of California Press.

Meyer, M., Steinhauer, K., Alter, K., Friederici, A. D., and von Cramon, D. Y. (2004). "Brain activity varies with modulation of dynamic pitch variance in sentence melody." *Brain and Language* 89(2): 277–289.

Mickelsen, W. C. (1977). *Hugo Riemann's Theory of Harmony*. Lincoln: University of Nebraska Press.

Miles, S. A., Rosen, D. S., Barry, S., Grunberg, D., and Grzywacz, N. (2021). "What to expect when the unexpected becomes expected: Harmonic surprise and preference over time in popular music." *Frontiers in Human Neuroscience* 15: 578644.

Miller, C. (2011). "Aesthetics and e-assessment: the interplay of emotional design and learner performance." *Distance Education* 32(3): 307–337.

Miller, G. A. (1956). "The magical number seven, plus or minus two: Some limits on our capacity for processing information." *Psychological Review* 63(2): 81.

Miller, G. A., and Heise, G. A. (1950). "The trill threshold." *Journal of the Acoustical Society of America* 22(5): 637–638.

Miller, S. E., Schlauch, R. S., and Watson, P. J. (2010). "The effects of fundamental frequency contour manipulations on speech intelligibility in background noise." *Journal of the Acoustical Society of America* 128(1): 435–443.

Milne, A. J., Laney, R., and Sharp, D. B. (2015). "A spectral pitch class model of the probe tone data and scalic tonality." *Music Perception* 32(4): 364–393.

Moisala, P. (1995). "Cognitive study of music as culture—basic premises for 'cognitive ethnomusicology.'" *Journal of New Music Research* 24(1): 8–20.

Möller, H., and Stephan, R., eds. (1991). *Die Musik des Mittelalters*. Lilienthal, Germany: Laaber.

Monson, B. B., Hunter, E. J., Lotto, A. J., and Story, B. H. (2014). "The perceptual significance of high-frequency energy in the human voice." *Frontiers in Psychology* 5: 587.

Monson, I. (1999). "Riffs, repetition, and theories of globalization." *Ethnomusicology* 43(1): 31–65.

Monzo, J. (n.d.). "Sonance. Encyclopedia of microtonal music theory." http://tonalsoft.com/enc/s/sonance.aspx

Mooney, K. (2000). "Hugo Riemann's debut as a music theorist." *Journal of Music Theory* 44(1): 81–99.

Moore, A. (1995). "The so-called 'flattened seventh' in rock." *Popular Music* 14(2): 185–201.

Moore, B. C. J. (1989). *An Introduction to the Psychology of Hearing*. 3rd ed. London, UK: Academic Press.

Moore, B. C. J. (1977). "Effects of relative phase of the components on the pitch of three-component complex tones." In *Psychophysics and Physiology of Hearing*, edited by E. F. Evans and J. P. Wilson, 349–358. London, GB: Academic Press.

Moore, B. C. J. (1990). "Auditory models: Pitch perception in silicon." *Nature* 343: 114–115.

Moore, B. C. J., and Glasberg, B. R. (1983). "Suggested formulae for calculating auditory-filter bandwidths and excitation patterns." *Journal of the Acoustical Society of America* 74(3): 750–753.

Moore, B. C. J., Glasberg, B. R., and Peters, R. W. (1985). "Relative dominance of individual partials in determining the pitch of complex tones." *Journal of the Acoustical Society of America* 77(5): 1853–1860.

Moore, B. C. J., Glasberg, B. R., and Peters, R. W. (1986). "Thresholds for hearing mistuned partials as separate tones in harmonic complexes." *Journal of the Acoustical Society of America* 80(2): 479–483.

Moore, B. C., Glasberg, B. R., Plack, C. J., and Biswas, A. K. (1988). "The shape of the ear's temporal window." *Journal of the Acoustical Society of America* 83(3): 1102–1116.

Moore, B. C. J., and Ohgushi, K. (1993). "Audibility of partials in inharmonic complex tones." *Journal of the Acoustical Society of America* 93(1): 452–461.

Moore, B. C. J., and Rosen, S. M. (1979). "Tune recognition with reduced pitch and interval information." *Quarterly Journal of Experimental Psychology* 31(2): 229–240.

Morgan, C. T., and Galambos, R. (1943). "A reinvestigation of the relation between pitch and intensity." *Journal of the Acoustical Society of America* 15(1): 77–77.

Morgan, R. P. (1978). "Schenker and the theoretical tradition: The concept of musical reduction." *College Music Symposium* 18(1): 72–96.

Morris, R. (1979). "A similarity index for pitch-class sets." *Perspectives of New Music* 18(1–2): 445–460.

Morrison, S. J. (2000). "Effect of melodic context, tuning behaviors, and experience on the intonation accuracy of wind players." *Journal of Research in Music Education* 48(1): 39–51.

Mos, L. P. (1988). "Cognitive representation: A psychology in search of mentality." In *Recent Trends in Theoretical Psychology*, edited by W. J. Baker, M. E. Hyland, R. Hezewijk, and S. Terwee, 145–156. New York, NY: Springer.

Motte-Haber, H. de la, and Kopiez, R. (1995). *Der Hörer als Interpret* [*The Listener as Performer*]. Frankfurt am Main: Peter Lang.

Mulholland, J., and Hojnacki, T. (2013). *The Berklee Book of Jazz Harmony*. Boston, MA: Berklee Press.

Mull, H. K. (1957). "The effect of repetition upon the enjoyment of modern music." *Journal of Psychology* 43(1): 155–162.

Mullennix, J. W., Pilot, K. M., Steeves, T. A., and Burns, J. C. (2018). "The effects of cognitive load on judgments of titled visual art." *Psychology of Aesthetics, Creativity, and the Arts* 12(2): 166.

Mummert, M. (1997). *Sprachcodierung durch Konturierung eines gehörangepassten Spektrogramms und ihre Anwendung zur Datenreduktion.* Doctoral dissertation, Technische Universität Munich. mediatum.ub.tum.de.

Nagy, S. I., Révész, G., Séra, L., Bandi, S. A., and Stachó, L. (2022). "Final-note expectancy and humor: An empirical investigation." *BMC Psychology* 10(1): 1–12.

Nakaseko, K. (1957). "Symbolism in ancient Chinese music theory." *Journal of Music Theory* 1(2): 147–180.

Narmour, E. (1990). *The Analysis and Cognition of Basic Melodic Structures: The Implication-Realization Model.* Chicago: University of Chicago Press.

Neubauer, S., and Hublin, J. J. (2012). "The evolution of human brain development." *Evolutionary Biology* 39(4): 568–586.

Nesse, R. M. (1990). "Evolutionary explanations of emotions." *Human Nature* 1(3): 261–289.

Nettl, B. (1999). "The institutionalization of musicology: Perspectives of a North American ethnomusicologist." In *Rethinking Music,* edited by N. Cook and M. Everest, 297–310. Oxford: Oxford University Press.

Nikjeh, D. A., Lister, J. J., and Frisch, S. A. (2008). "Hearing of note: An electrophysiologic and psychoacoustic comparison of pitch discrimination between vocal and instrumental musicians." *Psychophysiology* 45(6): 994–1007.

Nobile, D. (2016). "Harmonic function in rock music: A syntactical approach." *Journal of Music Theory* 60(2): 149–180.

North, A. C., and Hargreaves, D. J. (2007). "Lifestyle correlates of musical preference: 3. Travel, money, education, employment and health." *Psychology of Music* 35(3): 473–497.

Norton, R. (1984). *Tonality in Western Culture: A Critical and Historical Perspective.* University Park: Pennsylvania State University Press.

Notterman, J. M. (2000). "Note on reductionism in cognitive psychology: Reification of cognitive processes into mind, mind–brain equivalence, and brain–computer analogy." *Journal of Theoretical and Philosophical Psychology* 20(2): 116.

O'Keefe, V. (1975). "Psychological preference for harmonized musical passages in the just and equally-tempered systems." *Perception and Motor Skills* 40: 192–194.

Oakley, D. A., and Halligan, P. W. (2017). "Chasing the rainbow: The non-conscious nature of being." *Frontiers in Psychology* 8: 1924.

Ofosu, O. O., and Ofosu, J. S. (2020). "Tonal relationships in African music: A global phenomenon." *Icheke Journal* 18(4): 229–235.

Oppenheim, J. N., and Magnasco, M. O. (2013). "Human time-frequency acuity beats the Fourier uncertainty principle." *Physical Review Letters* 110(4): 044301.

Oram, N., and Cuddy, L. L. (1995). "Responsiveness of Western adults to pitch-distributional information in melodic sequences." *Psychological Research* 57(2): 103–118.

Oxenham, A. J. (2013). "Revisiting place and temporal theories of pitch." *Acoustical Science and Technology* 34(6): 388–396.

Oxenham, A. J., Bernstein, J. G., and Penagos, H. (2004). "Correct tonotopic representation is necessary for complex pitch perception." *Proceedings of the National Academy of Sciences* 101(5): 1421–1425.

Oxenham, A. J., and Plack, C. J. (1998). "Suppression and the upward spread of masking." *Journal of the Acoustical Society of America* 104(6): 3500–3510.

Ozmo, Ž. (2016). "Should (early) Baroque music be equally tempered? Vincenzo Galilei's 1584 Libro d'intavolatura di liuto and its wider implications for historical performance practice." *Early Music* 44(1): 119–124.

Páez, D., Rimé, B., Basabe, N., Wlodarczyk, A., and Zumeta, L. (2015). "Psychosocial effects of perceived emotional synchrony in collective gatherings." *Journal of Personality and Social Psychology* 108(5): 711.

Palmer, C. (1997). "Music performance." *Annual Review of Psychology* 48: 115–138.

Pamies, S. (2021). "Deconstructing modal jazz piano techniques: The relation between Debussy's piano works and the innovations of post-bop pianists." *Jazz Education in Research and Practice* 2(1): 76–105.

Pankhurst, T. (2008). *SchenkerGUIDE: A Brief Handbook and Website for Schenkerian Analysis*. New York, NY: Routledge.

Pankovski, T., and Pankovska, A. (2022). "Nonspecific Hebbian neural network model predicts musical scales discreteness and just intonation without using octave-equivalency mapping." *Scientific Reports* 12(1): 1–13.

Pantev, C., Hoke, M., Lütkenhöner, B., and Lehnertz, K. (1989). "Tonotopic organization of the auditory cortex: Pitch versus frequency representation." *Science* 246(4929): 486–488.

Papineau, D. (2013). "The rise of physicalism." In *Proper Ambition of Science*, edited by M. W. F. Stone and J. Wolff, 182–216. London, UK: Routledge.

Parks, R. S. (1976). "Voice leading and chromatic harmony in the music of Chopin." *Journal of Music Theory* 20(2): 189–214.

Parncutt, R. (1988). "Revision of Terhardt's psychoacoustical model of the root(s) of a musical chord." *Music Perception* 6: 65–94.

Parncutt, R. (1989). *Harmony: A Psychoacoustical Approach*. Berlin: Springer.

Parncutt, R. (1993). Pitch properties of chords of octave-spaced tones. *Contemporary Music Review* 9: 35–50.

Parncutt, R. (1994). "Template-matching models of musical pitch and rhythm perception." *Journal of New Music Research* 23: 145–168.

Parncutt, R. (1997). "A model of the perceptual root(s) of a chord accounting for voicing and prevailing tonality." In *Music, Gestalt, and Computing—Studies in Cognitive and Systematic Musicology*, edited by M. Leman, 181–199. Berlin: Springer.

Parncutt, R. (1999). "Systematic evaluation of the psychological effectiveness of non-conventional notations and keyboard tablatures." In *Music and Signs*, edited by I. Zannos, 146–174. Bratislava, Slovakia: ASCO Art & Science.

Parncutt, R. (2003). "Accents and expression in piano performance." In *Perspektiven und Methoden einer Systemischen Musikwissenschaft (Festschrift Fricke)*, edited by K. W. Niemöller, 163–185. Frankfurt: Peter Lang.

Parncutt, R. (2007). "Systematic musicology and the history and future of Western musical scholarship." *Journal of Interdisciplinary Music Studies* 1(1): 1–32.

Parncutt, R. (2009). "Tonal implications of harmonic and melodic Tn-types." In *Mathematics and Computing in Music*, edited by T. Klouche and T. Noll, 124–139. Berlin: Springer.

Parncutt, R. (2011). "The tonic as triad: Key profiles as pitch salience profiles of tonic triads." *Music Perception* 28: 333–365.

Parncutt, R., ed. (2012). "ESF Exploratory Workshop 'Cognition of Early Polyphony': Scientific report." https://homepage.uni-graz.at/de/richard.parncutt/research/books/

Parncutt, R. (2014). "The emotional connotations of major versus minor tonality: One or more origins?" *Musicae Scientiae* 18: 324–353.

Parncutt, R. (2019a). "Pitch-class prevalence in plainchant, scale-degree consonance, and the origin of the rising leading tone." *Journal of New Music Research* 48(5): 434–448.

Parncutt, R. (2019b). "Mother schema, obstetric dilemma, and the origin of behavioral modernity." *Behavioral Sciences* 9(12): 142.

Parncutt, R., and Bregman, A. S. (2000). "Tone profiles following short chord progressions: Top-down or bottom-up?" *Music Perception* 18(1): 25–57.

Parncutt, R., and Chuckrow, R. (2019). "Chuckrow's theory of the prenatal origin of music." *Musicae Scientiae* 23(4): 403–425.

Parncutt, R., and Cohen, A. J. (1995). "Identification of microtonal melodies: Effects of scale-step size, serial order, and training." *Perception & Psychophysics* 57: 835–846.

Parncutt, R., Engel, I., Zemmel, E., and Böhm, J. (2023). "Audible partials (spectral pitches) in common musical chords." *Psychological Topics* 32(1): 13–34.

Parncutt, R., and Hair, G. (2018). "A psychocultural theory of musical interval: Bye bye Pythagoras." *Music Perception* 35(4): 475–501.

Parncutt, R., and Radovanovic, L. (2023). "The missing fundamentals of harmonic theory: Chord roots and their ambiguity in arrangements of jazz standards." *Musicae Scientiae* 27(2): 366–389.

Parncutt, R., and Reisinger, D. (2021). "Octave-generalized analysis of chord progressions: Diatonic/fifth relations, missing fundamentals, completion tones." *Journal of New Music Research* 50(1): 52–73.

Parncutt, R., Reisinger, D., Fuchs, A., and Kaiser, F. (2019). "Consonance and prevalence of sonorities in Western polyphony: Roughness, harmonicity, familiarity, evenness, diatonicity." *Journal of New Music Research* 48(1): 1–20.

Parncutt, R., Sattmann, S., Gaich, A., and Seither-Preisler, A. (2019). "Tone profiles of isolated musical chords: Psychoacoustic versus cognitive models." *Music Perception* 36(4): 406–430.

Parncutt, R., and Stuckey, R. (1992). "Towards a standard alternative music notation and terminology based on the chromatic scale." *Musikometrika* 4: 117–143.

Parncutt, R., and Zemmel, E. (2022). "Psychoacoustic pitch analysis of common musical chords." Presented at the 14. Tagung der Österreichischen Gesellschaft für Psychologie [14th Conference of the Austrian Society for Psychology], Klagenfurt, Austria, September 4–7, 2022.

Patel, A. D., and Balaban, E. (2001). "Human pitch perception is reflected in the timing of stimulus-related cortical activity." *Nature Neuroscience* 4(8): 839–844.

Patel, A. D., Iversen, J. R., Wassenaar, M., and Hagoort, P. (2008). "Musical syntactic processing in agrammatic Broca's aphasia." *Aphasiology* 22(7–8): 776–789.

Patterson, R. D., Allerhand, M. H., and Giguere, C. (1995). "Time-domain modeling of peripheral auditory processing: A modular architecture and a software platform." *Journal of the Acoustical Society of America* 98(4): 1890–1894.

Patterson, R. D., Uppenkamp, S., Johnsrude, I. S., and Griffiths, T. D. (2002). "The processing of temporal pitch and melody information in auditory cortex." *Neuron* 36(4): 767–776.

Pearce, M. T., Ruiz, M. H., Kapasi, S., Wiggins, G. A., and Bhattacharya, J. (2010). "Unsupervised statistical learning underpins computational, behavioural, and neural manifestations of musical expectation." *NeuroImage* 50(1): 302–313.

Pegg, C. (1992). "Mongolian conceptualizations of overtone singing (xöömii)." *British Journal of Ethnomusicology* 1(1): 31–54.

Pellegrin, R. S. (2013). "On jazz analysis: Schenker, Salzer, and salience." Doctoral dissertation, University of Washington, Seattle, WA.

Pelofi, C., and Farbood, M. M. (2021). "Asymmetry in scales enhances learning of new musical structures." *Proceedings of the National Academy of Sciences* 118(31): e2014725118.

Peltola, H. R., and Vuoskoski, J. K. (2022). "'I hate this part right here': Embodied, subjective experiences of listening to aversive music." *Psychology of Music* 50(1): 159–174.

Peretz, I., Gaudreau, D., and Bonnel, A. M. (1998). "Exposure effects on music preference and recognition." *Memory & Cognition* 26(5): 884–902.

Perle, G. (1972). *Serial Composition and Atonality: An Introduction to the Music of Schoenberg, Berg, and Webern*. Berkeley: University of California Press.

Pesic, P. (2014). *Music and the Making of Modern Science*. Cambridge, MA: MIT Press.

Peters, R. W., Moore, B. C. J., and Glasberg, B. R. (1983). "Pitch of components of complex tones." *Journal of the Acoustical Society of America* 73(3): 924–929.

Pfordresher, P. Q., and Brown, S. (2016). "Vocal mistuning reveals the origin of musical scales." *Journal of Cognitive Psychology* 29(1): 35–52.

Pitt, M. A., Myung, I. J., and Zhang, S. (2002). "Toward a method of selecting among computational models of cognition." *Psychological Review* 109: 472–491.

Plack, C. J., and Oxenham, A. J. (2005). "The psychophysics of pitch." In *Pitch: Neural Coding and Perception*, edited by C. J. Plack, A. J. Oxenham, R. R. Fay, and A. N. Popper, 7–55. New York, NY: Springer.

Plack, C. J., Barker, D., and Hall, D. A. (2014). "Pitch coding and pitch processing in the human brain." *Hearing Research* 307: 53–64.

Plantinga, J., and Trehub, S. E. (2013). "Revisiting the innate preference for consonance." *Journal of Experimental Psychology* 40(1): 40–49.

Platt, J. R., and Racine, R. J. (1990). "Perceived pitch class of isolated musical triads." *Journal of Experimental Psychology: Human Perception and Performance* 16(2): 415.

Pléh, C. (2019). "The inspirational role of Chomsky in the cognitive turn of psychology." *Acta Linguistica Academica* 66(3): 397–428.

Plomp, R. (1964). "The ear as a frequency analyzer." *Journal of the Acoustical Society of America* 36(9): 1628–1636.

Plomp, R. (1965). "Detectability threshold for combination tones." *Journal of the Acoustical Society of America* 37(6): 1110–1123.

Plomp, R., and Levelt, W. J. M. (1965). "Tonal consonance and critical bandwidth." *Journal of the Acoustical Society of America* 38(4): 548–560.

Plomp, R., and Steeneken, H. J. (1969). "Effect of phase on the timbre of complex tones." *Journal of the Acoustical Society of America* 46(2B): 409–421.

Plomp, R., Wagenaar, W. A., and Mimpen, A. M. (1973). "Musical interval recognition with simultaneous tones." *Acustica* 29(2): 101–109.

Popescu, T., Neuser, M. P., Neuwirth, M., Bravo, F., Mende, W., Boneh, O., Moss, F. C. et al. (2019). "The pleasantness of sensory dissonance is mediated by musical style and expertise." *Scientific Reports* 9(1): 1–11.

Popper, K. (1934/1959). *The Logic of Scientific Discovery (Logik der Forschung).* London: Routledge.

Popper, K. R., and Eccles, J. C. (1977). *The Self and Its Brain: An Argument for Interactionism.* Berlin: Springer.

Powell, J., and Dibben, N. (2005). "Key-mood association: A self perpetuating myth." *Musicae Scientiae* 9(2): 289–311.

Powers, H. S. (1981). "Tonal types and modal categories in Renaissance polyphony." *Journal of the American Musicological Society* 34(3): 428–470.

Powers, H. S. (1998). "From psalmody to tonality." In *Tonal Structures in Early Music*, edited by C. C. Judd, 275–340. New York, NY: Routledge.

Prame, E. (1997). "Vibrato extent and intonation in professional Western lyric singing." *Journal of the Acoustical Society of America* 102(1): 616–621.

Preisler, A. (1993). "The influence of spectral composition of complex tones and of musical experience on the perceptibility of virtual pitch." *Perception & Psychophysics* 54(5): 589–603.

Pressing, J. (1978). "Towards an understanding of scales in jazz." *Jazz Research* 9: 25–35.

Presson, C. C., and Montello, D. R. (1988). "Points of reference in spatial cognition: Stalking the elusive landmark." *British Journal of Developmental Psychology* 6(4): 378–381.

Pritschet, H. (1992). *Wahrnehmung des Grundtons von Dreiklängen* [*Perception of the Root of Triads*]. Regensburg: Roderer.

Proust, D. (2009). "The harmony of the spheres from Pythagoras to voyager." *Proceedings of the International Astronomical Union* 5(S260): 358–367.

Raffman, D. (1993). *Language, Music, and Mind*. Cambridge, MA: MIT Press.

Rahn, J. (1980). *Basic Atonal Theory*. New York, NY: Schirmer.

Railsback, O. L. (1938). "A study of the tuning of pianos." *Journal of the Acoustical Society of America* 10(1): 86–86.

Rakowski, A. (1994). "Musicians' tendency to stretch larger-than-octave melodic intervals." *Journal of the Acoustical Society of America* 96(5): 3301–3301.

Rakowski, A., and Miśkiewicz, A. (1995). "Deviations from equal temperament in tuning isolated musical intervals." *Archives of Acoustics* 20(2): 95–104.

Rakowski, A., and Rogowski, P. (2007). "Experiments on long-term and short-term memory for pitch in musicians." *Archives of Acoustics* 32(4): 815–826.

Rameau, J.-P. (1722). *Traité de l'harmonie réduite à ses principes naturels*. Paris: Ballard.

Rameau, J.-P. (1737). *Génération harmonique, ou Traité de musique théorique et pratique*. Paris: Prault fils.

Ramsey, W. (2002). "Eliminative materialism." In *Stanford Encyclopedia of Philosophy*, edited by E. N. Zalta. https://plato.stanford.edu/

Randel, D. (1986). *The New Harvard Dictionary of Music*. Cambridge, MA: Harvard University Press.

Rasch, R. (2011). "Does 'well-tempered' mean 'equal-tempered'?" In *Bach*, edited by Y. Tomita, 511–528. London. Routledge.

Raskin, J. D. (2001). "On relativism in constructivist psychology." *Journal of Constructivist Psychology* 14(4): 285–313.

Read, G. (1987). *Source Book of Proposed Music Notation Reforms*. New York, NY: Greenwood.

Rehding, A. (2003). *Hugo Riemann and the Birth of Modern Musical Thought*. Cambridge, UK: Cambridge University Press.

Reichweger, G. (2010). "The perception of pitch salience in musical chords of different tone types." Master's thesis, University of Graz, Graz, Austria.

Révész, G. (1913). *Zur Grundlegung der Tonpsychologie*. Leipzig: Veit.

Richardson, M. J., Shockley, K., Fajen, B. R., Riley, M. A., and Turvey, M. T. (2008). "Ecological psychology: Six principles for an embodied–embedded approach to behavior." In *Handbook of Cognitive Science: An Embodied Approach*, edited by P. Calvo and T. Gomila, 159–187). San Diego, CA: Elsevier.

Riemann, H. (1872). "Musikalische Logik: Ein Beitrag zur Theorie der Musik." *Neue Zeitschrift für Musik* 68, 279–282, 287–288, 353–355, 363–364, 373–374 (as cited in K. Mooney 2000).

Riemann, H. (1882). *Die Natur der Harmonik*. Leipzig: Breitkopf & Härtel.

Riemann, H. (1893). *Vereinfachte Harmonielehre, oder die Lehre von den tonalen Funktionen der Akkorde*. London, UK: Augener.

Riemann, H. (1905). *Das Problem des harmonischen Dualismus*. Leipzig, Germany: Kahnt.

Riemann, H. (1914–1915). "Ideen zu einer 'Lehre von den Tonvorstellungen.'" *Jahrbuch der Musikbibliothek Peters* 21–22, 1–26 (as cited in E. Gollin and A. Rehding 2011).

Rings, S. (2011). *Tonality and transformation*. Oxford University Press.

Ritsma, R. J. (1967). "Frequencies dominant in the perception of the pitch of complex sounds." *Journal of the Acoustical Society of America* 42(1): 191–198.

Rivera, B. V. (1978). "The 'Isagoge' (1581) of Johannes Avianius: An early formulation of triadic theory." *Journal of Music Theory* 22(1): 43–64.

Rivera, B. V. (1984). "The seventeenth-century theory of triadic generation and invertibility and its application in contemporaneous rules of composition." *Music Theory Spectrum* 6: 63–78.

Rivera, B. V. (1989). "Studies in analysis and history of theory: The sixteenth and seventeenth centuries." *Music Theory Spectrum* 11(1): 24–28.

Roberts, B., and Brunstrom, J. M. (1998). "Perceptual segregation and pitch shifts of mistuned components in harmonic complexes and in regular inharmonic complexes." *Journal of the Acoustical Society of America* 104(4): 2326–2338.

Robinson, W. S. (2010). "Epiphenomenalism." *Wiley Interdisciplinary Reviews: Cognitive Science* 1(4): 539–547.

Rohrmeier, M. A., and Koelsch, S. (2012). "Predictive information processing in music cognition. A critical review." *International Journal of Psychophysiology* 83(2): 164–175.

Römer, H. (1998). "The sensory ecology of acoustic communication in insects." In *Comparative Hearing: Insects*, edited by J. H. Fullard, 63–96. New York, NY: Springer.

Rosch, E. (1975). "Cognitive reference points." *Cognitive Psychology* 7(4): 532–547.

Rosch, E., and Mervis, C. B. (1975). "Family resemblances: Studies in the internal structure of categories." *Cognitive Psychology* 7(4): 573–605.

Rosch, E., and Mervis, C. B. (1998). "Family resemblances: Studies in the internal structure of categories." In *Rethinking Intuition*, edited by M. R. DePaul and W. Ramsey, 17–44. Lanham, MD: Rowman & Littlefield.

Rosen, C., and Ricci, G. C. (2002). *Beethoven's Piano Sonatas: A Short Companion.* New Haven, CT: Yale University Press.

Ross, B., Eulitz, C., and Terhardt, E. (1996). "Binaural fusion and the representation of virtual pitch in the human auditory cortex." *Hearing Research* 100: 164–170.

Ross, B., and Knight, S. (2019). "Reports of equitonic scale systems in African musical traditions and their implications for cognitive models of pitch organization." *Musicae Scientiae* 23(4): 387–402.

Rothstein, W. (1991). "On implied tones." *Music Analysis* 10(3): 289–328.

Rowell, L. (2015). *Music and Musical Thought in Early India.* Chicago: University of Chicago Press.

Roy, D., and Reiter, E. (2005). "Connecting language to the world." *Artificial Intelligence* 167(1–2): 1–12.

Sachs, C. (1962). *The Wellsprings of Music.* Edited by J. Kunst. The Hague: Nijhoff.

Saddler, M. R., Gonzalez, R., and McDermott, J. H. (2021). "Deep neural network models reveal interplay of peripheral coding and stimulus statistics in pitch perception." *Nature Communications* 12(1): 1–25.

Saffran, J. R. (2003). "Statistical language learning: Mechanisms and constraints." *Current Directions in Psychological Science* 12(4): 110–114.

Salimpoor, V. N., Zald, D. H., Zatorre, R. J., Dagher, A., and McIntosh, A. R. (2015). "Predictions and the brain: How musical sounds become rewarding." *Trends in Cognitive Sciences* 19(2): 86–91.

Salzer, F. (1952). *Structural Hearing: Tonal Coherence in Music.* New York, NY: Charles Boni (as cited in S. Larson 1997).

Salzer, F. (1967). "Tonality in early medieval polyphony: Towards a history of tonality." *Music Forum* 1: 35–98.

Sanders, E. H. (2003). "'Are my ears on wrong?': Hearing polyphony, ca. 1375–ca. 1450." *Current Musicology* 75: 51–78.

Savage, P. E., Brown, S., Sakai, E., and Currie, T. E. (2015). "Statistical universals reveal the structures and functions of human music." *Proceedings of the National Academy of Sciences* 112(29): 8987–8992.

Schäfer, K., and Eerola, T. (2020). "How listening to music and engagement with other media provide a sense of belonging: An exploratory study of social surrogacy." *Psychology of Music* 48(2): 232–251.

Schäfer, T., and Sedlmeier, P. (2009). "From the functions of music to music preference." *Psychology of Music* 37(3): 279–300.

Schaffer, J. (2018). "Monism." In *Stanford Encyclopedia of Philosophy*, edited by E. N. Zalta. https://plato.stanford.edu/

Schelinski, S., and Kriegstein, K von. (2019). "The relation between vocal pitch and vocal emotion recognition abilities in people with autism spectrum disorder and typical development." *Journal of Autism and Development Disorders* 49: 68–82.

Schenker, H. (1906/1954). *Harmonielehre*. Stuttgart: Cotta. Translated by E. M. Borgese (1954) as *Harmony*. Chicago: Chicago University Press.

Schenker, H. (1935/1979). *Der freie Satz*. Vienna: Universal. Translated and Edited by E. Oster (1979) as *Free Composition*. New York, NY: Longman.

Schiavio, A., and van der Schyff, D. (2016). "Beyond musical qualia. Reflecting on the concept of experience." *Psychomusicology* 26(4): 366.

Schiller, P. H., and Carvey, C. E. (2005). "The Hermann grid illusion revisited." *Perception* 34: 1375–1397.

Schneider, A. (1996). "'Verschmelzung,' tonal fusion, and consonance: Carl Stumpf revisited." In *Music, Gestalt, and Computing*, edited by M. Leman, 115–143. Berlin: Springer.

Schneider, P., and Wengenroth, M. (2009). "The neural basis of individual holistic and spectral sound perception." *Contemporary Music Review* 28(3): 315–328.

Schoenberg, A. (1911). *Harmonielehre*. Vienna: Universal.

Schoenberg, A. (1975). *Style and Idea: Selected Writings*. Edited by L. Stein and Translated by L. Black. New York, NY (as cited in A. Forte 1978).

Schouten, J. F. (1940). "Five articles on the perception of sound (1938–1940)." Instituut voor Perceptic Caderzoek (as cited in Patterson and Wightman 1976).

Schouten, J. F., Ritsma, R. J., and Cardozo, B. L. (1962). "Pitch of the residue." *Journal of the Acoustical Society of America* 34(9B): 1418–1424.

Schön, D., Regnault, P., Ystad, S., and Besson, M. (2005). "Sensory consonance: An ERP study." *Music Perception* 23(2): 105–117.

Schubert, E. (2010). "Affective, evaluative, and collative responses to hated and loved music." *Psychology of Aesthetics, Creativity, and the Arts* 4(1): 36.

Schubert, E., and Wolfe, J. (2013). "The rise of fixed pitch systems and the slide of continuous pitch: A note for emotion in music research about portamento." *Journal of Interdisciplinary Music Studies* 7(1–2): 1–27.

Schubert, P. (1999). *Modal Counterpoint, Renaissance Style*. New York, NY: Oxford University Press.

Scruton, R. (1983). "Understanding music." In *The Aesthetic Understanding: Essays in the Philosophy of Art and Culture*, edited by R. Scruton, 77–100. Manchester, UK: Carcanet.

Sears, D. R., and Forrest, D. (2021). "Triadic patterns across classical and popular music corpora: Stylistic conventions, or characteristic idioms?" *Journal of Mathematics and Music* 15(2): 140–153.

Sears, D. R., Pearce, M. T., Caplin, W. E., and McAdams, S. (2018). "Simulating melodic and harmonic expectations for tonal cadences using probabilistic models." *Journal of New Music Research* 47(1): 29–52.

Seither-Preisler, A., Johnson, L., Krumbholz, K., Nobbe, A., Patterson, R., Seither, S., and Lütkenhöner, B. (2007). "Tone sequences with conflicting fundamental pitch and timbre changes are heard differently by musicians and nonmusicians." *Journal of Experimental Psychology: Human Perception and Performance* 33(3): 743.

Serrà, J., Corral, Á., Boguñá, M., Haro, M., and Arcos, J. L. (2012). "Measuring the evolution of contemporary Western popular music." *Scientific Reports* 2(1): 1–6.

Shackleton, T. M., and Carlyon, R. P. (1994). "The role of resolved and unresolved harmonics in pitch perception and frequency modulation discrimination." *Journal of the Acoustical Society of America* 95(6): 3529–3540.

Shamma, S., and Klein, D. (2000). "The case of the missing pitch templates: How harmonic templates emerge in the early auditory system." *Journal of the Acoustical Society of America* 107(5): 2631–2644.

Shepard, R. N. (1964). "Circularity in judgments of relative pitch." *Journal of the Acoustical Society of America* 36(12): 2346–2353.

Shepard, R. N. (1982). "Geometrical approximations to the structure of musical pitch." *Psychological Review* 89(4): 305–333.

Shepherd, D., and Sigg, N. (2015). "Music preference, social identity, and self-esteem." *Music Perception* 32(5): 507–514.

Shirlaw, M. (1917). *The Theory of Harmony: An Inquiry into the Natural Principles of Harmony, with an Examination of the Chief Systems of Harmony from Rameau to the Present Day*. London: Novello.

Shirlaw, M. (1951). "The music and tone-systems of ancient Greece." *Music & Letters* 32(2): 131–139.

Shirlaw, M. (1957). "The science of harmony." *Music Review* 18: 263–278.

Shonle, J. I., and Horan, K. E. (1976). "Trill threshold revisited." *Journal of the Acoustical Society of America* 59(2): 469–471.

Simms, B. (1975). "Choron, Fétis, and the theory of tonality." *Journal of Music Theory* 19: 112–139.

Skinner, B. F. (1938). *The Behavior of Organisms*. New York, NY: Appleton-Century-Crofts.

Skinner, B. F. (1957). *Verbal Behavior*. Brattleboro, Vermont: Echo Point.

Sloboda, J. A. (1991). "Music structure and emotional response: Some empirical findings." *Psychology of Music* 19(2): 110–120.

Smith, J. D., Wilson, M., and Reisberg, D. (1995). "The role of subvocalization in auditory imagery." *Neuropsychologia* 33(11): 1433–1454.

Smith, J. O., III (1996). "Physical modeling synthesis update." *Computer Music Journal* 20(2): 44–56.

Smith, J. R., Jr. (1967). "An investigation of the inherent qualities of musical intervals as definers of melodic tonality." Doctoral dissertation, Indiana University, Bloomington, IN.

Smith, J. W. (2019). "The salience, shapes, and functions of continuous processes in contemporary electronic dance music." Doctoral dissertation, University of Minnesota, Minneapolis, MN.

Smoorenburg, G. F. (1972). "Audibility region of combination tones." *Journal of the Acoustical Society of America* 52(2B): 603–614.

Snow, P. (1959). *The Two Cultures and the Scientific Revolution*. Cambridge, UK: Cambridge University Press.

Snow, W. B. (1936). "Change of pitch with loudness at low frequencies." *Journal of the Acoustical Society of America* 8(1): 14–19.

Sokal, A. D. (1996). "Transgressing the boundaries: Toward a transformative hermeneutics of quantum gravity." *Social Text* 46–47: 217–252.

Sokal, A. D. (2010). *Beyond the Hoax: Science, Philosophy and Culture*. Oxford: Oxford University Press.

Song, J. H., Skoe, E., Banai, K., and Kraus, N. (2011). "Perception of speech in noise: Neural correlates." *Journal of Cognitive Neuroscience* 23: 2268–2279.

Song, X., Osmanski, M. S., Guo, Y., and Wang, X. (2016). "Complex pitch perception mechanisms are shared by humans and a New World monkey." *Proceedings of the National Academy of Sciences* 113(3): 781–786.

Soon, C. S., Brass, M., Heinze, H. J., and Haynes, J. D. (2008). "Unconscious determinants of free decisions in the human brain." *Nature Neuroscience* 11(5): 543–545.

Spiller, K. (1995). "Diana Deutsch, founding editor (1983–1995)." *Music Perception* 12: 273–277.

Steege, B. (2010). "Helmholtz, music theory, and liberal-progressive history." *Journal of Music Theory* 54(2): 283–310.

Steege, B. (2011). "'The Nature of Harmony': A commentary." In *Oxford Handbook of Neo-Riemannian Music Theories*, edited by E. Gollin and A. Rehding, 55–91. New York, NY: Oxford University Press.

Steege, B. (2012). *Helmholtz and the Modern Listener*. Cambridge: Cambridge University Press.

Sterck, E. H., and Begeer, S. (2010). "Theory of mind: Specialized capacity or emergent property?" *European Journal of Developmental Psychology* 7(1): 1–16.

Stern, D. (1981). "Tonal organization in modal polyphony." *Theory and Practice* 6(2): 5–39.

Stevens, C. J. (2012). "Music perception and cognition: A review of recent cross-cultural research." *Topics in Cognitive Science* 4(4): 653–667.

Stevens, S. S. (1935). "The relation of pitch to intensity." *Journal of the Acoustical Society of America* 6(3): 150–154.

Stevens, S. S., Egan, J. P., and Miller, G. A. (1947). "Methods of measuring speech spectra." *Journal of the Acoustical Society of America* 19: 771–780 (as cited in B. B. Monson et al. 2014).

Stevens, S. S., and Galanter, E. H. (1957). "Ratio scales and category scales for a dozen perceptual continua." *Journal of Experimental Psychology* 54(6): 377.

Stevens, S. S., and Volkmann, J. (1940). "The relation of pitch to frequency: A revised scale." *American Journal of Psychology* 53(3): 329–353.

Stevens, S. S., Volkmann, J., and Newman, E. B. (1937). "A scale for the measurement of the psychological magnitude pitch." *Journal of the Acoustical Society of America* 8(3): 185–190.

Stieger, S., and Swami, V. (2015). "Time to let go? No automatic aesthetic preference for the golden ratio in art pictures." *Psychology of Aesthetics, Creativity, and the Arts* 9(1): 91.

Stock, J. P. (2007). "Alexander J. Ellis and his place in the history of ethnomusicology." *Ethnomusicology* 51(2): 306–325.

Stoll, G. (1982). "Spectral-pitch pattern." In *Music, Mind, and Brain*, edited by M. Clynes, 271–278. Boston, MA: Springer.

Stoll, G. (1984). "Pitch of vowels: Experimental and theoretical investigation of its dependence on vowel quality." *Speech Communication* 3(2): 137–147.

Stoll, G. (1985). "Pitch shift of pure and complex tones induced by masking noise." *Journal of the Acoustical Society of America* 77(1): 188–192.

Stoll, G., and Parncutt, R. (1987). "Harmonic relationship in similarity judgements of nonsimultaneous complex tones." *Acustica* 63(2): 111–119.

Straus, J. N. (2015). "Allen Forte's contribution to music theory." *Music Theory Spectrum* 37(1): 3–5.

Stravinsky, I. (1947). *Poetics of Music in the Form of Six Lessons*. Cambridge, MA: Harvard University Press.

Stumpf, C. (1883, 1890). *Tonpsychologie* (in 2 vols.). Leipzig: Hirtel.

Stumpf, C. (1911/2012). *The Origins of Music*. Translated by D. Trippett. Oxford: Oxford University Press.

Sukljan, N. (2018). "Zarlino's harpsichord: A contribution to the (pre) history of equal temperament." *Musicological Annual* 54(1): 5–22.

Sundberg, J. (1982). "In tune or not? A study of fundamental frequency in music practice." *Department of Speech, Music and Hearing: Quarterly Progress and Status Report (STL-QPSR)* 23(1): 49–78.

Swaminathan, S., and Schellenberg, E. G. (2015). "Current emotion research in music psychology." *Emotion Review* 7(2): 189–197.

Tagg, P. (2009). *Everyday Tonality: Towards a Tonal Theory of What Most People Hear*. Mass Media Scholar's Press. openlibrary.org.

Tartini, G. (1754). *Trattato di musica secondo la vera scienza dell'armonia*. Padua: Giovanni Manfrè.

Temperley, D. (1999). "The question of purpose in music theory." *Current Musicology* 66: 66–85.

Temperley, D. (2011). "Scalar shift in popular music." *Music Theory Online* 17(4).

Temperley, N. (1987). "Tonality and the Bourgeoisie." *Musical Times* 128(1738): 685–687.

Tendera, A., Rispoli, M., Senthilselvan, A., and Loucks, T. M. (2019). "Early speech rate development: A longitudinal study." *Journal of Speech, Language, and Hearing Research* 62(12): 4370–4381.

Tenney, J. (1988). *A History of "Consonance" and "Dissonance."* New York, NY: Excelsior.

Terhardt, E. (1968). "Über akustische Rauhigkeit und Schwankungsstarke [On acoustic roughness and fluctuation strength]." *Acustica* 20: 215–224.

Terhardt, E. (1972). "Zur Tonhöhenwahrnehmung von Klängen [On pitch perception in complex sounds] (in two parts)." *Acustica* 26: 173–199.

Terhardt, E. (1974a). "Pitch, consonance, and harmony." *Journal of the Acoustical Society of America* 55(5): 1061–1069.

Terhardt, E. (1975). "Influence of intensity on the pitch of complex tones." *Acustica* 33(5): 344–348.

Terhardt, E. (1976). "Concept and psychoacoustic foundations of musical consonance." *Acustica* 36(3): 121–137.

Terhardt, E. (1977). "The two-component theory of musical consonance." In *Psychophysics and Physiology of Hearing*, edited by E. F. Evans and J. P. Wilson, 381–390. London: Academic.

Terhardt, E. (1982). "Die psychoakustischen Grundlagen der musikalischen Akkordgrundtöne und deren algorithmische Bestimmung [The psychoacoustical foundations of musical chord-roots and their algorithmic determination]." In *Tiefenstruktur der Musik* [*Deep Structure of Music*], edited by C. Dahlhaus and M. Krause, 23–50. Berlin: Technical University of Berlin.

Terhardt, E. (1998). *Akustische Kommunikation: Grundlagen mit Hörbeispielen* [*Acoustic Communication: Foundations with Sound Examples*]. Berlin, Germany: Springer.

Terhardt, E., and Seewann, M. (1984). "Auditive und objektive Bestimmung der Schlagtonhöhe von historischen Kirchenglocken." *Acustica* 54(3): 129–144.

Terhardt, E., Stoll, G., Schermbach, R., and Parncutt, R. (1986). "Tonhöhenmehrdeutigkeit, Tonverwandtschaft und Identifikation von Sukzessivintervallen [Pitch ambiguity, tonal affinity, and melodic interval recognition]." *Acustica* 61(1): 57–66.

Terhardt, E., Stoll, G., and Seewann, M. (1982a). "Pitch of complex signals according to virtual-pitch theory: Tests, examples, and predictions." *Journal of the Acoustical Society of America* 71(3): 671–678.

Terhardt, E., Stoll, G., and Seewann, M. (1982b). "Algorithm for extraction of pitch and pitch salience from complex tonal signals." *Journal of the Acoustical Society of America* 71(3): 679–688.

Ternström, S., and Sundberg, J. (1988). "Intonation precision of choir singers." *Journal of the Acoustical Society of America* 84(1): 59–69.

Thelen, E., and Smith, L. B. (1994). *A Dynamic Systems Approach to the Development of Cognition and Action*. Cambridge, MA: MIT Press.

Thompson, D. E. (2011). "Speaking their language: Guitar tablature in the middle school classroom." *General Music Today* 24(3): 53–57.

Thompson, W. F., and Cuddy, L. L. (1989). "Sensitivity to key change in chorale sequences: A comparison of single voices and four-voice harmony." *Music Perception* 7(2): 151–168.

Thompson, W. F., and Parncutt, R. (1997). "Perceptual judgments of triads and dyads: Assessment of a psychoacoustic model." *Music Perception* 14: 263–280.

Thomson, W. (1958). "The problem of tonality in pre-Baroque and primitive music." *Journal of Music Theory* 2: 36–46.

Thomson, W. E. (1999). *Tonality in Music: A General Theory*. San Marino: Everett.

Tillmann, B. (2005). "Implicit investigations of tonal knowledge in nonmusician listeners." *Annals of the New York Academy of Sciences* 1060(1): 100–110.

Tillmann, B., and Bigand, E. (1996). "Does formal musical structure affect perception of musical expressiveness?" *Psychology of Music* 24(1): 3–17.

Tillmann, B., and Bigand, E. (2004). "The relative importance of local and global structures in music perception." *Journal of Aesthetics and Art Criticism* 62(2): 211–222.

Tillmann, B., Janata, P., Birk, J., and Bharucha, J. J. (2008). "Tonal centers and expectancy: Facilitation or inhibition of chords at the top of the harmonic hierarchy?" *Journal of Experimental Psychology: Human Perception and Performance* 34(4): 1031.

Tillmann, B., Koelsch, S., Escoffier, N., Bigand, E., Lalitte, P., Friederici, A. D., and von Cramon, D. Y. (2006). "Cognitive priming in sung and instrumental music: Activation of inferior frontal cortex." *NeuroImage* 31(4): 1771–1782.

Toiviainen, P., and Krumhansl, C. L. (2003). "Measuring and modeling real-time responses to music: The dynamics of tonality induction." *Perception* 32(6): 741–766.

Tokita, A. (2014). "Bi-musicality in modern Japanese culture." *International Journal of Bilingualism* 18(2): 159–174.

Trainor, L. J. (2015). "The origins of music in auditory scene analysis and the roles of evolution and culture in musical creation." *Philosophical Transactions of the Royal Society B: Biological Sciences* 370(1664): 20140089.

Trainor, L. J., Marie, C., Bruce, I. C., and Bidelman, G. M. (2014). "Explaining the high voice superiority effect in polyphonic music: Evidence from cortical evoked potentials and peripheral auditory models." *Hearing Research* 308: 60–70.

Trainor, L. J., and Trehub, S. E. (1994). "Key membership and implied harmony in Western tonal music: Developmental perspectives." *Perception & Psychophysics* 56(2): 125–132.

Tramo, M. J., Cariani, P. A., Delgutte, B., and Braida, L. D. (2001). "Neurobiological foundations for the theory of harmony in Western tonal music." *Annals of the New York Academy of Sciences* 930(1): 92–116.

Trehub, S. E. (2003). "The developmental origins of musicality." *Nature Neuroscience* 6(7): 669–673.

Trehub, S. E., Schellenberg, E. G., and Kamenetsky, S. B. (1999). "Infants' and adults' perception of scale structure." *Journal of Experimental Psychology: Human Perception and Performance* 25(4): 965.

Treitler, L. (1991). "Medieval improvisation." *World of Music* 33(3): 66–91.

Treitler, L. (1992). "The 'unwritten' and 'written transmission' of medieval chant and the start-up of musical notation." *Journal of Musicology* 10(2): 131–191.

Tsogli, V., Skouras, S., and Koelsch, S. (2022). "Brain-correlates of processing local dependencies within a statistical learning paradigm." *Scientific Reports* 12(1): 1–12.

Tymoczko, D. (2010). *A Geometry of Music: Harmony and Counterpoint in the Extended Common Practice*. New York, NY: Oxford University Press.

Uchida, R. (1990). "Tonal ambiguity in Debussy's piano works." MA thesis, School of Music, University of Oregon, Eugene, OR.

Ullal, S., Vanden Bosch der Nederlanden, C. M., Tichko, P., Lahav, A., and Hannon, E. E. (2013). "Linking prenatal experience to the emerging musical mind." *Frontiers in Systems Neuroscience* 7: 48.

Umpleby, S. A. (1997). "Cybernetics of conceptual systems." *Cybernetics & Systems* 28(8): 635–651.

Upadhyay, D., Shukla, R., and Chakraborty, A. (2016). "Factor structure of music preference scale and its relation to personality." *Journal of Indian Academy of Applied Psychology* 43(1): 104–113.

Utz, C. (2022). *Unerhöhte Klänge: Zur performativen Analyse und Wahrnehmung post-tonaler Musik und ihren historischen Voraussetzungen*. Hildesheim, Germany: Olms.

Väisälä, O. (2002). "Prolongation of harmonies related to the harmonic series in early post-tonal music." *Journal of Music Theory* 46(1–2): 207–283.

van der Werf, H. (1992). "Early Western polyphony." In *Companion to Medieval and Renaissance Music*, edited by T. Knighton and D. Fallows, 107–113. Berkeley: University of California Press.

Van Puyvelde, M., Vanfleteren, P., Loots, G., Deschuyffeleer, S., Vinck, B., Jacquet, W., and Verhelst, W. (2010). "Tonal synchrony in mother–infant interaction based on harmonic and pentatonic series." *Infant Behavior and Development* 33(4): 387–400.

van Waesberghe, J. S. (1951). "The musical notation of Guido of Arezzo." *Musica Disciplina* 5 (1951): 15–53.

Velmans, M. (2007). "Heterophenomenology versus critical phenomenology." *Phenomenology and the Cognitive Sciences* 6(1): 221–230.

Verhey, J. L., Dau, T., and Kollmeier, B. (1999). "Within-channel cues in comodulation masking release (CMR): Experiments and model predictions using a modulation-filterbank model." *Journal of the Acoustical Society of America* 106(5): 2733–2745.

Verschooten, E., Shamma, S., Oxenham, A. J., Moore, B. C., Joris, P. X., Heinz, M. G., and Plack, C. J. (2019). "The upper frequency limit for the use of phase locking to code temporal fine structure in humans: A compilation of viewpoints." *Hearing Research* 377: 109–121.

Verschuure, J., and Van Meeteren, A. A. (1975). "The effect of intensity on pitch." *Acustica* 32(1): 33–44.

Villegas, J., and Cohen, M. (2010). "Roughness minimization through automatic intonation adjustments." *Journal of New Music Research* 39(1): 75–92.

Vincent, H. J. (1894). *Ist unsere Harmonielehre wirklich eine Theorie?* [Is our theory of harmony really a theory?] Vienna: Rörich (as cited in R. W. Wason 1988).

Virtala, P., and Tervaniemi M. (2017). "Neurocognition of major–minor and consonance–dissonance." *Music Perception* 34(4): 387–404.

Vliegen, J., Moore, B. C. J., and Oxenham, A. J. (1999). "The role of spectral and periodicity cues in auditory stream segregation, measured using a temporal discrimination task." *Journal of the Acoustical Society of America* 106(2): 938–945.

Von Hoerner, S. (1976). "The definition of major scales, for chromatic scales of 12, 19 and 31 divisions per octave." *Psychology of Music* 4(1): 12–23.

Vos, J. (1986). "Purity ratings of tempered fifths and major thirds." *Music Perception* 3(3): 221–257.

Vos, P. G. (1999). "Key implications of ascending fourth and descending fifth openings." *Psychology of Music* 27(1): 4–17.

Vos, P. G. (2000). "Tonality induction: Theoretical problems and dilemmas." *Music Perception* 17(4): 403–416.

Vos, P. G., and Troost, J. M. (1989). "Ascending and descending melodic intervals: Statistical findings and their perceptual relevance." *Music Perception* 6(4): 383–396.

Vuoskoski, J. K., Thompson, W. F., McIlwain, D., and Eerola, T. (2011). "Who enjoys listening to sad music and why?" *Music Perception* 29(3): 311–317.

Vurma, A., Raju, M., and Kuuda, A. (2010). "Does timbre affect pitch? Estimations by musicians and non-musicians." *Psychology of Music* 39(3): 291–306.

Vyčinienė, D. (2012). "Relationships between Lithuanian and Balkan Schwebungs-Diaphonie: Interdisciplinary search key." In D. Despić et al. (Eds.), Musical Practices in the Balkans: Ethnomusicological Perspectives, 119–147. Belgrade, Serbia: Institute of Musicology. core.ac.uk.

Wagemans, J., Feldman, J., Gepshtein, S., Kimchi, R., Pomerantz, J. R., Van der Helm, P. A., and Van Leeuwen, C. (2012). "A century of Gestalt psychology in visual perception: II. Conceptual and theoretical foundations." *Psychological Bulletin* 138(6): 1218.

Walker, R. (1989). "Modes and pitch-class sets in Messiaen: A brief discussion of 'Première Communion de la Vierge.'" *Music Analysis* 8: 159–168.

Walliser, K. (1969). "Uber die Abhängigkeit der Tonhöhenempfindung von Sinustönen vom Schallpegel, von überlagertem drosselndem Störschall und von der Darbietungsdauer." *Acustica* 21: 211–221.

Wang, J. C. (2015). "Singing polyphony: An Asian experience." *Music Educators Journal* 101(4): 85–95.

Ward, W. D. (1954). "Subjective musical pitch." *Journal of the Acoustical Society of America* 26: 369–380.

Ward, W. D., and Martin, D. W. (1961). "Psychophysical comparison of just tuning and equal temperament in sequences of individual tones." *Journal of the Acoustical Society of America* 33(5): 586–588.

Wason, R. W. (1988). "Progressive harmonic theory in the mid-nineteenth century." *Journal of Musicological Research* 8(1–2): 55–90.

Weisberg, J. (n.d.). "The hard problem of consciousness." Internet Encyclopedia of Philosophy. https://iep.utm.edu/hard-problem-of-conciousness/#:~:text=The%20hard%20problem%20of%20consciousness%20is%20the%20problem,%E2%80%9Clight%20up%E2%80%9D%20and%20directly%20appear%20to%20the%20subject

Weiss, M. W., Cirelli, L. K., McDermott, J. H., and Trehub, S. E. (2020). "Development of consonance preferences in Western listeners." *Journal of Experimental Psychology: General* 149(4): 634–649.

Weiss, S. (2014). "Listening to the world but hearing ourselves: Hybridity and perceptions of authenticity in world music." *Ethnomusicology* 58(3): 506–525.

Wertheimer, M. (1938). "Gestalt theory." In *A Source Book of Gestalt Psychology*, edited and translated by W. D. Ellis, 1–11. London: Routledge.

White, C. W., and Quinn, I. (2018). "Chord context and harmonic function in tonal music." *Music Theory Spectrum* 40(2): 314–335O.

Wicks, S. A. (1989). "A belated salute to the 'old way' of 'snaking' the voice on its (ca.) 345th birthday." *Popular Music* 8(1): 59–96.

Wienpahl, R. W. (1971). "Modality, monality and tonality in the sixteenth and seventeenth centuries." *Music & Letters* 52(4): 407–417.

Wiggins, G. A., Müllensiefen, D., and Pearce, M. T. (2010). "On the non-existence of music: Why music theory is a figment of the imagination." *Musicae Scientiae* 14(1_suppl): 231–255.

Wiliford, L. (1999). "Overtone singing: The key to 'past life melodies.'" *Choral Journal* 40(3): 43.

Will, U. (1997). "Two types of octave relationships in central Australian vocal music?" *Musicology Australia* 20(1): 6–14.

Willimek, B., and Willimek, D. (2017). "Feelings which strike a chord, and chords which strike a feeling." *Open Journal of Acoustics* 7(1): 10.

Wilson, M. (2002). "Six views of embodied cognition." *Psychonomic Bulletin & Review* 9(4): 625–636.

Windsor, W. L., and De Bézenac, C. (2012). "Music and affordances." *Musicae Scientiae* 16(1): 102–120.

Withagen, R. (2018). "Towards an ecological approach to emotions and the individual differences therein." *New Ideas in Psychology* 51: 21–26.

Wittgenstein, L. (1953). *Philosophical Investigations*. New York, NY: Macmillian (as cited in E. Rosch and C. B. Mervis 1998).

Wolf, W., and Bernhart, W., eds. (2006). *Framing Borders in Literature and Other Media*. Amsterdam, Netherlands: Rodopi.

Wright, J. K., and Bregman, A. S. (1987). "Auditory stream segregation and the control of dissonance in polyphonic music." *Contemporary Music Review* 2: 63–92.

Yih, A. K. (2000). "Analysing Debussy: Tonality, motivic sets and the referential pitch-class specific collection." *Music Analysis* 19(2): 203–229.

Zarlino, G. (1558). *Le institutioni harmoniche*. Venetia: Senese.

Zatorre, R. J. (1988). "Pitch perception of complex tones and human temporal-lobe function." *Journal of the Acoustical Society of America* 84(2): 566–572.

Zhmud, L. (2012). "Aristoxenus and the Pythagoreans." In *Aristoxenus of Tarentum: Texts and Discussion*, edited by C. A. Huffman, 223–249. Abingdon, Oxfordshire, UK: Transaction.

Zweifel, P. F. (1994). "Just tuning and the unavoidable discrepancies." *Indiana Theory Review* 15(2): 89–120.

Zweifel, P. F. (1996). "Generalized diatonic and pentatonic scales: A group-theoretic approach." *Perspectives of New Music* 34: 140–161.

Zwicker, E. (1954). "Die Verdeckung von Schmallbandgeräuschen durch Sinustöne." *Acustica* 4: 415–420 (as cited in R. Plomp 1976).

Zwicker, E. (1961). "Subdivision of the audible frequency range into critical bands (Frequenzgruppen)." *Journal of the Acoustical Society of America* 33(2): 248.

Zwicker, E., and Feldtkeller, R. (1967). *Das Ohr als Nachrichtuenempfänger*. Stuttgart: Hirzel. Translated by E. Zwicker and R. Feldtkeller (1999) as *The Ear as a Communication Receiver*. Woodbury, NY: Acoustical Society of America.

Index